A Relaxation-Based Approach to Optimal Control of Hybrid and Switched Systems

A Relaxation-Based Approach to Optimal Control of Hybrid and Switched Systems

A Practical Guide for Engineers

Vadim Azhmyakov

Department of Mathematical Sciences,
Universidad EAFIT,
Medellin, Republic of Colombia

Butterworth-Heinemann
An imprint of Elsevier

Butterworth-Heinemann is an imprint of Elsevier
The Boulevard, Langford Lane, Kidlington, Oxford OX5 1GB, United Kingdom
50 Hampshire Street, 5th Floor, Cambridge, MA 02139, United States

Library of Congress Cataloging-in-Publication Data
A catalog record for this book is available from the Library of Congress

British Library Cataloguing-in-Publication Data
A catalogue record for this book is available from the British Library

ISBN: 978-0-12-814788-7

For information on all Butterworth-Heinemann publications
visit our website at https://www.elsevier.com/books-and-journals

Publisher: Mara Conner
Acquisition Editor: Sonnini R. Yura
Editorial Project Manager: Joshua Mearns
Production Project Manager: Bharatwaj Varatharajan
Designer: Victoria Pearson

Typeset by VTeX

To my students from the North and South Americas

Contents

Preface

This book is written as an application-oriented extension of the conventional relaxation theory to hybrid and switched optimal control problems. Modern optimal control theory is concerned with the analysis and design of sophisticated dynamical systems, where the aim is at steering such a system from a given configuration to some desired target by minimizing a suitable performance. Nowadays this theory constitutes a powerful design methodology for the computer-oriented development of several types of high-performance controllers. General optimal control problems associated with various types of advanced control systems have been comprehensively studied due to their natural engineering applications. The material discussed in this research monograph is the result of the author's work at the Ernst-Moritz-Arndt University of Greifswald (Greifswald, Germany), the Centro de Investigacion y de Estudios Avanzados del Instituto Politecnico Nacional (Mexico City, Mexico), and the University of Medellin (Medellin, Colombia). The main purpose of this book is to propose a unified approach to effective and numerically tractable relaxation schemes for optimal control of hybrid and switched systems. We study several generic classes of hybrid and switched dynamic models and corresponding optimal control problems in the context of suitable relaxation schemes.

"Relaxing the initial problem" has various meanings in applied mathematics, depending on the areas where it is defined, depending also on what one relaxes (a functional, the underlying space, etc.). In the context of an optimal control problem, when dealing with the minimization of an objective functional, the most common way of looking at relaxation is to consider the lower semicontinuous hull of this functional determined on a convexification of the set of admissible controls. The concept of relaxed controls was introduced by L.C. Young in 1937 under the name of generalized curves and surfaces. It has been used extensively in the professional literature for the study of diverse optimal control problems. It is common knowledge that a real-world optimal control problem does not always have a (mathematical) solution. On the other hand, the corresponding relaxed problem has an optimal solution under some mild assumptions. In practice, this solution can be considered as a suitable approximation for the sophisticated initial problem. In the absence of the so-called "relaxation gap," the generalized problem is of prime interest for the initial optimal control problem. In this case, the minimal value of the objective functional in the initial problem coincides with the minimum of the objective functional in the relaxed problem. Therefore, in that situation, an adequately relaxed problem can be used as a theoretical fundament for adequate numerical solution algorithms for the initial problem. When solving optimal control problems with ordinary differential equations, we deal with functions and systems which, except in very special cases, are to be replaced by numerically tractable approximations. In contrast to the conventional optimal control problems an effective implementation of adequate computational schemes for hybrid/switched system optimization is predominantly based on the relaxed controls. Therefore, our aim is to consider the relaxations of the hybrid and switched optimal control problems in a close methodological relationship to the corresponding numerical methods and possible engineering applications.

Recall that various types of hybrid and switched control systems and the related optimal control problems have been comprehensively studied in the past several years due to their important engineering applications. Let us mention here some real-world applications from the mobile robot technology, intelligent automotive control, modern telecommunications, process control, and data science. We first give an extensive overview of the existing (conventional and newly developed) relaxation techniques associated with the "conventional" systems described by ordinary differential equations. Next we construct a self-contained relaxation theory for optimal control processes governed by various types (subclasses) of general hybrid and switched systems. Note that due to the extreme complexity of hybrid/switched dynamic systems this "construction" is a challenging analytic and computational problem and cannot be considered as a simple "theory/fact transfer" from the conventional optimal control to hybrid and switched cases. Let us also note that the book we propose contains all mathematical tools that are necessary for an adequate understanding and use of the sophisticated relaxation techniques. All in all, this manuscript follows the "engineering" and "numerical" concepts. However, it can also be considered as a mathematical "compendium" that contains all the necessary formal results and some important algorithms related to the modern relaxation theory. This fact makes it possible to use this book in systems engineering (specifically in electrical, aerospace, and financial engineering) and in practical systems optimization.

The target audience of this book includes but is not restricted to academic researchers and PhD candidates from Electrical Engineering and/or Applied Mathematics faculties (technical and regular universities, academic research centers) and R&D departments of electrical/electronic companies (research engineers, developers). This book can also be useful for economic schools and mathematically oriented economists. Parts of this book can also be used for advanced courses such as Dynamic Optimization, Optimal Control of Modern Dynamic Systems, and Advanced Mathematics for Engineers. The book can also be included into PhD qualification programs in control engineering, applied mathematics, advanced computer science, and mathematical economics. We expect that this monograph will be useful to the interested graduate students and some undergraduate students with sufficient knowledge of functional analysis, mathematical optimization and dynamic systems. The book can also be considered as a complementary text to graduate courses in applied mathematics. We assume that the reader has knowledge of analysis and linear algebra, while a little more is presupposed from nonlinear analysis, optimization theory, and optimal control. We decided to use a style with detailed and often transparent proofs of the significant results. Of course, the book only claims to present an (extended) introduction to the relaxation theory for hybrid/switched optimal control problems and possible applications. We have made an attempt to unify, simplify, and relate many scattered results in the literature. Some of the topics discussed here are new, others are not. Therefore the book is not a collection of research papers, but it is a monograph to present recent developments of the theory that could be the foundations for further developments.

Many people have influenced the contents and final presentation of this book, and I am grateful to all of them. I would like to thank Professor W.H. Schmidt (Ernst-Moritz-Arndt University of Greifswald, Greifswald, Germany), Professor M.V. Basin (Universidad Autonoma de Nuevo Leon, Monterrey, Mexico), and Professor A. Poznyak (Centro de Investigacion y de Estudios Avanzados del Instituto Politecnico Nacional, Mexico City, Mexico). I also would like to express my warm gratitude to Professor M. Egerstedt, Professor Y. Wardi, and Professor E. Verriest (Georgia Institute of Technology, Atlanta, USA). Many methodological aspects of the book are finally improved due to several professional discussions with my Master and PhD students from Germany, Mexico, and Colombia.

I wish to thank Professor M.V. Basin (Universidad Autonoma de Nuevo Leon, Monterrey, Mexico), Professor St. Pickl (University of Bundeswehr Munich, Munich, Germany), and Professor M. Shamsi (Amirkabir University of Technology, Tehran, Iran) for reading the first version of this monograph.

Finally, I wish to express my appreciation to Elsevier for their excellent and accomplished handling of the manuscript, their understanding, and their patience.

CHAPTER 1

Introduction and Motivation

1.1 Optimal Control of Hybrid and Switched Dynamic Systems

Modern optimal control theory (OCT) is concerned with the analysis and design of sophisticated dynamical systems, aiming at steering such a system from a given configuration to some desired target by minimizing a suitable performance. Nowadays this theory constitutes a powerful design methodology for the computer-oriented development of several types of high-performance controllers. General optimal control problems (OCPs) associated with various types of advanced control systems have been comprehensively studied due to their natural engineering applications.

Optimization of sophisticated constrained dynamic models (including hybrid systems [HSs] and switched systems [SSs]) is, nowadays, not only a mathematical theory buy also a mature and relative simple design methodology for the practical development of several types of modern controllers. Consideration of the general OCPs usually incorporates two conceptual sophisticated aspects: the presence of various additional constraints and the existence of an optimal solution of an OCP under consideration. These two basic aspects determine the basic facets in optimization of the dynamics systems described by ordinary differential equations (ODEs). It is easy to understand that in the context of a hybrid optimal control problem (HOCP)/switched optimal control problem (SOCP) the same theoretic points, namely, constraints and existence of an optimal solution, obtain an increasing complexity. This fact is a simple consequence of the following observation: the general HSs and SSs constitute a formal generalization of the ODEs involved in control systems. The same is evidently true for the corresponding OCPs.

It is readily appreciated that the real-world control systems have a corresponding set of constraints; for example, inputs always have maximum and minimum values, and states are usually required to lie within certain ranges. The second problem mentioned above, namely, the existence of an optimal solution, can in fact be denoted as a "millennium problem" from the classic OCT. A constructive consideration of this sophisticated problem was finally realized in the framework of the relaxation theory (RT). Of course, in a concrete OCP one could proceed by ignoring the theoretically complicated existence question, ignore various state-control constraints, and hope that no serious consequences result from this approach. This simple procedure may be sufficient at times. It is generally true that optimal levels of a suitable performance are associated with operating on, or near, constraint boundaries. Thus, a control

A Relaxation-Based Approach to Optimal Control of Hybrid and Switched Systems
https://doi.org/10.1016/B978-0-12-814788-7.00007-2

engineer really cannot ignore constraints without incurring a performance penalty. Since a relaxed OCP usually possesses some newly determined admissible sets of constraints, the above observation is also true for relaxed OCPs.

For the classic as well as hybrid OCPs the suitable relaxation procedures also imply new numerical methodologies. Relaxed OCPs involve specific computational treatment and the corresponding practical algorithms. Recently, the problem of effective numerical methods for constrained systems optimization has attracted a lot of attention, thus both theoretical results and applications were developed. The handling constraints in an original or relaxed form in practical systems design are an important issue in most, if not all, real-world applications. Our book deals with some specific classes of (finite) hybrid and switched control systems and with the corresponding OCPs. We give in fact a "reformulation" of the conventional and extended RT for the case of these OCPs. Taking into consideration the wide use of the relaxation schemes in numerical treatment of the classic OCPs, we are also interested to elaborate consistent computational algorithms for HOCPs/SOCPs (OCP) in the presence of constraints. We study classes of OCPs with convex (or convexified) costs functionals. The structure of the admissible control functions we consider in this book is mainly motivated by some important practical control applications as well as by the widely applicable modern quantization procedure associated with the original system dynamics. We refer to the interesting self-closed results on the convex-like and generalized dynamics. For example, the formal treatment of linear quadratic OCPs was based on the backward solutions of the Riccati differential equations, and the optimum had to be recomputed for each new final state. Computation of nonlinear gains using the Hamilton–Jacobi–Bellman (HJB) equation and the convex optimization techniques has also been performed.

On the other hand, the existing optimization approaches to the relaxed dynamics are not sufficiently advanced for the hybrid and switched types of OCPs. In our book, we propose analytic as well as numerical methods based on a combination of the convex-like relaxation schemes and the first-order projection approach. Also, it should be noted already at this point that a computational algorithm we propose can be effectively used in a concrete control synthesis phase associated with a practical engineering design of hybrid and switched dynamic systems. Recall that the general HSs and SSs constitute a class of mathematical abstractions where two types of dynamics are present, i.e., continuous and discrete event dynamic behavior. In order to understand how these systems can be operated efficiently, both aspects of the actual dynamics have to be taken into account during the optimization and the corresponding control design procedure. The nonlinear systems we study can be interpreted as a particular family of the general systems with the state-driven (HSs) and time-driven (SSs) location transitions.

Dynamic processes described by HSs/SSs constitute an adequate and more detailed modeling approach to many important real-world problems in engineering, social science, and

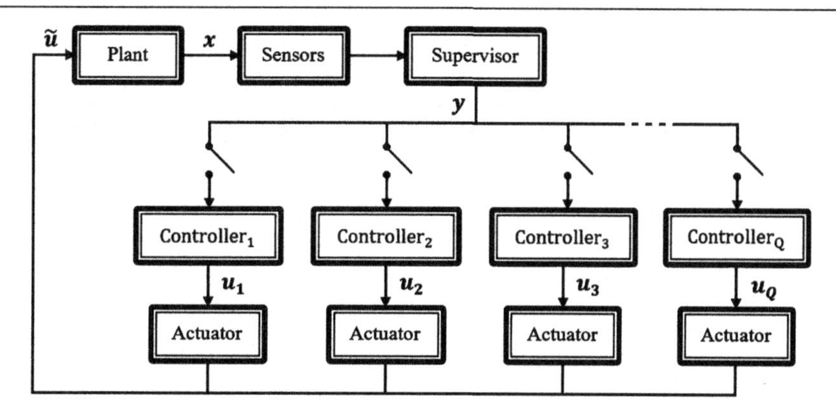

Figure 1.1: A general structure of an HS.

bioscience. Variety of suitable mathematical abstraction for control and measurements (engineering instrumentation) can be characterized by this novel modeling framework. These include, for example, the cost of hardware implementation, measured for example not only by computational requirements such as speed and memory, but also by communication requirements of complex and costly systems. In general, a sophisticated interplay of complex nonlinear dynamic objects, such as autonomous robots/vehicles, airplanes, and satellites, in the presence of the associated communication networks is a prevailing attribute of many modern applicable control systems. Let us also note that the switched discrete-continuous structure of the generic HSs and SSs makes it possible to use these dynamic models in the (optimal) decision science. Note that the necessary analysis, optimization, and adequate design procedures for these systems have been recognized as major problems in modern control engineering.

Let us give a qualitative illustration (the structural scheme) of a generic HS (Fig. 1.1).

The presented feedback-type structural scheme contains $Q \in \mathbb{N}$ (a natural number) controllers and actuators that are triggered by a supervisor. We later give a formal definition of a wide class of HSs under consideration.

Recently a vast body of research on mathematical models for hybrid and switched control systems has been produced, drawing its motivation from the fact that many modern application domains involve complex systems, in which subsystem interconnections, mode transitions, and heterogeneous computational devices are presented. A common example of an interconnected (hybrid) dynamic system is given by the following qualitative model of a batch reactor (see Fig. 1.2).

An adequate detailed model of the abovementioned complexly interconnected technical objects is usually characterized by switched or hybrid dynamics and, moreover, includes an as-

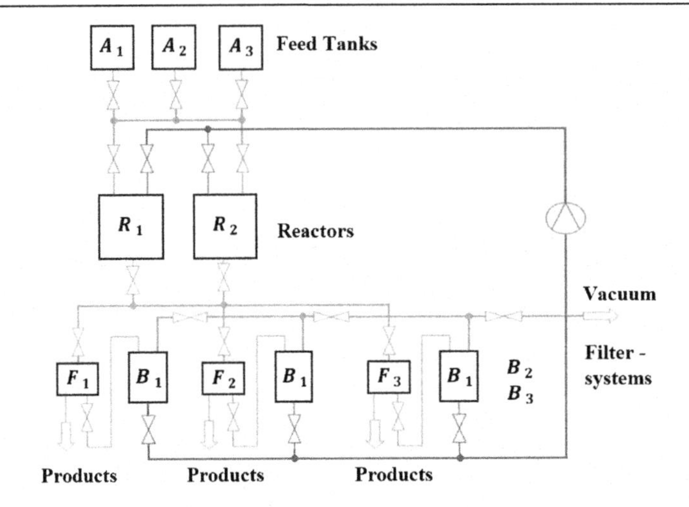

Figure 1.2: Structural scheme of a batch reactor.

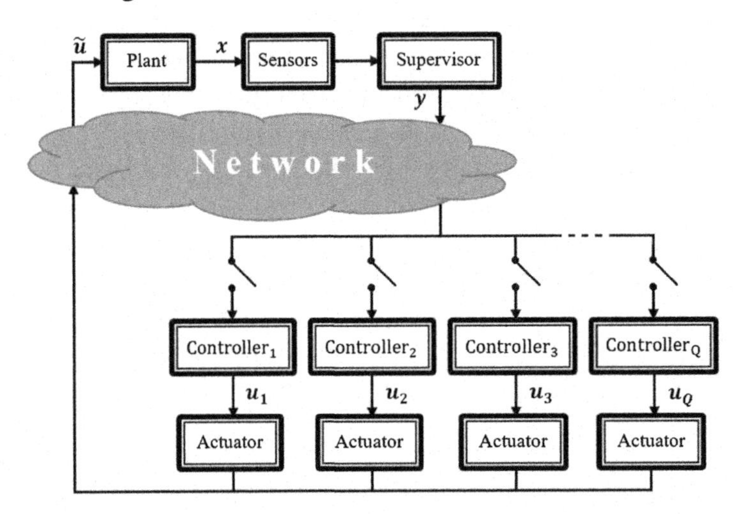

Figure 1.3: An HS with the communication network.

sociated (local or global) communication network. Note that in the case of the interconnected control-communication systems the real discrete transitions of the state vectors (switches) can be triggered not only by simple discrete control commands but also autonomously (for example, by an autonomous location transition mechanism). An illustration of an HS that contains a communication network is given in Fig. 1.3.

Evidently, an optimal control synthesis for systems presented in Figs. 1.1–1.3 is a very sophisticated task. However, we are optimistic as regards the use of the relaxation bases control design schemes. This optimism is due to the solid mathematical and computational foundations of the classic optimization methodologies in complex (real Banach) spaces and also to the clear engineering interpretation of the switched and hybrid dynamical models, their reliability, and the existence of well-established examples and simulation results. Evidently, a possible generalization of the relaxation approach for HOCPs and SOCPs is an extremely interesting theoretic and numeric problem which has a high potential in engineering applications. The aim of our book is to extend the theoretic and numeric aspects of the classic RT to the powerful and effective computational schemes associated with the abovementioned classes of optimal control systems.

Let us now pay attention to some formal concepts. This brief issue illustrates some conceptual aspects mentioned above. We start by introducing a variant of a useful definition of an HS under consideration.

Definition 1.1. *An HS is a 7-tuple* $\{\mathcal{Q}, M, U, F, \mathcal{U}, I, \mathcal{S}\}$, *where*

- $\mathcal{Q}$ *is a finite set of discrete states (called locations);*
- $M = \{M_q\}_{q\in\mathcal{Q}} \subset \mathbb{R}^n$ *is a family of smooth manifolds, indexed by* $\mathcal{Q}$*;*
- $U \subseteq \mathbb{R}^m$ *is a set of admissible control input values (called control set);*
- $F = \{f_q\},\ q \in \mathcal{Q}$, *is a family of maps*

$$f_q : [0, t_f] \times M_q \times U \to TM_q,$$

  where TM_q *is the tangent bundle of* M_q*;*
- $\mathcal{U}$ *is the set of all admissible control functions;*
- $I = \{I_q\}$ *is a family of adjoint subintervals of* $[0, t_f]$ *such that*

$$\sum_{q\in\mathcal{Q}} |I_q| = t_f;$$

- $\mathcal{S}$ *is a subset of* Ξ, *where*

$$\Xi := \{(q, x, q', x') \ :\ q, q' \in \mathcal{Q}, x \in M_q, x' \in M_{q'}\}.$$

An HS from Definition 1.1 is determined on the finite time interval $[0, t_f]$. Note that in contrast to the general definition of an HS, the control set U here is the same for all locations. Moreover, in the sense of this definition the set $\mathcal{U}$ is also independent of a location. We generally assume that U is a compact set. Let us assume here that

$$\mathcal{U} := \{u(\cdot) \in \mathbb{L}^\infty_m([0, t_f]) \ :\ u(t) \in U \text{ a.e. on } [0, t_f]\},$$

where $\mathbb{L}^\infty_m([0, t_f])$ is the Lebesgue space of all measurable, essentially bounded functions $u : [0, t_f] \to \mathbb{R}^m$. Note that $\mathbb{L}^\infty_m([0, t_f])$ is assumed to be equipped with the usual sup-norm. We now introduce the following generic hypothesis associated with the vector fields $f_q,\ q \in \mathcal{Q}$:

- all functions $f_q(t, \cdot, \cdot)$ from F are differentiable;
- $$f_q,\ \partial f_q/\partial x,\ \partial f_q/\partial u$$

  are continuous and there exist constants $C_q < \infty$ such that

  $$||\frac{\partial}{\partial x} f_q(t, x, u)|| \leq C_q,\ \ q \in \mathcal{Q},\ (t, x, u) \in [0, t_f] \times M_q \times U.$$

In particular, we require that the maps f_q from F are Lipschitz on

$$[0, t_f] \times M_q \times U.$$

Note that this requirement is a formal technical condition that implies the differentiability of the objective functional we will introduce below. For $q, q' \in \mathcal{Q}$ one can also define the switching set

$$S_{q,q'} := \{(x, x') \in M_q \times M_{q'}\ :\ (q, x, q'x') \in \mathcal{S}\}$$

from location q to location q'. The intervals $I_q,\ q \in \mathcal{Q}$, indicate the lengths of time intervals on which the system can stay in location q. We say that a location switching from q to q' occurs at a *switching time* $t^{switch} \in [0, t_f]$.

We next consider an HS with $r \in \mathbb{N}$ switching times $\{t_i\}$, $i = 1, ..., r$, where

$$0 = t_0 < t_1 < ... < t_r < t_{r+1} = t_f.$$

Note that the sequence of switching times $\{t_i\}$ is not defined a priori. A hybrid control system remains in location, i.e.,

$$q_i \in \mathcal{Q}\ \forall t \in [t_{i-1}, t_i[,\ i = 1, ..., r+1.$$

Let $[t_{i-1}, t_i] \subset I_{q_i}$ for all $i = 1, ..., r+1$.

Definition 1.2. *Let $u(\cdot) \in \mathcal{U}$ be an admissible control for an HS. Then a "continuous" trajectory of HSs is an absolutely continuous function*

$$x : [0, t_f] \to \bigcup_{q \in \mathcal{Q}} M_q$$

such that

$$x(0) = x_0 \in M_{q_1}$$

and

- $$\dot{x}(t) = f_{q_i}(t, x(t), u(t))$$

  for almost all $t \in [t_{i-1}, t_i]$ *and all* $i = 1, ..., r+1$;
- *the switching condition*

  $$(x(t_i), x(t_{i+1})) \in S_{q_i, q_{i+1}}$$

  holds if $i = 1, ..., r$.

The vector

$$\mathcal{R}_{r+1} := (q_1, ... q_{r+1})$$

is called a "discrete trajectory" of the hybrid control system.

Definition 1.2 describes dynamics of an HS. Since $x(\cdot)$ is an absolutely continuous function, Definition 1.2 describes a class of HSs without impulse components of the continuous trajectories. Therefore, the corresponding switching sets $S_{q,q'}$ (and $S_{q_i,q_{i+1}}$) are defined for

$$x(t_i) = x(t_{i+1}).$$

Under the above assumptions for the given family of vector fields F, for each admissible control $u(\cdot) \in \mathcal{U}$ and for every interval $[t_{i-1}, t_i]$ (for every location $q_i \in \mathcal{R}$) there exists a unique absolutely continuous solution of the corresponding differential equation. This means that for each $u(\cdot) \in \mathcal{U}$ we have a unique absolute continuous trajectory of the HS. Moreover, the switching times $\{t_i\}$ and the discrete trajectory $\mathcal{R}$ for an HS are also uniquely defined. Therefore, it is reasonable to introduce the following concept.

Definition 1.3. *Let an HS be defined as above. For an admissible control* $u(\cdot) \in \mathcal{U}$, *the triplet*

$$\mathcal{X}^u := (\tau, x(\cdot), \mathcal{R}),$$

where τ *is the set of the corresponding switching times* $\{t_i\}$ *and* $x(\cdot)$ *and* $\mathcal{R}$ *are the corresponding continuous and discrete trajectories, respectively, is called a hybrid trajectory of the HS.*

Let $\phi : \mathbb{R}^n \to \mathbb{R}$ be a continuously differentiable function. Given an HS we now consider a "discrete" trajectory $\mathcal{R}_{r+1}$ from Definition 1.2. Finally we are ready to formulate the following Mayer-type HOCP:

$$\begin{aligned}
&\text{minimize } \phi(x(t_f)) \\
&\text{subject to } \dot{x}(t) = f_{q_i}(t, x(t), u(t)) \text{ a.e. on } [t_{i-1}, t_i] \\
&q_i \in \mathcal{Q}\; i = 1, ..., r+1, \;\; x(0) = x_0 \in M_{q_1}, \;\; u(\cdot) \in \mathcal{U}.
\end{aligned} \tag{1.1}$$

Note that the main dynamic optimization problem, namely, problem (1.1), constitutes a useful abstract framework for the concrete engineering systems optimization.

Analogously to Definition 1.1 one can introduce a concept of an SS. In this section we only give a definition of an affine SS (ASS). The general concept will be studied in Chapter 6.

Definition 1.4. *An ASS is a 7-tuple* $\{\mathcal{Q}, \mathcal{X}, U, \mathcal{A}, \mathcal{B}, \mathcal{U}, \Psi\}$, *where*

- $\mathcal{Q}$ *is a finite set of indices;*
- $$\mathcal{X} = \{\mathcal{X}_q\}, q \in \mathcal{Q},$$
  is a family of state spaces such that $\mathcal{X}_q \subseteq \mathbb{R}^n$;
- $U \subseteq \mathbb{R}^m$ *is a set of admissible control input values (called control set);*
- $$\mathcal{A} = \{a_q(\cdot,\cdot)\},\ \mathcal{B} = \{b_q(\cdot,\cdot)\},\ q \in \mathcal{Q},$$
  are families of uniformly bounded on an open set $(0, t_f) \times \mathcal{R}$ *Carathéodory functions*
  $$a_q : (0, t_f) \times \mathcal{R} \to \mathbb{R}^n,\ b_q : (0, t_f) \times \mathcal{R} \to \mathbb{R}^{n \times m};$$
- $\mathcal{U}$ *is the set of admissible control functions introduced above;*
- Ψ *is a subset of* Ξ, *where*
  $$\Xi := \left\{(x, x') \mid x \in \mathcal{X}_q,\ x' \in \mathcal{X}_{q'},\ q, q' \in \mathcal{Q}\right\}.$$

An admissible trajectory associated with an ASS is an absolutely continuous function $x(\cdot)$ *such that*

- $$x_i(\cdot) = x(\cdot)|_{(t_{i-1}, t_i)}$$
  is an absolutely continuous function on (t_{i-1}, t_i) *continuously prolongable to* $[t_{i-1}, t_i]$, $i = 1, \dots, r$;
- $$\dot{x}_i(t) = a_{q_i}(t, x_i(t)) + b_{q_i}(t, x_i(t))u_i(t)$$
  for almost all times $t \in [t_{i-1}, t_i]$, *where* $u_i(\cdot)$ *is a restriction of the chosen control function* $u(\cdot)$ *on the time interval* $[t_{i-1}, t_i]$.

Using Definition 1.4 we can formulate an OCP involving an SS.

$$
\begin{aligned}
&\text{minimize } \phi(x(t_f)) \\
&\text{subject to } \dot{x}(t) = f_{q_i}(t, x(t), u(t)) \text{ a.e. on } [t_{i-1}, t_i] \\
&q_i \in \mathcal{Q}\ i = 1, ..., r+1, \ x(0) = x_0, \ u(\cdot) \in \mathcal{U}.
\end{aligned} \tag{1.2}
$$

In spite of a "visual identity" of the OCPs (1.1) and (1.2), these two problems are conceptually different. In Section 1.2 we will discuss the conceptual difference between problems (1.1) and (1.2).

The generic HOCP (1.1) and SOCP (1.2) constitute very sophisticated (constrained) dynamic optimization problems in an abstract functional space of all absolutely continuous functions. Even in the case of a one-element set $\mathcal{Q}$, namely, in the case of a conventional OCP, the existence of an optimal solution is a very complicated question. As mentioned above, this existence problem is in fact solved (for the conventional case) in the framework of the classic RT. Additionally to this abstract result (existence), the celebrated RT provides various constructive numerical approaches to the initial and generalized OCPs. Summarizing, we can say that the possible theoretic extension of the conventional RT to the new classes of systems, namely, to hybrid and switched dynamic models, can be useful in the practical (optimal) control design. As a main result of the presented book, one can indicate the creation of the analytic basis for practically relaxed optimal control strategies to be associated with the wide classes of modern hybrid/switched dynamical systems.

1.2 Questions Relaxation Theory Can Answer

Consider an HS from Definition 1.1. We immediately observe that the switching times $\{t_i\}$, $i = 1, ..., r$ are determined by the switching set $S_{q_i, q_{i+1}}$, where $i = 1, ..., r$ (see above). The corresponding switching manifolds are M_q, where $q \in \mathcal{Q}$ is a subset of the state space $\mathbb{R}^n$. This fact makes it clear that the switching times

$$\tau := \{t_i\}, \ i = 1, ..., r$$

introduced above are in fact functions of the system state $t_i = t_i(x(\cdot))$, where $x(\cdot)$ is the continuous trajectory of the HS introduced in Definition 1.2. That means the switching times for a general HS depend on the a posteriori information about the system and cannot be determined before the definition of the complete dynamics. From the formal point of view this situation is similar to the feedback control philosophy (state-dependent control inputs). The theory of hybrid dynamic systems contains many important and "heavy" consequences of this state-dependent structure of the switching times $\tau = \{t_i\}$. For example, the formal proof of the main optimality tool in OCT of HSs, namely, the celebrated hybrid Pontryagin maximum principle

(HPMP), is technically more complex compared with the classic case. The same is also true with respect to possible generalizations of the usual techniques of the RT in the framework of HSs.

On a conceptual level the switching times τ in an HS are a part of the state output ("hybrid trajectory") $\mathcal{X}^u$ introduced in Definition 1.3. Consider now an HS from Definition 1.1. An admissible control function $u(\cdot) \in \mathcal{U}$ generates the corresponding complete hybrid trajectory $\mathcal{X}^u$. For every interval $[t_{i-1}, t_i]$, where $t_i \in \tau$, we next define the characteristic function

$$\beta_{[t_{i-1},t_i)}(t) = \begin{cases} 1 & \text{if } t \in [t_{i-1}, t_i), \\ 0 & \text{otherwise.} \end{cases}$$

Using these introduced characteristic functions, we can rewrite the differential equations from Definition 1.2 in the following compact form:

$$\dot{x}(t) = \sum_{i=1}^{r+1} \beta_{[t_{i-1},t_i)}(t) f_{q_i}(t, x(t), u(t)), \tag{1.3}$$

where $x(0) = x_0$. Under the basic technical assumptions for the family of vector fields F (see Definition 1.1), the right-hand side of the obtained differential equation (1.3) satisfies the conditions of the extended Carathéodory theorem. Therefore, there exists a unique (absolutely continuous) solution $x^u(\cdot)$ of (1.3). Taking into consideration the above observations we deduce that for an HS the characteristic functions introduced above also (significantly) depend on the state $x^u(\cdot)$, i.e.,

$$\beta_{[t_{i-1}(x^u(\cdot)),t_i(x^u(\cdot)))}(t).$$

Evidently, this dependence is highly nonlinear and all the usual first-order techniques from the classic control/systems theory are affected by this fact. For example, a simple linearization technique is being converted to a mathematically nontrivial procedure. Let us use the following compact notation:

$$\beta(\cdot) := (\beta_{[t_0,t_1)}(\cdot), \ldots \beta_{[t_r,t_{r+1})}(\cdot))^T.$$

Following the main Definition 1.1 we conclude that $\beta(\cdot)$ constitutes an additional "state" of the HS under consideration.

On the other side the SS from Definition 1.4 does not possess any additional "state." An SS can formally also be presented in the form (1.3). However, the characteristic function

$$\beta_{[t_{i-1},t_i)}(\cdot), \; i = 1, \ldots, r+1,$$

does not depend on the trajectory $x^u(\cdot)$ and needs to be interpreted as additional "systems input." Summarizing we can conclude that the conceptual difference between HSs and SSs consists in a formal determination of the corresponding switching mechanism. In (1.3) this switching mechanism (determined by the set of switching times τ) is formally described by the characteristic functions consolidated in the vector $\beta(\cdot)$. For HSs $\beta(\cdot)$ represents a posteriori information, namely, the system state, and in the case of the SSs this vector is an additional control input (a priori information). As mentioned in Section 1.1, HSs and SSs have a similarity from the point of view of the "visibility" of the formal representation (for example, by (1.3)) but have strong conceptual differences.

Recall that the classic RT for OCPs is usually used for two principal tasks. The first is establishment of the existence of an optimal solution to the given (usually sophisticated) OCP. This use of the RT is more known for the experts in optimization theory as well as for research control engineers, computer scientists, and researchers in mathematical economy. The second benefit which the conventional RT provides is related to the constructive computational schemes for OCPs. Note that this second application possibility of the RT is far less known for the experts and researchers from control engineering and practical optimization.

Questions RT can answer in the context of HSs/SSs and the corresponding OCPs are in fact the same. We will discuss in this book the generic existence questions for hybrid and switched OCPs as well as generalize the relaxation-based numerical approaches. This "knowledge transfer" from classic OCT to hybrid and switched cases cannot be considered as a simple formal "transfer." The conceptually new dynamic aspects of HSs and SSs in comparison to the conventional ODEs involving control systems imply some mathematical challenges and the necessity of additional theoretical development and effort for a successful knowledge transfer mentioned above.

1.3 A Short Historical Remark

It is common knowledge that an OCP does not always have a solution (see, e.g., [159,321, 154]). On the other hand, the corresponding relaxed problem has, under mild assumptions, an optimal solution [127,167,279]. This solution can be considered, in practice, for constructing an approximating solution for the initial problem. In the absence of the so-called "relaxation gap" (see, e.g., [246,247,121]) the relaxed problem is of primary interest for the initial OCP. In this case the minimal value of the objective functional in the initial OCP coincides with the minimum of the objective functional in the relaxed problem. Therefore, in this situation a solution of the relaxed problem can be used as a basis for constructing a minimizing sequence for the initial problem [302,154].

Extensions in problems of variational calculus, beginning with the idea of Hilbert, were realized many times for various purposes. Let us consider the following abstract regular variational problem:

$$\int_{\Upsilon} \mathcal{L}(s, \omega(s), \dot{\omega}(s))ds \rightarrow \min,$$
$$\omega = \varsigma \text{ on } \partial\Upsilon.$$

In the 20th problem of his Paris lecture, Hilbert asked the following question: "Is it true that the presented variational problem has a solution $\omega(\cdot)$ in case we generalize the notion of solution in an appropriate sense?" Today we know the answer is positive. Under some natural assumption (see, e.g., [328]) for a bounded region

$$\Upsilon \in \mathbb{R}^n$$

and for the smooth functions $\mathcal{L}(\cdot, \cdot, \cdot)$, $\varsigma(\cdot)$ there exists a solution $\omega(\cdot)$ from the Sobolev space

$$\mathbb{W}^1_l(\Upsilon), \ 1 < l < \infty.$$

The first constructive investigation on this subject was a work of N.N. Bogoljubov [94]. The concept of relaxed controls was introduced by L.C. Young in 1937 under the name of generalized curves and surfaces [326]. It has been used extensively in the literature for the study of diverse OCPs [321,127,173,192,154,279]. As Clarke points out in [130], a relaxed problem is, in general, the only one for which existence theorems can be proved and, for this reason, there are many who deem it the only reasonable problem to consider in practice. However, though relaxation of a problem is important in order to prove existence theorems, one is also interested in proving convergence of numerical approximation algorithms.

Relaxations in the theory of HSs and SSs and the corresponding OCPs are studied in [81] and [320]. Let us note that these works constitute initial studies in RT for HOCPs/SOCPs. The first paper does not give a self-closed theoretical methodology related to applications of relaxations for HSs and SSs. The second work is mostly dedicated to particular algorithmic aspects of the RT in the context of HSs/SSs. The critical methodological analysis of these two basic publications dedicated to applications of the RT in hybrid/switched dynamic models implies the necessity of a systematic compendium devoted to this topic.

1.4 Outline of the Book

In this section we briefly discuss the breakdown of the book into chapters. The book consists of nine chapters.

The mathematical systems theory and the RT in OCT are based on the solid formal fundament of the modern functional analysis and general topology. Therefore Chapter 2 constitutes in fact a necessary collection of the analytic and geometric (topological) facts and results. It also contains a short introduction into the convex analysis and elements of the approximation theory.

Using some facts presented in Chapter 2, we next give a compact overview of the main results of the convex optimization. Recall that many relaxation schemes of the OCPs associated with the systems described by ODEs finally lead to a so called "convexified" generalization of an initial OCP. This fact motivates the presence of Chapter 3 "convex programming." The existence theorems from a convex minimization problem in a real Hilbert space are in fact "prototypes" for the future existence results in the case of the relaxed HOCPs. The same is true with respect to the corresponding first-order optimality conditions. Chapter 3 also contains some related questions, namely, on well-posedness in optimal control, and a short discussion of the numerical approaches in convex programming.

Chapter 4 contains some necessary results from the theory of continuous time control systems. We follow the approach based on the differential inclusions technique. In this chapter we try to "minimize" a sophisticated mathematical formalism that is naturally included into the theory of differential inclusions. The classic fundamental material is complemented by an important applied topic related to the constructive approximations of the general differential inclusions. These approximations are in fact the expected "bridges" between the highly theoretical part of the established RT and possible engineering applications. In our opinion concrete engineering applications of an abstract theory or sophisticated mathematical model are usually implemented by the way of adequate numerical approximations.

Since the RT for HOCPs and SOCPs is in fact a natural extension of the classical relaxation schemes, Chapter 5 deals with some known generalization approaches from the optimal control processes governed by ODEs. We consider various classic relaxation methods as well as some approximative approaches to the convex relaxation in optimal control. The question here is the resulting "relaxation gap" of a concrete generalization scheme. This specific "gap" describes an adequateness of the selected relaxation in the context of the initially given (non-relaxed) OCP. Let us note that a concrete quantitative characterization of the relaxation gap is in general a very sophisticated theoretical question. However, for some particular cases one can compare the corresponding relaxation gaps qualitatively. For example, the Rubio relaxation discussed in Chapter 5 is broader than the Gamkrelidze–Tikhomirov relaxation.

Chapter 6 represents some types of HOCPs and SOCPs. We discuss the existence questions and we are mostly interested in conceptual computational methodologies that can be applied to these nontrivial dynamic optimization problems. An eventual RT-based effective numerical treatment of the relaxed HOCPs and SOCPs is the main aim of our consideration.

Chapter 7 is an immediate logical continuation of Chapter 5 and Chapter 6. It studies the application of some relaxation schemes from the classic OCT to HOCPs and SOCPs. In particular we consider the "hybrid" version of the celebrated Gamkrelidze–Tikhomirov generalization. Moreover, we also apply the approximative β-relaxations to the HOCPs and SOCPs. The next novel concept we propose is the so-called "weak relaxation." We study this new relaxation idea in the context of regular HOCPs and, moreover, apply it to the singular OCPs of hybrid and switched nature. Chapter 7 also contains some critical remarks in connection with the newly developed Bengea–DeCarlo relaxation approach.

Chapter 8 of our book is devoted to the necessary numerical part of the extended RT. We propose some conceptual numerical approaches to the HOCPs and SOCPs under consideration. In fact, this chapter constitutes an introduction to the numerical analysis of the obtained relaxation schemes. The conceptual algorithms we propose can be easily extended to the concrete numerical solution procedures and implementable algorithms. We close this chapter by considering some illustrative examples.

Chapter 9 summarizes our book.

The presented book constitutes in fact an initial work dedicated to the relaxation techniques in hybrid and switched optimal control. We hope that the resulting theoretical and computational schemes we obtain will be complimented in the future by real-world examples and engineering applications.

1.5 Notes

We now indicate some useful references for this chapter. The classic theory of optimal control can be found in [5,17,28,29,37,66,78,86,92,93,96,127,129–131,154,161,167,173,192,236,269, 285,286,300,302,305,321,277,326,332]. Classic and modern applications of this theory are discussed in [90,110,166,177,206,229,230,258,267,284,295]. We also refer to some results related to the HSs and SSs in [19,33–35,38–41,43,44,46,47,125,150,151]. Some interesting applications of various types of HSs and SSs can be found in [9,48,51,59,60,62,69,97–100, 103–105,124,152,174,231,232,296,306,311,316,323–325,333].

CHAPTER 2

Mathematical Background

2.1 Necessary Results and Facts From Topology and Functional Analysis

In this book we consider only real functional (normed vector) spaces. This selection is in fact motivated by the concrete applications to the real-world dynamic systems.

Definition 2.1. *A set W of elements, called vectors, is said to be a* real vector space *if*

a) *the operations of addition "$+$", and multiplication by areal number, denoted by "$\cdot$" (and usually omitted), are defined on W;*
b) *with respect to addition the following hold:*
  (b.1) *for any $x, y \in W$, the sum $x+y$ is a unique element in W,*
  (b.2) *for any $x, y \in W$, $x+y=y+x$,*
  (b.3) *for any $x, y \in W$,*

$$(x+y)+z=x+(y+z),$$

  (b.4) *there is a zero element $0 \in W$ such that, for all $x \in W$,*

$$x+0=x,$$

  (b.5) *for every $x \in W$, there exists an element $-x \in W$ such that*

$$x+(-x)=0;$$

c) *with respect to multiplication by a real number, the following hold:*
  (c.1) *for any $\alpha \in \mathbb{R}$ and $x \in W$, αx is a unique element of W,*
  (c.2) *for any $\alpha, \beta \in \mathbb{R}$ and $x \in W$,*

$$(\alpha\beta)x=\alpha(\beta x),$$

  (c.3) *for any $\alpha \in \mathbb{R}$ and $x, y \in W$,*

$$\alpha(x+y)=\alpha x+\alpha y,$$

  (c.4) *for any $\alpha, \beta \in \mathbb{R}$ and $x \in W$,*

$$(\alpha+\beta)x=\alpha x+\beta x,$$

  (c.5) *for any $x \in W$, $1 \cdot x=x$.*

A Relaxation-Based Approach to Optimal Control of Hybrid and Switched Systems
https://doi.org/10.1016/B978-0-12-814788-7.00008-4

Definition 2.2. *Given a real vector space W, we say that a function $x \to ||x||_W$, from W into $\mathbb{R}$, is a norm on W, if*

a) for all $x \in W$, $||x||_W \geq 0$ and $||x||_W = 0$ iff $x = 0$;
b) for any $\alpha \in \mathbb{R}$ and any $x \in W$,

$$||\alpha x||_W = |\alpha| ||x||_W;$$

c) for any $x, y \in W$,

$$||x + y||_W \leq ||x||_W + ||y||_W.$$

A real vector space W equipped with a norm $||\cdot||_W$ is called a real normed space and denoted $\{W, ||\cdot||_W\}$ (or simply W when no confusion should arise).

The fundamental concepts of the real inner product (pre-Hilbert) space, Banach space, and Hilbert space are of great importance in applied functional analysis and, in particular, in relaxation theory. Basic analytic properties of the Banach spaces are extremely useful in theoretical engineering and applied science.

Definition 2.3. *Given a real vector space W, we will say that a bilinear (i.e., linear in each argument separately) function $(x, y) \to \langle x, y\rangle_W$ from $W \times W$ into $\mathbb{R}$ is an* inner product *on W, if*

a) for any $x, y \in W$,

$$\langle x, y\rangle_W = \langle y, x\rangle_W;$$

b) for any $x, y, z \in W$,

$$\langle x + y, z\rangle_W = \langle x + z\rangle_W + \langle y, z\rangle_W;$$

c) for any $x, y \in W$ and any $\alpha \in \mathbb{R}$,

$$\langle \alpha x, y\rangle_W = \alpha\langle x, y\rangle_W;$$

d) for any $x \in W$, $\langle x, x\rangle \geq 0$ and $\langle x, x\rangle_W = 0$ iff $x = 0_W$, the zero element in W.

Theorem 2.1. *Suppose that W is a real vector space and that $\langle\cdot, \cdot\rangle_W$ is an inner product on W. Then, for all $x, y \in W$, the* Schwartz inequality *holds, i.e.,*

$$|\langle x, y\rangle_W|^2 \leq \langle x, x\rangle_W \langle y, y\rangle_W.$$

The function $x \to ||x||_W$, from W into the reals defined by

$$||x||_W := \sqrt{\langle x, x\rangle},$$

is a norm on W.

The basic types of convergence play a decisive role in system theory and, in particular, are very useful for various relaxation schemes in optimal control.

Definition 2.4. *Let $\{W, ||\cdot||_W\}$ be a real normed space. A sequence $\{x_i\}$, $i \in \mathbb{N}$, in $\{W, ||\cdot||_W\}$ is the image of $\mathbb{N}$ in $\{W, ||\cdot||_W\}$ under a mapping from $\mathbb{N}$ into $\{W, ||\cdot||_W\}$. A sequence $\{x_i\}$* converges *to a point x^*, which will be indicated by writing*

$$x_i \to x^*,$$

as $i \to \infty$, if

$$\lim_{i\to\infty} ||x - x^*|| = 0.$$

The point x^* is called the limit point of $\{x_i\}$. A sequence $\{x_i\}$ is said to be Cauchy sequence *if for any $\delta > 0$ there exists an i_δ such that, if $\imath > i_\delta$ and $j > i_\delta$,*

$$||x_i - x_j||_W < \delta.$$

We call a real normed space $\{W, ||\cdot||_W\}$ a real Banach space if every Cauchy sequence in $\{W, ||\cdot||_W\}$ converges to a point in $\{W, ||\cdot||_W\}$. If $\{W, ||\cdot||_W\}$ is a real Banach space with an inner product $\langle\cdot,\cdot\rangle_W$ such that

$$||x||_W := \sqrt{\langle x, x\rangle_W},$$

then $\{W, ||\cdot||_W\}$ is a real Hilbert space.

Let $\{W, ||\cdot||_W\}$ be a real Banach space. A functional $|\cdot| : H \to \mathbb{R}$ is called *seminorm* if it has all properties of a norm except for $|x| > 0$ for all $x \neq 0$. A subset of a vector space is called a *vector subspace* if it is a vector space in its own right under the induced operations. The *span* of a subset $E \subset W$ is the smallest vector subspace including it. A function (operator)

$$\Psi : W_1 \to W_2$$

between two vector spaces is linear if it satisfies

$$\Psi(\alpha x + \beta y) = \alpha\Psi(x) + \beta\Psi(y)$$

for every $x, y \in W_1$ and $\alpha, \beta \in \mathbb{R}$. Linear functions between two vector spaces are usually called linear operators.

Definition 2.5. *A Banach space $\{X, ||\cdot||_X\}$ is* strictly convex *if for two elements $x_1, x_2 \in X$ which are linear independent,*

$$||x_1 + x_2||_X < ||x_1||_X + ||x_2||_X.$$

The following references are very useful for a detailed study of the convex and strictly convex Banach spaces: [211,297].

We next recall some basic topological concepts and facts.

Definition 2.6. *A* topology τ *on a set* Ξ *is a collection of subsets of* Ξ *satisfying*

a) $\emptyset,\ \Xi \in \tau$*;*
b) τ *is closed under finite intersections;*
c) τ *is closed under arbitrary intersection.*

A nonempty set Ξ *equipped with a topology* τ *is called a* topological space *and is denoted* $\{\Xi, \tau\}$ *(or simply* Ξ *when no confusion should arise). We call a member of* τ *an open set in* Ξ*. The complement of an open set is a closed set.*

An open cover of a set K is a collection of open sets whose union includes K. A subset K of a topological space is *compact* if every open cover of K includes a finite subcover. A topological space is called a *compact space* if it is a compact set. A subset of a topological space is called *relatively compact* if its closure is compact. Note that $\emptyset$ and Ξ are open and closed. A *semimetric* $\rho(\cdot,\cdot)$ on a space Ξ is a real-valued function on $\Xi \times \Xi$ that is nonnegative and symmetric, satisfies

$$\rho(\xi, \xi) = 0$$

for every $\xi \in \Xi$, and in addition satisfies the triangle inequality,

$$\rho(\xi_1, \xi_3) \leq \rho(\xi_1, \xi_2) + \rho(\xi_2, \xi_3).$$

Recall that a *metric* is a semimetric (determined above) with the additional property: $\rho(\xi_1, \xi_2) = 0$ implies $\xi_1 = \xi_2$. A pair $\{\Xi, \rho(\cdot,\cdot)\}$, where $\rho(\cdot,\cdot)$ is a metric on Ξ, is called a *metric space*. A topological space Ξ is *metrizable* if there exists a metric $\rho(\cdot,\cdot)$ on Ξ generating the topology on Ξ. In the case of a normed space $\{W, ||\cdot||_W\}$, the *norm topology* is the metrizable topology generated by the metric

$$\rho(x, y) := ||x - y||_W.$$

A topology τ on a vector space W is called a *linear topology* if the vector operations addition and (scalar) multiplication are τ-continuous. The pair $\{W, \tau\}$ is called a *topological vector space*.

Definition 2.7. *Let* $\Psi : W_1 \to W_2$ *be an operator between two topological vector spaces. By* $R(\Psi) \subseteq W_2$ *and* $D(\Psi) \subseteq W_1$ *we denote the rank and domain of definition of* Ψ*, respectively.*

We say that the operator Ψ is invertible, *if for every $y \in R(\Psi)$ the equation $\Psi(x) = y$ has a unique solution. The operator*

$$\Psi^{-1} : R(\Psi) \to D(\Psi), \ \Psi^{-1}(y) = x$$

is an inverse operator.

Let Ψ be an invertible linear operator. Then Ψ^{-1} is also a linear operator.

Definition 2.8. *A vector space W is an* algebra *if an additional operation of multiplication is defined on W such that*

a) $(xy)z = x(yz)$;
b) $x(y+z) = xy + xz$, $(y+z)x = yx + zx$;
c) $\alpha(xy) = (\alpha x)y = x(\alpha y)$, $\alpha \in \mathbb{R}$,

for all $x, y \in W$. If there exists an element $e \in W$ such that

$$ex = xe = x$$

for all $x \in W$, then W is an algebra with unit. If $xy = yx$ for all $x, y \in W$, then W is a commutative algebra. In the case of a Banach *space W, the algebra W is called* Banach algebra.

Definition 2.9. *An element x of an algebra W is* invertible *if there exists an element $x^{-1} \in W$ such that*

$$xx^{-1} = x^{-1}x = e.$$

Otherwise, x is a noninvertible *element of W.*

Theorem 2.2. *The set of all invertible elements of a Banach algebra is open in the norm topology. The set of all noninvertible elements of a* Banach *algebra is closed.*

We refer to [208,209] for further theoretical details.

The very important concept of a continuous function is presented below.

Definition 2.10. *A function $\Upsilon : \Xi_1 \to \Xi_2$ between two topological spaces is* continuous *if $\Upsilon^{-1}(U)$ is open for each open set U. We say that Υ is* continuous at the point ξ, *if $\Upsilon^{-1}(U)$ is a neighborhood of ξ whenever U is a neighborhood of $\Upsilon(\xi)$.*

Note that in a metric space, continuity at a point reduces to the usual ϵ/δ definition. Recall that a function

$$\Psi : W_1 \to W_2$$

between two normed spaces is *uniformly continuous* if for each $\epsilon > 0$ there exists some $\delta > 0$ (depending only on ϵ) such that

$$||x - y||_{W_1} < \delta$$

implies

$$||\Psi(x) - \Psi(y)||_{W_2} < \epsilon, \; x, y \in W_1.$$

It is common knowledge that the norm of a Banach space is a uniformly continuous function. The above function Ψ is called *Lipschitz continuous* if there is some real number $L > 0$ such that for every $x, y \in W_1$,

$$||\Psi(x) - \Psi(y)||_{W_2} \leq L||x - y||_{W_1}.$$

If for every point $x \in W_1$ there exists a neighborhood of x where Ψ is Lipschitz continuous, then we call Ψ *locally Lipschitz.*

Two topological spaces Ξ_1 and Ξ_2 are called *homeomorphic* if there is a one-to-one continuous function Υ from Ξ_1 onto Ξ_2 such that Υ^{-1} is continuous, too. The function Υ is called a *homeomorphism.* A mapping

$$\Upsilon : \Xi_1 \to \Xi_2$$

between two topological spaces is an *embedding* if $\Upsilon : \Xi_1 \to \Upsilon(\Xi_1)$ is a homeomorphism.

Definition 2.11. *A normed space W_1 is called* continuously embedded *into a normed space W_2 if $W_1 \subset W_2$ and*

$$||x||_{W_2} \leq C||x||_{W_1} \; \forall x \in W_1,$$

where C is a fixed constant.

Obviously, an operator which performs this embedding (the *embedding operator*) is linear and continuous.

The convergence theory in topological spaces constitutes the main analytic tool of the relaxation theory. We actively use this abstract concept in Chapters 5, 6, and 7. Assume

$$\{F_i : S \to \{\Xi_i, \tau_i\}\}, \; i \in I \subseteq \mathbb{N},$$

is a collection of functions from the nonempty set S into the topological spaces Ξ_i. The *weak topology* on S generated by the family of functions $\{F_i\}$ is the weakest topology on S that makes all the functions F_i continuous. It is the topology generated by the family of sets

$$\{F_i^{-1}(U) \; : i \in I \subset \mathbb{N}, \; U \in \tau_i\}.$$

A subset V of a topological space Ξ is *dense* if every nonempty open subset of Ξ contains a point in V. A topological space is *separable* if it includes a countable dense set.

Let $\{H, \langle\cdot,\cdot\rangle_H\}$ be a real Hilbert space. A sequence $\{h_i\}$, $i \in \mathbb{N}$, converges weakly to $h \in H$, if for every $y \in H$,

$$\lim_{i\to\infty} \langle h_i, y\rangle_H = \langle h, y\rangle_H = 0.$$

Theorem 2.3. *Let $\{H, \langle\cdot,\cdot\rangle_H\}$ be a separable Hilbert space. If*

$$\{h_i\},\ i \in \mathbb{N},$$

is a bounded sequence in H, then $\{h_i\}$ contains a subsequence that converges weakly.

A subset of a real Banach space $\Xi \subset Y$ is called *norm bounded* if there is a constant $C \in \mathbb{R}_+$ such that

$$||y||_Y \leq C$$

for all $y \in \Xi$. It is common knowledge that a closed in the weak topology and norm bounded subset of a normed space is compact in the weak topology *weakly compact set* (*weakly compact*).

Definition 2.12. *A topology τ on a topological space Ξ is called* Hausdorff (*or separated*) *if any two distinct points can be separated by disjoint neighborhoods of the points. That is, for each pair $\xi_1, \xi_2 \in \Xi$ with $\xi_1 \neq \xi_2$ there exist neighborhoods $\Gamma(\xi_1)$ and $\Gamma(\xi_2)$ such that*

$$\Gamma(\xi_1) \bigcap \Gamma(\xi_2) = \emptyset.$$

The space $\{\Xi, \tau\}$ is called a Hausdorff space.

Recall that the *graph*

$$\mathrm{Gr}_\Upsilon$$

of a function $\Upsilon : \Xi_1 \to \Xi_2$ is the set

$$\mathrm{Gr}_\Upsilon = \{(\xi_1, \xi_2) \in \Xi_1 \times \Xi_2 \ :\ \xi_2 = \Upsilon(\xi_1)\}.$$

Theorem 2.4 (Closed graph theorem). *A function from a topological space into a compact* Hausdorff space *is continuous if and only if its graph is closed.*

The terms mapping, function, and operator will be next used synonymously.

Definition 2.13. *The* operator norm *of a linear operator* $\Psi : W_1 \to W_2$ *between normed spaces is the nonnegative extended real number* $||\Psi||$ *defined by*

$$||\Psi|| := \sup_{||x||_{W_1} \leq 1} ||\Psi x||_{W_2} = \min\{M \geq 0 \,:\, ||\Psi x||_{W_2} \leq M||x||_{W_1} \; \forall x \in W_1\}.$$

If $||\Psi|| = \infty$*, we say that* Ψ *is a linear unbounded operator, while in the case*

$$||\Psi|| < \infty,$$

we say that Ψ *is a linear bounded operator.*

It is well known that a linear operator between topological vector spaces is continuous if and only if it is continuous in zero. Moreover, a linear operator between normed spaces is continuous if and only if it is bounded.

Let X, Y be real Banach spaces. The vector space $L(X, Y)$ of all bounded linear operators from X into Y is a Banach space (with the introduced operator norm). Moreover, $L(X, X)$ is a noncommutative Banach algebra with unit (a unit operator). The multiplication in $L(X, X)$ is defined as

$$\mathcal{C} := \mathcal{A}_1\mathcal{A}_2, \;\; \mathcal{C}x := \mathcal{A}_1(\mathcal{A}_2 x), \; \forall x \in X,$$

for

$$\mathcal{A}_1, \mathcal{A}_2 \in L(X, X).$$

Evidently, $\mathcal{C} \in L(X, X)$ and

$$||\mathcal{C}|| \leq ||\mathcal{A}_1|| \cdot ||\mathcal{A}_2||.$$

A family of operators

$$\mathcal{A}_i \in L(X, Y), \; i \in \mathbb{N},$$

is *pointwise bounded* if for each $x \in X$ there exists some $M_x > 0$ such that $||\mathcal{A}_i x||_Y \leq M_x$ for each $i \in \mathbb{N}$. Recall that an *open mapping* is one that carries open sets to open sets.

Theorem 2.5 (Banach open mapping theorem). *A bounded linear operator* $\mathcal{A}$ *from a* Banach *space onto another* Banach *space is an open mapping. Consequently, if it is also one-to-one, then there exists the bounded inverse operator* $\mathcal{A}^{-1}$.

Definition 2.14. *The space* W^* *of all continuous linear functionals on a topological vector space* W *is called the* topological dual *of* W*. A dual system is a pair* $\{W, W^*\}$ *of vector spaces together with a function*

$$(x, x^*) \to \langle x, x^* \rangle_{(W^*, W)}$$

(called pairing or duality of the pair), satisfying:

a) *the mapping*

$$x^* \to \langle x, x^* \rangle_{(W^*, W)}$$

is linear for each $x \in W$*;*

b) *the mapping* $x \to \langle x, x^* \rangle_{(W^*, W)}$ *is linear for each* $x^* \in W^*$*;*

c) *if*

$$\langle x, x^* \rangle_{(W^*, W)} = 0$$

for each $x^* \in W^*$*, then* $x = 0_W$*;*

d) *if*

$$\langle x, x^* \rangle_{(W^*, W)} = 0$$

for each $x \in W$*, then* $x^* = 0_{W^*}$*.*

The (norm) dual of a normed space $\{W, ||\cdot||_W\}$ *is the vector space* $\{W^*, ||\cdot||_{W^*}\}$ *consisting of all (norm) continuous linear functionals on* W*, equipped with the operator norm* $||\cdot||_{W^*}$*. The norm dual* W^{**} *of* W^* *is called the* second dual.

The following result and a reflexivity concept of Banach spaces will be used in Chapters 5, 6, and 7.

Theorem 2.6. *The norm dual of a normed space is a* Banach *space.*

Definition 2.15. A Banach *space* W *is called* reflexive *if*

$$W = W^{**}.$$

Note that W is reflexive if and only if W^* is reflexive. Moreover, the reflexivity of W is equivalent to the compactness of the closed unit ball

$$V_1(0) := \{x \in W \ : \ ||x||_W \leq 1\}$$

in the weak topology.

Theorem 2.7 (Riesz theorem)**.** *Let* H *be a real* Hilbert *space and* H^* *be the dual to* H*. For any functional* $l \in H^*$ *there exists a unique element* $\tilde{h} \in H$ *such that*

$$l(h) = \langle \tilde{h}, h \rangle_H \ \forall h \in H.$$

Let H be a Hilbert space. Consider a closed subspace K of H. The set

$$\{h \in H \ : \ \langle h, h_K \rangle_H = 0 \, \forall h_K \in K\}$$

is called an orthogonal complement of K and is denoted by $K^{\perp}$. For the concept of the *generalized orthogonal complement* consult [331].

We next suppose the knowledge of the differentiability aspects of functions in normed spaces and frequently use the Fréchet derivatives. Let

$$A : X \to Y$$

and $x_0 \in X$. If there is a continuous linear mapping $A'(x_0) : X \to Y$ with the property

$$\lim_{||\Delta||_X \to \infty} \frac{||A(x_0 + \Delta) - A(x_0) - A'(x_0)\Delta||_Y}{||\Delta||_X} = 0,$$

then $A'(x_0)$ is called the *Fréchet derivative* of A at x_0 and the operator A is called *Fréchet differentiable at* x_0. According to this definition we obtain for Fréchet derivatives

$$A(x_0 + \Delta) = A(x_0) + A'(x_0)\Delta + o(||\Delta||_X),$$

where the expression $o(||\Delta||_X)$ of this Taylor series has the property

$$\lim_{||\Delta||_X \to 0} \frac{o(||\Delta||_X)}{||\Delta||_X} = 0.$$

Theorem 2.8. *Let W be a real normed space. Consider a functional $J : W \to \mathbb{R}$. If $J(\cdot)$ is once continuous* Fréchet *differentiable, then for any $x, y \in W$,*

$$J(x) - J(y) = J_x(x + \lambda(y - x))(y - x)$$

for some $\lambda \in [0, 1]$, where $J_x(\cdot)$ is the Fréchet *derivative. If W is a real* Hilbert *space, then*

$$J(x) - J(y) = \langle J_x(x + \lambda(y - x)), (y - x) \rangle_H$$

for some $\lambda \in [0, 1]$.

Theorem 2.9 (Classic implicit function theorem)**.** *Let X, Y, Z be real Banach spaces, V be a neighborhood in $X \times Y$, and*

$$\Psi : V \to Z$$

be a continuously Fréchet *differentiable mapping. Assume that*

a) $\Psi(x_1, y_1) = 0$;

b) there exists an inverse operator

$$[\Psi'_y(x_1, y_1)]^{-1} \in L(Z, Y),$$

where Ψ'_y *is the Fréchet derivative of* $\Psi(x, \cdot)$.

Then there exist $\epsilon, \delta > 0$ *and there exists a continuously* Fréchet *differentiable mapping*

$$\psi : B(x_1, \delta) \to Y,$$

where

$$B(x_1, \delta) := \{x \in X \ : \ ||x - x_1|| < \delta\},$$

such that

a) $\psi(x_1) = y_1$;
b) *from* $||x - x_1||_X < \delta$ *follows*

$$||\psi(x) - y_1||_Y < \epsilon$$

and $\Psi(x, \psi(x)) = 0$;
c) *if*

$$(x, y) \in B(x_1, \delta) \times B(y_1, \epsilon),$$

where

$$B(y_1, \epsilon) := \{x \in X \ : \ ||x - x_1|| < \epsilon\},$$

then $\Psi(x, y) = 0$ *implies* $y = \psi(x)$;
d) *the mapping* $\psi(\cdot)$ *is Fréchet differentiable and*

$$\psi'_x(x) = -[\Psi'_y(x, \psi(x))]^{-1}\Psi'_x(x, \psi(x)),$$

where Ψ'_x *is the Fréchet derivative of* $\Psi(\cdot, y)$ *and* ψ'_x *is the Fréchet derivative of* $\psi(\cdot)$.

Definition 2.16. *A continuous operator* $A : W_1 \to W_2$ *between two normed spaces is a nonexpansive operator, if*

$$||A(x_1) - A(x_2)||_{W_2} \leq ||x_1 - x_2||_{W_1}$$

for all $x_1, x_2 \in W_1$. *This operator is called contractive if there exists a nonnegative number* $r > 0$ *with the property that*

$$||A(x_1) - A(x_2)||_{W_2} \leq r||x_1 - x_2||_{W_1}$$

for all $x_1, x_2 \in W_1$.

Theorem 2.10 (Banach fixed point theorem). *Let X be a Banach space and let A be a contraction of X into itself. Then A has a unique fixed point, in the sense that $A(x) = x$ for some $x \in X$.*

Let W be a vector space. A functional

$$J : W \to \bar{\mathbb{R}} := \mathbb{R} \bigcup \{+\infty\}$$

is convex if, for any $x_1, x_2 \in W$ and $\lambda \in [0, 1]$,

$$J(\lambda x_1 + (1-\lambda)x_2) \leq \lambda J(x_1) + (1-\lambda)J(x_2).$$

This functional is called *strictly convex* if

$$J(\lambda x_1 + (1-\lambda)x_2) < \lambda J(x_1) + (1-\lambda)J(x_2).$$

For J the set

$$\text{epi}J := \{(\alpha, x) \in \mathbb{R} \times W \ : \ \alpha \geq J(x)\}$$

is said to be an epigraph of $J(\cdot)$. Usually, for a convex functional

$$J(\cdot) : W \to \bar{\mathbb{R}},$$

$J(\cdot) \neq \infty$, the notation of a *proper convex* functional is used. The set

$$\text{dom}J := \{x \in W \ : \ J(x) < \infty\}$$

is called an *effective domain* of a functional $J(\cdot)$. Recall that a set of all interior points of a convex set S is called the *interior* of S and denoted by $\text{int}\{S\}$.

Definition 2.17. *A functional $J : W \to \bar{\mathbb{R}}$ is said to be* lower semicontinuous *if* epi$J$ *is a closed set.*

Theorem 2.11 (Mean value theorem). *Let H be a real* Hilbert *space and*

$$J : H \to \bar{\mathbb{R}}$$

is a lower semicontinuous proper functional. Suppose J is Gâteaux *differentiable on an open neighborhood that contains the line segment*

$$[h_1, h_2] := \{th_1 + (1-t)h_2 \ : \ 0 \leq t \leq 1\},$$

where $h_1, h_1 \in H$. Then there exists

$$h_3 \in [h_1, h_2]$$

such that

$$J(h_1) - J(h_2) = \langle J'_G(h_3), h_1 - h_2 \rangle_H,$$

where $J'_G(h_3)$ *is the Gâteaux derivative of* J *at the point* h_3.

Consider now a functional $J : X \to \bar{\mathbb{R}}$, where X is a real Banach space. Let $U \subseteq X$. A sequence

$$\{x_i\},\ i \in \mathbb{N},$$

in X is a *minimizing sequence* if

$$\lim_{i \to \infty} J(x_i) = \inf_{x \in U} J(x).$$

For some further facts from the convex analysis see Section 2.2. The following references may be useful for a detailed study of the basic facts: on the theory of *convergence spaces* (or *Fréchet spaces*) [282,6], and on the theory of *Sobolev spaces* [1,226].

We now recall some necessary definitions and facts related to the linear operators in a real Banach space.

A linear operator $\mathcal{A} : X \to Y$ between real Banach (or normed) spaces is called a *linear homeomorphism* if

$$\mathcal{A} : X \to R(\mathcal{A})$$

is a *homeomorphism*, or, equivalently, if there exist positive constants d and M such that

$$d||x||_X \leq ||\mathcal{A}x||_Y \leq M||x||_X$$

for each $x \in X$. This concept is a natural consequence of the Banach open mapping theorem. A linear operator $\mathcal{A} : X \to Y$ that satisfies

$$||\mathcal{A}x||_Y = ||x||_X$$

for all $x \in X$ is a *linear isometry* [6].

Definition 2.18. *Let* V *be an open unit ball of a real Banach space* X. *An operator* $\mathcal{A} \in L(X, Y)$ *is called* compact operator, *if the set* $\mathcal{A}(V)$ *is relatively compact* (*i.e., the closure of the set* $\mathcal{A}(V)$ *is compact*).

Let us now examine the case $X = Y = H$ (a real Hilbert space). Consider a linear continuous operator $\mathcal{B} : H \to H$, where H is a Hilbert space (not necessarily real). By I we denote the identity operator on H. If we deal with a compact operator B, then we can formulate the following very useful *Fredholm alternative* (see, e.g., [272])

Theorem 2.12. *Let $\mathcal{B} \in L(H, H)$ be a compact operator. Then there exists the inverse operator $(I - \mathcal{B})^{-1}$, or, alternatively, the equation*

$$\mathcal{B}h = h, \ h \in H,$$

has a solution.

2.2 Elements of Convex Analysis and Approximation Theory

Let X be a real reflexive Banach space and $Q \subseteq X$ be a nonempty closed convex set. Assume

$$J : X \to (-\infty, \infty]$$

is a proper convex and lower semicontinuous functional and consider the abstract convex optimization problem

$$\begin{aligned} &\text{minimize } J(x) \\ &\text{subject to } x \in Q. \end{aligned} \tag{2.1}$$

Recall that a convex functional is proper if it never assumes the value $-\infty$ and its effective domain

$$\text{dom}\psi := \{z \in Z \ : \psi(z) < \infty\}$$

is nonempty. We next omit the word "proper," because only such convex functionals will be considered in this book. In addition, we suppose that $J(\cdot)$ is bounded on

$$Q + \epsilon B.$$

Here B is an open unit ball of X and $\epsilon > 0$. Clearly,

$$Q + \epsilon B \subset \text{int dom}\{J(\cdot)\}.$$

Since $J(\cdot)$ is bounded on $Q + \epsilon B$, it follows that this functional is also continuous on $Q + \epsilon B$ (see, e.g., [192]). Note that $J(\cdot)$ is also Lipschitz continuous on Q (see [275], Theorem 10.4). Let us also note that a convex function on a convex subset of an infinite-dimensional topological vector space does not need to be continuous on the interior of its domain. For instance, any discontinuous linear functional on an infinite-dimensional topological vector space provides such an example.

Let W be a vector space and $Q \subset W$ be a convex set. Recall that a function

$$f : Q \to (-\infty, \infty]$$

is convex on W if for any $w_1, w_2 \in W$ and $\alpha \in [0, 1]$,

$$f(\alpha w_1 + (1-\alpha)w_2) \leq \alpha f(w_1) + (1-\alpha) f(w_2).$$

In the case of a strong inequality

$$f(\alpha w_1 + (1-\alpha)w_2) < \alpha f(w_1) + (1-\alpha) f(w_2)$$

the function $f(\cdot)$ is called *strictly convex*. We now consider some useful properties of convex functions.

Theorem 2.13. *Let Q be a convex subset of a vector space W and let*

$$f : Q \to (-\infty, \infty]$$

be a convex function on W. Then, for any $c \in \mathbb{R}$,

$$G_c := \{w \in W \ : \ f(x) \leq c\}$$

is convex.

Theorem 2.14. *Let Q be a convex subset of a vector space W, let f, g be convex functions on W, and let $c \in \mathbb{R}_+$. Then two functions $f + g$ and cf defined by*

$$(f+g)(w) = f(w) + g(w), \ (cf)(w) = cf(w)$$

for all $w \in W$ are convex on W.

Definition 2.19. *We say that a functional $g : \Gamma \subset \mathbb{R}^n \to \mathbb{R}$ is* monotonically nondecreasing *if*

$$g(\xi) \geq g(\zeta)$$

for all $\xi, \zeta \in \Gamma$ such that

$$\xi_k \geq \zeta_k, \ k = 1, ..., n.$$

The presented monotonicity concept can be expressed by introducing a *positive cone* $\mathbb{R}^n_{\geq 0}$ (the positive orthant). For a similar monotonicity concept see, for example, [180]. A general example of a monotonically nondecreasing functional can be deduced from the mean value theorem, namely, a differentiable functional

$$g : \mathbb{R}^n \to \mathbb{R}$$

with

$$\partial g(\xi)/\partial \xi_k \geq 0$$

for all $\xi \in \mathbb{R}^n$, $k = 1, ..., n$, where

$$\partial g(\xi)\partial \xi_k$$

is the kth component of the gradient of g, is monotonically nondecreasing.

It is a well-known fact that the composition of two convex functionals is not necessarily convex. In the following we will need two basic results providing conditions that ensure convexity of the composition (see, e.g., [275]).

Theorem 2.15. *Let $g^1 : \mathbb{R}^n \to \mathbb{R}$ be a monotonically nondecreasing, convex functional. Assume that for every $k = 1, ..., n$ the functionals*

$$g_k^2 : \mathbb{R}^m \to \mathbb{R}$$

are convex. Then the functional

$$g : \mathbb{R}^m \to \mathbb{R}, \ g(\cdot) := g^1(g^2(\cdot)),$$

where

$$g^2(\xi) := (g_1^2(\xi), ..., g_n^2(\xi))^T, \ \xi \in \mathbb{R}^m,$$

is convex.

Theorem 2.16. *Let $g^1 : \mathbb{R}^n \to \mathbb{R}$ be a convex functional and $g^2 : H \to \mathbb{R}^n$ be a linear function on a real Hilbert space H. Then the functional*

$$g : H \to \mathbb{R}, \ g(\cdot) := g^1(g^2(\cdot))$$

is convex.

For the generic optimization problem (2.1) we formulate the following fundamental existence result [153].

Theorem 2.17. *Let the set Q be bounded. Then there exists a solution x^{opt} of* (2.1). *If J is strictly convex, then* (2.1) *has a unique solution.*

Note that the boundedness condition in Theorem 2.17 can be replaced by the condition J to be *coercive*, i.e.,

$$\lim_{||x||_X \to \infty} J(x) = \infty$$

for $x \in Q$. Moreover, the solution set of (2.1) is a closed convex set [153].

Suppose X, Y are real Banach spaces and let $W \subset X$ be an open set. A function $f : X \to Y$ is said to be directionally differentiable at a point $w \in W$ in the direction $x \in X$ if there exists the following limit:

$$D(w, x) := \lim_{h \downarrow 0} \frac{f(w + hx) - f(w)}{h}.$$

In that case $D(w, x)$ is called a *directional derivative* of function $f(\cdot)$ in the point $w \in W$ in the direction $x \in X$. Assume that for every $x \in X$ there exists a directional derivative $D(w, x)$. The mapping

$$\delta f(w, \cdot) : X \to Y,$$

where $\delta f(w, x) := D(w, x)$, is called the *first variation* of function $f(\cdot)$ at the point $w \in W$. We sometimes say that function $f(\cdot)$ possesses a first variation. Note that a symmetry case, namely, the condition

$$\delta f(w, -x) = -\delta f(w, x)$$

for all $x \in X$, is equivalent to the existence of the mutual limit

$$D(w, x) := \lim_{h \to 0} \frac{f(w + hx) - f(w)}{h}.$$

In that specific case the map $x \to D(w, x)$ is called *Lagrange variation*. Finally, if a Lagrange variation is a linear operator

$$D(w, x) = \Lambda x, \ \forall x \in X,$$

then $D(w, x)$ constitutes the celebrated *Gâteaux derivative* and the function under consideration is called *Gâteaux differentiable*.

Theorem 2.18. *Let in addition to the assumptions of* Theorem 2.17 *the objective functional J be Gâteaux differentiable. The element $x^{opt} \in Q$ is a solution of* (2.1) *if and only if*

$$\langle \nabla_G J(x^{opt}), x - x^{opt} \rangle_{(X^*, X)} \geq 0 \ \forall x \in X,$$

where $\nabla_G J(x^{opt})$ denotes the Gâteaux *derivative of J at x^{opt}.*

One of the central results of the general optimization theory describes the first-order necessary optimality conditions for the smooth finite-dimensional version of the following nonlinear problem:

$$\begin{aligned} &\text{minimize } J(x) \\ &\text{subject to } x \in S \\ &g_i(x) \leq 0, \ i = 1, ..., m \ , \ h(x) = 0, \end{aligned} \tag{2.2}$$

where $S \subseteq \mathbb{R}^n$ is an open set, $J : \mathbb{R}^n \to \mathbb{R}$ is Fréchet differentiable, $g_i : \mathbb{R}^n \to \mathbb{R}$ for all $i = 1, ..., m$, and

$$h : \mathbb{R}^n \to \mathbb{R}^r.$$

Let $x^{opt} \in \mathbb{R}^n$ be a local minimizer of (2.2). We denote the set of indices of the *active* inequality constraints by

$$I(x^{opt}) := \{i \ : \ g_i(x^{opt}) = 0\}.$$

Let the given functions g_i, $i = 1, ..., m$ be continuous and Fréchet differentiable at the point x^{opt}. The function h is assumed to be strictly differentiable, with the surjective gradient, at x^{opt}. Suppose that there is a direction $p \in \mathbb{R}^n$ satisfying

$$\langle \nabla g_i(x^{opt}), p\rangle < 0 \, \forall i \in I(x^{opt}). \tag{2.3}$$

The introduced conditions (2.3) are the *Mangasarian–Fromovitz regularity conditions* (constraint qualifications).

Theorem 2.19 (Karush–Kuhn–Tucker). *If the above assumptions and the above Mangasarian–Fromovitz constraint qualifications hold, then there exist the Lagrange multipliers*

$$\lambda_i \in \mathbb{R}_+$$

(for $i \in I(x^{opt})$) and $\mu \in \mathbb{R}^n$ satisfying

$$\nabla J(x^{opt}) + \sum_{i \in I(x^{opt})} \lambda_i \nabla g_i(x^{opt}) + \langle \nabla h(x^{opt}), \mu\rangle = 0.$$

Next we present the celebrated Ekeland variational principle in Banach spaces. Note that the corresponding result can also be proved for complete metric spaces [297].

Theorem 2.20 (Ekeland variational principle). *Let X be a real* Banach *space and let*

$$J : X \to (-\infty, \infty]$$

be a proper, bounded below, and lower semicontinuous functional. Then, for any $\epsilon > 0$ and $\tilde{x} \in X$ with

$$J(\tilde{x}) \leq \inf_{x \in X} J(x) + \epsilon,$$

there exists $z \in X$ satisfying the following conditions:

$$J(z) \leq J(\tilde{x}), \ \ ||\tilde{x} - z||_X \leq 1,$$
$$J(\xi) > J(z) - \epsilon||z - \xi||_X \, \forall \xi \in X, \ \xi \neq z.$$

Let X be a real Banach space and $J : X \to (-\infty, \infty]$ be a functional on X. An element $u^* \in X^*$ is called a *subgradient* of J at $u \in X$ if $J(u) \neq \pm\infty$ and

$$J(x) \geq J(u) + \langle u^*, x - u \rangle_{(X^*,X)} \ \forall x \in X.$$

The set of all subgradients of J at $u \in X$ is called a subdifferential $\partial_S J(x)$ of J at $x \in X$. For a proper convex functional J with $J(x) < \infty$ we have

$$\partial_S J(x) \neq \emptyset$$

for all $x \in X$ and the set $\partial_S J(x)$ is convex. The definition of the generalized Jacobian was introduced in [129]. We also refer to [129,130,137,277] for analytical details and other concepts from the nonsmooth analysis. Using the generalized Jacobian of a mapping

$$G : D \subseteq \mathbb{R}^n \to \mathbb{R},$$

we formulate the next well-known and useful result.

Theorem 2.21 (Inverse function theorem). *Let $D \subseteq \mathbb{R}^n$ be an open set,*

$$G : D \to \mathbb{R}^n$$

let be a locally Lipschitz *mapping, and let $x_0 \in D$. Assume that all linear operators from the generalized Jacobian*

$$\partial G(x_0)$$

are invertible. Put $G(x_0) = y_0$. Then there exist

$$\eta > 0, \ \delta > 0,$$

and a Lipschitz *mapping G^{-1} defined on an open ball*

$$\text{int}\{V_\eta(y_0)\}$$

such that

$$G(G^{-1}(y)) = y \ \forall y \in \text{int}\{V_\eta(y_0)\},$$
$$G^{-1}(G(x)) = x \ \forall x \in \text{int}\{V_\delta(x_0)\}.$$

Let us now discuss a fundamental principle of the linear functional analysis and its two consequences.

Theorem 2.22 (Hahn–Banach theorem). *Let $\mathcal{M}$ be a subspace of a normed space W, and let Λ_0 be a continuous linear functional on $\mathcal{M}$. Then Λ_0 can be extended to a continuous linear functional Λ_1 defined on the whole space W such that*

$$||\Lambda_0||_W = ||\Lambda_1||_W.$$

Theorem 2.23 (Separation theorem). *Let W be a normed space and let C be a nonempty closed convex subset of W. If x is a vector not in C, then there exists a continuous linear functional $\Lambda \in W^*$ such that*

$$\Lambda(x) < \inf_{y \in W} \Lambda(y).$$

Theorem 2.24. *Let X be a real* Banach *space and let*

$$J : X \to (-\infty, \infty]$$

be a proper convex and lower semicontinuous functional. Then there are $e \in \mathbb{R}$ and $x^* \in X^*$ such that for all $x \in X$

$$J(x) \geq e + \langle x^*, x \rangle_{(X^*, X)}.$$

If $S \subseteq X$ is a closed convex set such that $J : S \to \mathbb{R}$, then J is continuous on the interior $\text{int}\{S\}$.

The celebrated *Liusternik theorem* is a basic of some proofs presented in this book. In connection with this we formulate two classical results and refer to [107] for some generalizations of Liusternik's result.

Theorem 2.25 (Liusternik theorem). *Let X, Y be real* Banach *spaces and let*

$$A : X \to Y$$

be a strictly Fréchet differentiable operator at $x_0 \in X$. Assume that

$$M := \{x \in X \ : \ A(x) = 0\} \neq \emptyset.$$

Then there exists a tangent space to the set M at the point $x_0 \in M$. If the operator $A'(x_0)$ is surjective, then

$$T_z(M) = T_z^+(M) = \text{Ker}(A'(z))$$

for all $z \in M$, where

$$\text{Ker}(A'(z)) := \{x \in X \ : \ A'(z)x = 0\}$$

is the kernel of $A'(z)$.

Theorem 2.26 (Graves–Liusternik theorem). *Let H_1, H_2 be real* Hilbert *spaces, W be a metric space, $U \subseteq H_1$, and*

$$F : H_1 \times W \to H_2$$

be a partial Fréchet *differentiable mapping. Let $(h_0, \omega_0) \in U \times W$ be a point satisfying*

$$F(h_0, \omega_0) = 0,$$

and suppose that $F_h'(h_0, \omega_0)$ is onto:

$$R(F_h'(h_0, \omega_0)) = H_2.$$

Let Ω be a neighborhood of h_0. Then for some $\delta > 0$, for all (h, ω) sufficiently near (h_0, ω_0), we have

$$\text{dist}(\Phi(\omega), h) \leq \frac{||F(h, \omega)||_{H_2}}{\delta},$$

where

$$\Phi(\omega) := \{z \in \Omega \ : \ F(z, \omega) = 0\}.$$

Let X be a real Banach space and $S \subset X$ be closed and convex. We consider the following *variational inequality:*

$$\langle f(\omega, \tilde{x}), x - \tilde{x} \rangle_{(X^*, X)} \geq 0 \ \forall x \in S, \tag{2.4}$$

where $f : W \times X \to X^*$ and W is a normed space (a parameter set). Let us introduce the *normal cone* to S at $x \in X$, i.e.,

$$N_S(x) := \begin{cases} \{v \in X \ : \ \langle v, x - \tilde{x} \rangle_{(X^*, X)} \leq 0 \ \forall x \in S\}, & \text{if } \tilde{x} \in S, \\ \emptyset, & \text{otherwise.} \end{cases}$$

We next assume that the given variational inequality (2.4) has a solution $x_1 \in X$ for a fixed parameter $\omega_1 \in W$. Let $f(\cdot, x_1)$ and $f_x'(\cdot, x_1)$ be continuous at ω_1 and $f(\omega, \cdot)$ be Fréchet differentiable. Let $f_x'(\omega, \cdot)$ be continuous uniformly in $\omega \in W$. The next result is of importance in explaining the regularity conditions in the general optimization theory.

Theorem 2.27 (Robinson implicit function theorem). *Let the conditions given above hold. Assume that there exist neighborhoods Ξ of a vector $x_1 \in X$, and Θ of the origin 0_X such that the mapping*

$$x \to [f(\omega_1, x_1) + f_x'(\omega_1, x_1)(\cdot - x_1) + N_S(\cdot)](x) \bigcap \Xi$$

is single-valued and Lipschitz *continuous in* Θ *with a* Lipschitz *constant L. Then for every* $\mu > 0$ *there exist neighborhoods* $\mathcal{V}_\mu$ *of* x_1 *and* $\mathcal{U}_\mu$ *and a single-valued function*

$$\omega \to x(\omega)$$

from $\mathcal{U}_\mu$ *to* $\mathcal{V}_\mu$ *such that, for any*

$$\omega \in \mathcal{U}_\mu,$$

$x(\mu)$ *is a unique solution in* $\mathcal{V}_\mu$ *of* (2.4) *and, moreover, for any* $\omega, \tilde{\omega} \in \mathcal{V}_\mu$ *one has*

$$||x(\omega) - x(\tilde{\omega})||_X \leq (L + \mu)||f(\omega, x(\tilde{\omega})) - f(\tilde{\omega}, x(\tilde{\omega}))||_{X^*}.$$

Consider a proper convex and lower semicontinuous functional

$$J : X \to (-\infty, \infty]$$

on the real Banach space X. The *conjugate functional* is given by

$$J^*(p) := \sup_{x \in X} \langle p, x \rangle_{X^*, X}.$$

Now recall a *Tikhonov well-posedness* concept in abstract optimization. *Tikhonov regularization* [303,304] and the well-posedness definition [304,308] are of crucial importance in the context of ill-posed problems in functional analysis and abstract optimization theory. The problems of Tikhonov regularization and well-posedness are only touched upon in this book. We refer to [303,308,307,13] for basic results of the Tikhonov theory. Let H_1, H_2 be real Hilbert spaces. Consider a linear problem (linear operator equation)

$$\mathcal{B}h = b, \ \ \mathcal{B} \in L(H_1, H_2), \ \ b \in H_2.$$

Instead of this original linear problem one can consider a perturbed equation

$$\mathcal{B}h = b_\epsilon, \ \ \mathcal{B} \in L(H_1, H_2), \ \ b_\epsilon \in H_2,$$

where

$$||b - b_\epsilon||_{H_2} \leq \epsilon.$$

The Tikhonov regularization scheme is given by the following auxiliary optimization problem:

$$\text{minimize} \ \frac{1}{2}||\mathcal{B}h - b_\epsilon||^2_{H_2} + \alpha\Gamma(h),$$

where

$$\alpha = \alpha(\delta) > 0$$

and $\Gamma : H_1 \to \mathbb{R}$ is a Gâteaux differentiable functional. Assume that the above minimization problem has an optimal solution. A functional Γ is called a *stabilizer* if

a) $\Gamma(h) \leq 0 \; \forall h \in H_1$;
b) the set

$$\mathcal{D}_r := \{h \in H_1 \; : \; \Gamma(h) \leq r\}$$

is a compact subset of H_1 for any $r > 0$.

A minimizer h_α for the Tikhonov optimization problem is given by a solution of the following equation:

$$\alpha \nabla_G \Gamma(h_\alpha) + \mathcal{B}^* \mathcal{B} h_\alpha = \mathcal{B}^* b_\delta.$$

Note that in the special case

$$\Gamma(h) = \frac{1}{2} ||h||^2_{H_1}$$

we easily obtain the following equation:

$$\alpha h + \mathcal{B}^* \mathcal{B} h = \mathcal{B}^* b,$$

where $b^* \in H_2$ and B^* is the adjoint operator (to B). The aim of the Tikhonov theory consists in proving that for a suitable choice of

$$\alpha = \alpha(\delta),$$

the sequence $\{h_\alpha\}$ generated by the auxiliary optimization problem converges (in some sense) as $\delta \downarrow 0$. The next result establishes the relationship between Tikhonov well-posedness and differentiability of the conjugate functional [18].

Theorem 2.28 (Asplund–Rockafellar theorem). *Let X be a real Banach space. The problem*

$$\begin{aligned} &\text{minimize } J(x) \\ &\text{subject to } x \in X \end{aligned}$$

is Tikhonov well-posed with respect to the strong convergence if and only if J^* is Fréchet differentiable at 0.

Theorem 2.29. *Let Q be a closed convex subset of a Banach space X and let f be a convex function of Q into $(-\infty, \infty]$. Then f is lower semicontinuous in the norm topology if and only if f is lower semicontinuous in the weak topology.*

We formulate here an important existence result for the basic convex optimization problem (2.1) in the form of a theorem.

Theorem 2.30. *Let Q be a closed convex subset of a reflexive* Banach *space X. Let $J(\cdot)$ be a proper convex lower semicontinuous function of Q into $(-\infty, \infty]$ and suppose that*

$$\lim_{k\to\infty} J(x_k) = \infty$$

as

$$\lim_{k\to\infty} ||x_k||_X = \infty.$$

Then there exists an element $x_0 \in D(J)$ such that

$$J(x_0) = \inf_{x\in Q} J(x).$$

For the aim of relaxations we also need some useful property of the strongly convex functions. Let D be a nonempty convex set in $\mathbb{R}^n$. Recall that a function

$$\Gamma : D \subseteq \mathbb{R}^n \to \mathbb{R}$$

is called strongly convex on D with modulus $\gamma > 0$ if

$$\Gamma(\alpha x_1 + (1-\alpha)x_2) \le \alpha\Gamma(x_1) + (1-\alpha)\Gamma(x_2) - \frac{\gamma}{2}\alpha(1-\alpha)||x_1 - x_2||^2$$

for all $x_1, x_2 \in D$ and all $\alpha \in]0, 1[$. Note that the function Γ is strongly convex on D with modulus γ if and only if the function

$$\Gamma(\cdot) - \frac{\gamma}{2}||\cdot||^2$$

is convex on D.

Theorem 2.31. *Let Γ be a function differentiable on an open set $\Omega \subset \mathbb{R}^n$, and let D be a convex subset of Ω. Then Γ is strongly convex with modulus γ on D if and only if, for all $x_1, x_2 \in D$,*

$$\Gamma(x_1) - \Gamma(x_2) \ge \langle \nabla\Gamma(x_1), x_1 - x_2\rangle + \frac{\gamma}{2}||x_1 - x_2||^2.$$

The gradient of a convex function is a monotone operator and the gradient of the strongly convex function Γ is strongly monotone (strongly stable), i.e.,

$$\langle \nabla\Gamma(x_1) - \nabla\Gamma(x_2), x_1 - x_2\rangle \ge \gamma||x_1 - x_2||^2.$$

The stability concept of an operator is given below.

Definition 2.20. *An operator* $A : X \to Y$ *is called* stable *if*

$$||A(x_1) - A(x_2)||_Y \geq g(||x_1 - x_2||_X) \; \forall x_1, x_2 \in X,$$

where

$$g : \mathbb{R}_+ \to \mathbb{R}_+$$

is a strictly monotone increasing and continuous function with

$$g(0) = 0, \quad \lim_{t \to +\infty} g(t) = +\infty.$$

The function $g(\cdot)$ *is called a* stabilizing function *of the operator* A.

Evidently, $\nabla\Gamma(\cdot)$ introduced above is a *stable* operator.

In the context of the main convex optimization problem (2.1) we now introduce the following *proximal mapping:*

$$\mathcal{P}_{f,Q,\chi} : \alpha \to \text{Argmin}_{x \in Q}[J(x) + \frac{\chi}{2}||x - \alpha||^2], \; \chi > 0, \; \alpha \in X,$$

and define the classical *proximal point* method [275,276,202,203]

$$x_{cl}^{i+1} \approx \mathcal{P}_{f,Q,\chi_i}(x_{cl}^i), \quad x_{cl}^0 \in Q, \; i = 0, 1, \dots ,$$

where $\{\chi_i\}$ is a given sequence with

$$0 < \chi_i \leq C < \infty.$$

Thus the original convex minimization problem (2.1) can be replaced by a sequence of the (strongly convex) auxiliary problems

$$f(x) + \frac{\chi_i}{2}||x - x_{cl}^i||^2 \to \min, \; x \in Q, \; i = 0, 1, \dots,$$

with strong convex objective functionals. Recall that a functional $f : X \to \mathbb{R}$ is called strongly convex with a parameter $r > 0$ if the following inequality holds for all points x, y in its domain:

$$f(\lambda x + (\lambda - 1)y) \leq \lambda f(x) + (1 - \lambda) f(y) - \frac{1}{2}\lambda(1 - \lambda)||x - y||^2$$

with $\lambda \in [0, 1]$. Evidently, for a constructive treatment of the given problem of convex minimization the proximal point method must be combined with an effective numerical procedure for the auxiliary problems.

Suppose now that the approximation $\{x_{cl}^i\}$, $i=0, 1, ...$, satisfies the following condition:

$$||x_{cl}^{i+1} - \mathcal{P}_{f,Q,\chi_i}(x_{cl}^i)|| \leq \epsilon_i, \ \sum_{i=0}^{\infty} \frac{\epsilon_i}{\chi_i} < \infty,$$

for all $i = 0, 1, ...$ Under these conditions the sequence $\{x_{cl}^i\}$ converges in the weak topology to an optimal solution x^{opt} of (2.1) [202]. Moreover, $\{x_{cl}^i\}$, $i = 0, 1, ...$, is a minimizing sequence. Let us note that in some specific cases one can establish the strong convergence of the corresponding minimizing sequences (see, e.g., [27,28]).

Let $\{X, \ ||\cdot||_X\}$, $\{Y, \ ||\cdot||_Y\}$ be real Banach spaces. The basic inspiration for studying operators is the general operator equation of the form

$$A(x) = a, \tag{2.5}$$

where $A : X \to Y$ is, in general, a nonlinear operator and $x \in X, \ a \in Y$. The symbol $A : X \to Y$ will mean a single-valued mapping whose domain of definition is X and whose range is contained in Y, that is, for every $x \in X$ the mapping A assigns a unique element $A(x) \in Y$. Recall that the *range* of the operator A is the set

$$R(A) := \{y \in Y \ : \ y = A(x), \ x \in X\}$$

of all image elements. For further notations, we refer to the corresponding sections.

Along with the original operator equation (2.5) a sequence of the approximate equations

$$\begin{aligned} &A^n(x^n) = a^n, \\ &A^n : X^n \to Y^n, \ x^n \in X^n, \ a^n \in Y^n, \ n \in \mathbb{N}, \end{aligned}$$

is of frequent use in numerical functional analysis. Here,

$$\begin{aligned} &X^1 \subset X^2 \subset ... \subset X^n \subset ... \subset X, \\ &Y^1 \subset Y^2 \subset ... \subset Y^n \subset ... \subset Y \end{aligned}$$

are sequences of subspaces of the space X and of the space Y, respectively. Suppose that

$$A(\hat{x}) = a.$$

If we assume that $Y = X$ and $\{X^n\}$ is a sequence of finite-dimensional subspaces of the space X, then approximations

$$\hat{x}^n, \ n \in \mathbb{N},$$

of the solution $\hat{x}$ of Eq. (2.5) are often obtained by the following procedure, called the *Galerkin method* (see, e.g., [307]):

$$\hat{x}^n = P^n A(x^n), \; x^n \in X^n,$$

where $P^n : X \to X^n$ are *projection operators*. Each equation of this type can be treated as a finite system of equations with scalar variables.

For a sequence $\{\Psi_s\}$, $s \in \mathbb{N}$, of operators $\Psi_s : X \to Y$ one can consider various convergence concepts. We discuss shortly the *uniform convergence* and the *pointwise convergence* and recall the corresponding definitions. An operator

$$\Psi : X \to Y$$

is the *uniform limit* of $\{\Psi_s\}$ if, for every $\epsilon > 0$, there exists a number $N(\epsilon) \in \mathbb{N}$ (depending on ϵ) such that $s \geq N(\epsilon)$ implies

$$||\Psi_s(x) - \Psi(x)||_Y \leq \epsilon$$

for every $x \in X$. Note that $\Psi : X \to Y$ is called the *pointwise limit* of $\{\Psi_s\}$ if, for each $x \in X$ and for every $\epsilon > 0$, there exists a number

$$N(x, \epsilon) \in \mathbb{N}$$

(depending on x and ϵ) such that

$$||\Psi_s(x) - \Psi(x)||_Y \leq \epsilon$$

for all $s \geq N(x, \epsilon)$.

From Definition 2.20 it follows that the solution $\hat{x}$ of (2.5) with a stable A is "stable" in the following sense: for each $\epsilon > 0$ there exists a number $\delta(\epsilon) > 0$ such that

$$||a_1 - a_2||_Y < \delta(\epsilon),$$

where $a_1, a_2 \in R(A)$ always implies that

$$||\hat{x}_1 - \hat{x}_2||_X < \epsilon$$

for the corresponding solution $\hat{x}_1, \hat{x}_2 \in X$ of the problems

$$A(x) = a_1$$

and $A(x) = a_2$, respectively. We now formulate the main theorem on strongly stable operators in separable Hilbert spaces [328–331].

Theorem 2.32. *Let H be a real separable* Hilbert *space with* $\dim H = \infty$ *and* $\{e_i\}$ *be a complete orthonormal system in H. Let*

a) $H^n = \text{span}\{e_1, ..., e_n\}$*;*
b) the operator $B : H \to H$ *be continuous and strongly stable;*
c) $P^n : H \to H^n$ *be the orthogonal projection operator from H onto H^n, i.e.,*

$$P^n h := \sum_{i=1}^{n} \langle e_i, h \rangle_H e_i;$$

d) $E^n : H^n \to H$ *be the embedding operator corresponding to* $H^n \subseteq H$*.*

Then, for each $b \in H$, the equation $B(h) = b,\ b \in H$, has a unique solution $\hat{h} \in H$. Moreover, the approximate equation

$$P^n B(h^n) = P^n b,\ h^n \in H^n,$$

has a unique solution $\hat{h}^n$ and

$$\lim_{n\to\infty} ||E^n \hat{h}^n - \hat{h}||_H = 0.$$

In fact, Theorem 2.32 presents solvability conditions for a special case of the operator equations (2.5), namely, for the equations with *strongly stable* operators and for the corresponding approximate equations.

Note that one can consider *linear expanding* operators. Consider a strongly stable operator $B : H \to H$, where H is a real Hilbert space. The Schwarz inequality implies that

$$||B(h_1) - B(h_2)||_H \geq c||h_1 - h_2||_H.$$

Let

$$\aleph(X, Y) \subset L(X, Y)$$

be the set of all expanding linear continuous surjective operators $\mathcal{A}$ which have an inverse linear continuous operator $\mathcal{A}^{-1}$. We now suggest the following general definition.

Definition 2.21. *An operator $A : X \to Y$ is called* expanding *if there is a number $d > 0$ such that*

$$||A(x_1) - A(x_2)||_Y \geq d||x_1 - x_2||_X\ \forall x_1, x_2 \in X.$$

It is evident that an expanding operator A is a stable operator with the stabilizing function

$$g(t) = d \cdot t, \ t \geq 0$$

(see Definition 2.20). Motivated by the above concepts, we now present an important consequence of the Banach open mapping theorem.

Theorem 2.33. *Let* $\mathcal{A} \in \aleph(X, Y)$,

$$||\mathcal{A}x_1 - \mathcal{A}x_2||_Y \geq d||x_1 - x_2||_X,$$

and $\mathcal{D} \in L(X, Y)$ *such that*

$$||\mathcal{D}||_{L(X,Y)} < \frac{1}{||\mathcal{A}^{-1}||_{L(Y,X)}} \text{ and } ||\mathcal{D}||_{L(X,Y)} < \tilde{d} < d.$$

Then the operator $(\mathcal{A} + \mathcal{D})$ *is expanding and invertible. Moreover,*

$$(\mathcal{A} + \mathcal{D})^{-1}$$

is a linear continuous operator, i.e.,

$$(\mathcal{A} + \mathcal{D}) \in \aleph(X, Y).$$

As we can see, Theorem 2.33 shows that $\aleph(X, Y)$ is an open (in the usual norm topology) subset of $L(X, Y)$.

Recall that a closed subspace F of a Banach space Y is *a complemented subspace* if there exists another closed subspace $\tilde{F}$ such that

$$Y = F \oplus \tilde{F},$$

where $\oplus$ denotes the *direct sum* of two vector subspaces F and $\tilde{F}$. This means that every $y \in Y$ has a unique decomposition, i.e.,

$$y = y_1 + y_2,$$

with $y_1 \in F$ and $y_2 \in \tilde{F}$. The closed subspace $\tilde{F}$ is called a *complement* of F.

We now take a look at the linear variant of the operator equation (2.5)

$$\mathcal{A}(x) = a, \ x \in X, \tag{2.6}$$

where $\mathcal{A} : X \to Y$ is a linear operator and X and Y are real Banach spaces. This equation has a solution for each $a \in Y$ iff

$$R(\mathcal{A}) = Y,$$

i.e., if and only if $\mathcal{A}$ is surjective. Let

$$R(\mathcal{A})^{\perp}$$

be a *generalized orthogonal complement* to the set $R(\mathcal{A})$. The next theorem (see [330]) exhibits remarkable properties of $R(\mathcal{A})$.

Theorem 2.34. *Let $\mathcal{A} : X \to Y$ be a linear operator, where X and Y are Banach spaces. Then we have*

$$R(\mathcal{A}) = Y$$

if and only if the following two conditions are satisfied:

a) $R(\mathcal{A})$ *is closed;*
b) $R(\mathcal{A})^{\perp} = \{0\}$.

An important generalization of the notion of the general continuity is the concept of the *graph closed operator.*

Definition 2.22. *An operator $A : X \to Y$, where X and Y are* Banach *spaces, is called graph closed if the graph*

$$G_A := \{(x, Ax) \in X \times Y \; : \; x \in X\}$$

is a closed set, i.e., for each sequence $\{x_i\}$ in X, it follows from

$$\lim_{i \to \infty} ||x - x_i||_X = 0, \;\; \lim_{i \to \infty} ||a - Ax_i||_Y = 0$$

that $Ax = a$.

Using Theorem 2.34 and the concept of graph closed operators one can prove the following solvability result for (2.5) (see [330]).

Theorem 2.35. *Let $\mathcal{A} : X \to Y$ be a linear expanding and graph closed operator, where X and Y are real* Banach *spaces. Suppose that*

$$R(\mathcal{A})^{\perp} = \{0\}.$$

Then, for each $a \in Y$, Eq. (2.5) *has a unique solution.*

Note that the linear version (2.6) of the general nonlinear operator equation (2.5) is in fact a theoretical fundament of the extremely important "system linearization" operation that is widely used in control theory and in particular in this book.

We continue by considering an operator $\mathcal{B} \in L(H, H)$, where H is a real Hilbert space.

Definition 2.23. *Let $\mathcal{B} : H \to H$ be a bounded linear operator mapping a real* Hilbert *space H into itself. Then a complex number λ is called an eigenvalue of B if there exists an element*

$$h \in H, \ h \neq 0,$$

such that

$$\mathcal{B}h = \lambda h.$$

This element h is called an eigenelement of $\mathcal{B}$. A complex number λ is called a regular value *of $\mathcal{B}$ if*

$$(\lambda I - \mathcal{B})^{-1}$$

exists and is bounded. Here I is the identity operator. The set of all regular values λ is called the resolvent set $\rho(\mathcal{B})$ and

$$\mathrm{Res}(\lambda; \mathcal{B}) := (\lambda I - \mathcal{B})^{-1}$$

is called the resolvent. *The complement of $\rho(\mathcal{B})$ is called the* spectrum $\sigma(\mathcal{B})$ *and*

$$r(\mathcal{B}) := \sup_{\lambda \in \sigma(\mathcal{B})} |\lambda|$$

is called the spectral radius *of $\mathcal{B}$.*

We now consider a linear *symmetric operator* $\mathcal{B} : H \to H$, where H is a Hilbert space. Thus for every $h_1, h_2 \in H$ we have

$$\langle \mathcal{B}h_1, h_2 \rangle_H = \langle \mathcal{B}h_2, h_1 \rangle_H.$$

Recall that a symmetric operator $\mathcal{B} \in L(H, H)$ has only real eigenvalues [282].

Theorem 2.36. *A number $\lambda \in \mathbb{R}$ is a regular value of a symmetric operator*

$$\mathcal{B} \in L(H, H)$$

if and only if

$$(\lambda I - \mathcal{B})$$

is an expanding operator, i.e.,

$$||(\lambda I - \mathcal{B})h||_H \geq d||h||_H \ \forall h \in H.$$

Every expanding (stable) operator is evidently injective. That is, the expanding operator

$$(\lambda I - \mathcal{B})$$

is bijective. By the *Banach open mapping theorem*, we deduce that there exists an inverse continuous operator

$$\mathrm{Res}(\lambda; \mathcal{B}) : H \to H.$$

Hence λ is a regular value.

Note that if $\lambda \in \mathbb{R}$ is a regular value of a symmetric operator $\mathcal{B} \in L(H, H)$, then the resolvent

$$\mathrm{Res}(\lambda; \mathcal{B})$$

is a Lipschitz continuous mapping.

Recall that a mapping $B : H \to H$ (real Hilbert spaces) is said to be a monotone operator if

$$\langle B(h_1) - B(h_2), h_1 - h_2 \rangle_H \geq 0$$

for all $h_1, h_2 \in H$. We now present some concepts and results from the abstract optimization theory in Hilbert spaces.

Definition 2.24. *Let $B : H \to H$ be an operator on the real* Hilbert *space H. Then*

a) *B is called* maximal monotone operator *if B is monotone and the graph*

$$G_B := \{(y, h) \in H \times H \ : \ y = B(h)\}$$

is not properly contained in the graph of any other monotone operator

$$\tilde{B} : H \to H;$$

b) *B is called* accretive operator *if*

$$(I + \mu B) : H \to H$$

is injective and

$$(I + \mu B)^{-1}$$

is nonexpansive for all $\mu > 0$;

c) *B is called* maximal accretive operator *if B is accretive and*

$$(I + \mu B)^{-1}$$

exists on H for all $\mu > 0$.

Note that a linear operator $\mathcal{B} : H \to H$ is accretive if

$$||\mathcal{A}x + \mu x||_Y \geq \mu ||x||_X$$

for all $x \in X$ and all $\mu > 0$.

Theorem 2.37. *Let $B : H \to H$ be an operator on the real* Hilbert *space H. Then the following are equivalent:*

a) B is monotone and $(I + B)$ is surjective;
b) B is maximal monotone;
c) B is maximal accretive.

Note that the classic *Minty theorem* [242] states that a monotone operator B is maximal if and only if

$$(I + B)$$

is surjective.

Consider an operator $\mathcal{B} \in L(H, H)$. Let $\lambda = -1$ be a regular value of $\mathcal{B}$. That is, the equation

$$\mathcal{B}h = -h$$

has a unique solution. We call the operator $\mathcal{B}$ *positive*, if

$$\langle h, \mathcal{B}h \rangle_H \geq 0$$

for every $h \in H$. It is evident that for a linear operator $\mathcal{B}$ the monotonicity condition is equivalent to the simpler positivity condition. We establish the following result.

Theorem 2.38. *Let $\mathcal{B} \in L(H, H)$ be a positive symmetric operator. Then $\lambda = -1$ is a regular value of the operator $\mathcal{B}$. If $\mathcal{B}$ is surjective, then this operator is maximal monotone.*

Finally, we note that symmetric operators play a large role not only in analysis and applied mathematics, but also in physics. They are of prime importance in quantum theory.

2.3 Notes

Let us mention here some classical and modern additional works related to functional analysis and topology [1,6,12,15,18,20,84,94,137,139,168,182,185,187,194,196,197,200,201,208,209, 211,213–217,226–228,241,265,272–274,277,282,283,297,328–331].

Useful elements of convex analysis and approximation theory can be found in [4,8,11,14,21, 23–25,70,71,75,76,79,83,87,102,109,115–117,119,132,133,153,178,190–193,195,198,220, 221,237,239,246–250,252,256,259,261–264,267,271,275,276,278,307–309,315,327,336].

CHAPTER 3

Convex Programming

3.1 Problem Formulation

We complement here the theoretic aspects related to the basic convex optimization problem (2.1) formulated in Section 2.2 and consider practically oriented concepts of the convex programming. These "practically oriented concepts" are focused on the various useful convergence concepts and results.

Definition 3.1. *A sequence $\{x_k\}$, $k \in \mathbb{N}$, is called a minimizing sequence for the optimization problem (2.1) if*

$$\lim_{k \to \infty} = J(x^{opt}) \equiv \min_{Q} J(x).$$

Note that in a real Hilbert space the "functional convergence" indicated in Definition 3.1 does not generally imply any "argument convergence," i.e.,

$$\lim_{k \to \infty} x_k = x^{opt}.$$

However, if this is the case $\{x_k\}$, $k \in \mathbb{N}$, is called a convergent minimizing sequence. Due to the various convergence meanings in H we usually consider strongly and weakly convergent minimizing sequences.

Note that from the engineering or applied point of view an element $x^{\hat{k}}$ of $\{x_k\}$, $k \in \mathbb{N}$, for a sufficiently big number $\hat{k} \in \mathbb{N}$ can be interpreted as a "practical solution" to the main problem (2.1). We are usually interested to obtain convergent (in some adequate sense) minimizing sequences as a result of a suitable computational procedure applied to (2.1).

Definition 3.2. *A functional $J(\cdot)$ is (strictly) quasiconvex on a convex set Q whenever*

$$J((\alpha x_1 + (1-\alpha)x_2)) < max\{J(x_1), J(x_2)\} \; \forall x_1, x_2 \in Q, \; 0 < \alpha < 1.$$

An equivalent characterization of a quasiconvex function is the following well-known fact: if for every $\alpha > 0$ the set

$$\{x \in Q \; : \; J(x) \leq \alpha\}$$

A Relaxation-Based Approach to Optimal Control of Hybrid and Switched Systems
https://doi.org/10.1016/B978-0-12-814788-7.00009-6

is convex, then $J(\cdot)$ is quasiconvex. Evidently, a strictly convex functional is quasiconvex. If

$$J : U \to \mathbb{R}$$

is continuously differentiable on the open subset U of a linear space, then $J(\cdot)$ is quasiconvex iff for every $x_1, x_2 \in U$,

$$J(x_2) < J(x_1)$$

means that

$$\nabla J(x)[y - x] < 0.$$

The next result guarantees a strong convergence of a minimizing sequence.

Theorem 3.1. *If $J(\cdot)$ is continuous and strictly quasiconvex on a compact convex set Q, then any minimizing sequence for (2.1) converges in H-norm to a unique minimum.*

Definition 3.3. *An operator $P_Q : H \to Q$ is called a projection operator associated with a point $x \in H$ (projection onto the convex set Q sometimes called a projection mapping) if*

$$\min_{y \in Q} ||x - y|| = ||x - P_Q(x)||.$$

We also will use the following notation:

$$P(x) = \text{Argmin}_{y \in Q} ||x - y||^2.$$

Recall some useful properties of the operator $P_Q(\cdot)$.

Theorem 3.2. *Let $P_Q(\cdot)$ be the projection operator onto a closed convex set $Q \subset H$. Then for $x \in H$ and $y \in Q$,*

$$\langle x - y, y \rangle_H \geq \langle x - y, q \rangle \; \forall q \in Q$$

if and only if $y = P_Q(x)$.

Let

$$C^{\perp} := \{x | \langle x, y \rangle_H = 0 \; \forall y \in Q\}.$$

The general Theorem 3.1 has two important specific cases.

Theorem 3.3. *Let P_Q be the projection operator onto a closed convex set Q.*

(a) If Q is a subspace of H, then for $x \in H$ and $y \in Q$,

$$(x - y) \in Q^{\perp}$$

if and only if $y = P_Q(x)$.

(b) If Q is a cone, then for $x \in H$ and $y \in Q$,

$$\langle x - y, y \rangle_H = 0,$$
$$\langle x - y, q \rangle_H \leq 0 \; \forall q \in Q$$

if and only if

$$y = P_Q(x).$$

Evidently, $P_Q(\cdot)$ is well defined (see Theorem 2.17) and Lipschitz continuous, and assigns to a given point in H its closest point in a convex subset Q. Assume $J(\cdot)$ in (2.1) is Fréchet differentiable and $J'(x, h)$ (also sometimes denoted by $\nabla J(x)h$) is the corresponding derivative. A point $\hat{x}$ in Q will be called stationary if the linear functional

$$\nabla J(x)h = J'(x, h)$$

achieves a minimum on Q at $\hat{x}$. In that case we evidently have

$$P_Q(\hat{x} - \rho \nabla J(\hat{x})) = \hat{x} \; \forall \rho > 0.$$

The last relation shows that a stationary point is a fixed point of the projection operator.

We next discuss the nonexpansivity property of the projection operator onto a closed convex set Q. The following nice property is sometimes called a variational characterization of the projection operator.

Theorem 3.4. *Let P_Q be a projection operator onto a closed convex set $Q \subset H$ and $x_1, x_2 \in H$. Then*

$$\langle x_1 - P_Q(x_1), x - P_Q(x_1) \rangle_H \leq 0 \; \forall x \in Q.$$

Proof. Since

$$P_Q(x_2) \in Q$$

we deduce

$$\langle x_1 - P_Q(x_1), P_Q(x_2) - P(x_1) \rangle_H \leq 0.$$

Similarly,

$$\langle x_2 - P - Q(x-2), P_Q(x_1) - P_Q(x_2) \le 0,$$
$$\langle P - Q(x_2) - x_2, P_Q(x_2) - P_Q(x_1)\rangle_H \le 0.$$

We now apply the Cauchy–Schwarz inequality and obtain

$$\langle P_Q(x_2) - P_Q(x_1), P_Q(x_2) - P_Q(x_1)\rangle_H \le$$
$$\langle x_2 - x_1, P_Q(x_2) - P_Q(x_1)\rangle_H \le ||x_2 - x_1|| \times ||P_Q(x_2) - P_Q(x_1)||.$$

Thus

$$||P_Q(x_2) - P_Q(x_1)||^2 \le$$
$$||x_2 - x_1|| \times ||P_Q(x_2) - P_Q(x_1)|| \times ||P_Q(x_2) - P_Q(x_1)|| \le ||x_2 - x_1||.$$

The proof is completed. □

Finally we give an interesting convergence result. Let

$$J : H \to \mathbb{R}$$

be a real-valued function on a real Hilbert space H and $x_0 \in Q \subset H$, where Q is assumed to be convex (see problem (2.1)). Let S denote the level set

$$\{x \in Q \ : \ J(x) \le J(x_0)\},$$

and let $\hat{S}$ be any open set containing the convex hull of S.

Theorem 3.5. *Assume $J(\cdot)$ is bounded. For each*

$$x \in \hat{S}, \ h \in H,$$

and $\rho > 0$ assume that there exists the Fréchet derivative $J'(x, h)$ (also sometimes denoted by $\nabla J(x)h$) and a Gâteaux derivative $J''(x, h, h)$. Moreover, let

$$||J''(x, h, h)|| \le \frac{1}{\rho}||h||^2.$$

Choose σ and ρ_k satisfying

$$0 < \sigma \le \rho_0$$

and

$$\sigma \le \rho_k \le 2\rho_0 - \sigma.$$

Define the following iterations:

$$x_{k+1} = P_Q(x_k - \rho_k \nabla J(x_k)),$$

where $P_Q(\cdot)$ is a projection operator onto Q.

Then:

- *The sequence $\{x_k\},\ k \in \mathbb{N}$, belongs to S,*

$$(x_{k+i} - xk)$$

  converges to 0, and $J(x_k)$ converges downward to a limit L.
- *If S is compact, z is a cluster point of*

$$\{x_k\},\ k \in \mathbb{N},$$

  and $\nabla J(\cdot)$ is continuous in a neighborhood of z, then z is a stationary point. If z is unique, $\{x_k\},\ k \in \mathbb{N}$, converges to z, and z minimizes $J(\cdot)$ on Q.
- *If S is convex and*

$$J''(x, h, h) \geq \mu ||h||^2$$

  for each $x \in S$, $h \in H$, and $\mu \geq 0$, then

$$L = \inf\{J(x)\ :\ x \in C\}.$$

- *Additionally assume that S is bounded. Then the weak cluster points of $\{x_k\},\ k \in \mathbb{N}$, are solutions to the optimization problem (2.1).*
- *Assume $\mu > 0$ and $\nabla J(\cdot)$ bounded on S. Then*

$$J(z) = L$$

  for a $z \in S$ and $\{x_k\},\ k \in \mathbb{N}$, converges to (unique) z.

Theorem 3.1 is a useful tool that makes it possible to establish the convergence properties of a minimizing sequence associated with (2.1).

3.2 Existence Theorems

The general existence result was mentioned in Section 2.2 (Theorem 2.17). The corresponding geometrical characterization of the set $\mathcal{F}$ of all optimal solutions to (2.1) is also well known (see, e.g., [198,275]).

Theorem 3.6. *Let Q be a nonempty convex subset of a real Hilbert space H. For every quasiconvex functional $J : Q \to \mathbb{R}$ the set of minimal points of $J(\cdot)$ on Q is convex.*

Evidently, Theorem 3.6 can also be applied to problem (2.1). Theorem 3.6 can be easily proved by a direct calculation. It is necessary to underline that the result of Theorem 2.17 is an immediate consequence of the following general existence result for real Banach spaces.

Theorem 3.7. *Let Q be a nonempty, convex, closed, and bounded subset of a reflexive real Banach space, and let $J : Q \to \mathbb{R}$ be a continuous quasiconvex functional. Then problem $J(\cdot)$ has at least one minimal point on Q.*

Recall (see Chapter 2) that if we eliminate the boundedness hypothesis from Theorem 3.7 we need to make some additional assumptions in order to guarantee the existence of an optimal solution in problem (2.1). Evidently, Theorem 2.30 makes it possible to establish the existence of an optimal solution to (2.1) in simple terms of a minimizing sequence.

The next fundamental result constitutes in some sense a helpful tool for the existence proofs (see, e.g., [6]). Let us note that this result is generally true in a reflexive Banach space.

Theorem 3.8 (Eberlein–Smulyan theorem). *Let H be a real Hilbert space and*

$$B_0 := \{x \in H \ : \ ||x||_H \leq 1\}$$

be the closed unit ball. Then the following are equivalent:

- *B_0 is weakly compact;*
- *B_0 is sequentially weakly compact;*
- *every bounded sequence in H has a weakly convergent subsequence.*

As mentioned in Section 3.1 a practically oriented (engineering) "solution" to the general convex minimization problem (2.1) is represented by a minimizing sequence. In a "worst case" the minimizing sequence generated by a numerical optimization method does not possess any convergence property. Since the concept of a minimizing sequence can be considered as a new numerically oriented solution concept, we are interested in the existence of a convergent minimizing sequence for the original problem (2.1). So we also understand in this book an "existence question" from the point of view of numerical analysis and study certain criteria under which a minimizing sequence must converge in norm. Such conditions must be known before the iterative and extrapolative numerical solution procedure and can be used in the computational step for the approximate minima search. Note that the class of applied problems which can be effectively interpreted as minimization problems is growing rapidly and the numerical solution concept, namely, a generation of a numerically stable minimizing sequence, constitutes nowadays a challenging engineering and mathematical problem.

Taking into consideration the above motivating arguments we next present some general results related to the convergent minimizing sequences. First of all let us study a counterexample.

Example 3.1. *Consider the specific case $H = l^2$ and a closed positive cone C in H. Then the set $Q := C \bigcap B_0$, where B_0 is a closed unit ball centered at zero, is closed and convex. By Theorem 3.8 it is also weakly compact. Let*

$$e_1 = (1, 0, O, ...),\ e_2 = (0, 1, O, ...),\ ...$$

be elements of the canonical orthonormal Schauder basis. That means every $x \in l^2$ can be expressed as

$$x = \sum_{i=1}^{\infty} \alpha_i e_i. \tag{3.1}$$

Consider

$$y = (1, 1/2, 1/3, ...) \in l^2$$

and define a linear functional

$$J(x) := \sum_{i=1}^{\infty} \alpha_i (1/i)$$

for $x \in l^2$ defined by (3.1). Since $J(\cdot)$ has a unique minimum at $x = 0$, the sequence

$$\{e_i\},\ i \in \mathbb{N},$$

is a minimizing sequence for problem (2.1). We have

$$J(e_i) = 1/i \to 0 = J(0).$$

However,

$$||e_i - e_j||_{l^2} = 2^{1/2},\ i \neq j,$$

so that $\{e_i\}$, $i \in \mathbb{N}$, contains no subsequences which converge to 0 in norm.

Note that Example 3.1 shows that a minimizing sequence to the problem (2.1) does not necessarily contain any norm-convergent subsequence.

We next will see that the usual first-order numerical optimization methods generate a weakly convergent minimizing sequence (Section 3.6). The next general result illustrates a possibility to obtain a strongly convergent sequence from a weakly convergent one. This analytic result is also true in general normed spaces (see, e.g., [192]).

Theorem 3.9 (Mazur lemma). *Assume H is a real Hilbert space, and assume $\{x_k\}$, $k \in \mathbb{N}$, is a sequence converging weakly to $x \in H$. Then there is a sequence*

$$\{x_k\}, \ k \in \mathbb{N},$$

of convex combinations of $\{x_k\}$, $k \in \mathbb{N}$, which converges strongly to $x \in H$.

Let us also recall that the extremal set of a compact convex subset of a reflexive Banach space X need not be compact.

Theorem 3.10 (Krein–Milman). *Let X be a locally convex Hausdorff topological vector space and let $Q \subset X$ be a nonempty compact convex set. Then Q is the closed convex hull of its extremal points $Ext(Q)$, i.e.,*

$$Q = c\bar{o}nv(Ext(Q)).$$

In particular, Q admits an extremal point, i.e., $Ext(Q) \neq \emptyset$.

Assume now X is strictly convex, i.e., for all $x, y \in X$,

$$||x + y||_X = 2||x||_X = 2||y||_X \Rightarrow x = y.$$

Recall that a real Hilbert space H is a reflexive and strictly convex Banach space [20]. The closed unit ball B_0 of H is weakly compact by Theorem 3.8 and its extremal set is the unit sphere $S(B_0)$. Thus the extremal set is not weakly compact and B_0 is the weak closure of its extremal set.

We finally discuss the convergence results for minimizing sequences in convex optimization. If $J(\cdot)$ has certain analytical properties then convergence in H-norm is guaranteed for any minimizing sequence. It is well known that in the case $J(\cdot)$ is strictly quasiconvex and continuous on a compact convex subset Q of a real Hilbert space H, $J(\cdot)$ is uniformly quasiconvex on Q. A direct consequence for problem (2.1) of this geometrical fact can be expressed as follows (see [131]).

Theorem 3.11. *If $J(\cdot)$ is continuous and strictly quasiconvex on a compact convex set Q, then any minimizing sequence for (2.1) converges in H-norm to a unique minimum.*

We next need a relative fine analytic concept, namely, the concept of a dentable set [276]. Let

$$\text{epi}\{J(\cdot)\}$$

be the epigraph of $J(\cdot)$ on Q.

Definition 3.4. *We will say that $J(\cdot)$ is dentable on Q at x_0, whenever, for $\epsilon > 0$, the point $(x_0, J(x_0))$ does not belong to the closed convex hull of*

$$\text{epi}\{J(\cdot)\} \setminus B_\epsilon,$$

where B_ϵ is the ϵ-ball centered on $(x_o, J(x_0))$.

Theorem 3.12. *Let $J(\cdot)$ be a lower semicontinuous convex functional on a weakly compact convex subset of a Hilbert space H with a unique minimum $x_0 \in Q$. If $J(\cdot)$ is dentable at x_0, then any minimizing sequence*

$$\{x_k\} \subset Q, \; k \in \mathbb{N},$$

converges in H-norm to x_0.

Note that the proof of this result is based on the generic properties of the projection operator and also uses the fundamental separation principle.

Theorem 3.13. *Assume $J(\cdot)$ is strictly quasiconvex and lower semicontinuous on a weakly compact convex set $Q \subset H$ and is dentable at its unique minimum. Then any minimizing sequence*

$$\{x_k\}, \; k \in \mathbb{N},$$

for (2.1) converges in H-norm to the minimum.

One can show that in Example 3.1

$$\inf\{J(x) \; : \; x \in Q\} = 0$$

but the point $(0, 0)$ is not a denting point of epi$\{J(\cdot)\}$. The necessary conditions for a convergence of a minimizing sequence are also closely related to the concept of dentable sets (Definition 3.4).

Theorem 3.14. *Assume $J(\cdot)$ is a lower semicontinuous quasiconvex functional on a closed convex $Q \subset H$ and suppose that (2.1) has a unique solution*

$$x_0 \in Q.$$

If every minimizing sequence of (2.1) converges to $x_0 \in Q$ in H-norm, then $J(\cdot)$ is dentable at x_0.

We finally give the following general result.

Theorem 3.15. *Let $J(\cdot)$ be a lower semicontinuous strictly quasiconvex functional on a weakly compact convex subset Q of a Hilbert space H. Then every minimizing sequence for (2.1) converges in H-norm to a unique minimum $x_0 \in Q$ iff $J(\cdot)$ is dentable at x_0.*

Evidently, a convex and closed set Q in a real Hilbert space H is weakly compact (see Theorem 3.8 above). Without the weak compactness there may be no minimum in (2.1). Of course, even in that case a minimizing sequence

$$\{x_k\}, \ k \in \mathbb{N},$$

for (2.1) can exist.

Example 3.2 (K. Weierstrass). *Let*

$$Q := \{g(\cdot) \in \mathbb{L}^2([t_0, t_f], R) \ : \ g(-1) = -1, \ g(1) = 1\} \bigcap \mathbb{C}^1(-1, 1),$$

where $\mathbb{L}^2([t_0, t_f], R)$ is the Lebesgue space of all square integrable functions from $[t_0, t_f]$ to $\mathbb{R}$.

Let $J : Q \to \mathbb{R}$ with

$$J(g(\cdot)) := \int_{-1}^{1} t^2 [g(t)]^2 dt, \tag{3.2}$$

where $g(\cdot)$ is a derivative of the function $g(\cdot) \in$. The infimum of $J(\cdot)$ on the convex set Q is equal to 0. Since the integrand in (3.2) is nonnegative, we get

$$J(g(\cdot)) > 0$$

for every $g(\cdot) \in Q$. Thus there is no minimum of $J(\cdot)$ on Q. However,

$$g_k(t) := \arctan(kt)/\arctan(k)$$

defines a minimizing sequence for the resulting problem (2.1);

$$J(g_k(\cdot)) < \int_{-1}^{1} (t^2 + (1/k))(g(t))^2 dt = 2/k \arctan(k) \to 0$$

for $k \to \infty$.

In the above example the convex set Q is nonclosed (in the sense of the $\mathbb{L}^2([t_0, t_f], R)$-norm) and is nonweakly compact.

3.3 Optimality Conditions

The necessary optimality conditions for convex optimization problems (2.1) and (2.2) are in fact discussed in Theorem 2.18 and Theorem 2.19 (see Chapter 2). We present here a useful generalization of the Kuhn–Tucker theorem (Theorem 2.19), namely, a "subdifferential" form of this fundamental result. Let X and Y be Banach spaces and

$$\Lambda : X \to Y$$

be a continuous linear operator. Consider the convex functions

$$f_i : X \to \mathbb{R}, \; i = 0, 1, ..., m,$$

and let $a \in \mathbb{R}^m$, $b \in Y$. Moreover, assume that A is a convex subset of X. Consider the specific convex programming problem

$$\begin{aligned} &\text{minimize } f_0(x) \\ &\text{subject to } f_i(x) < a_i, \; i = l, ..., m, \\ &\Lambda x = b, \; x \in A. \end{aligned} \tag{3.3}$$

Let

$$B := \{x \in X \mid \Lambda x = b\}.$$

Introduce the following Lagrange function associated with the optimization problem (3.3)

$$L(x, p, \lambda, \lambda_0) := \lambda_0 f_0(x) + \sum_{i=1}^{m} \lambda_i (f_i(x) - a;) + \langle p, Ax - b \rangle_Y.$$

Here $p \in Y*$ (the dual space to Y). The following characterization of an optimal solution $x^{opt} \in A$ to problem (3.3) is useful in optimal control.

Theorem 3.16. *Let $x^{opt} \in A$ be an optimal solution to the convex problem (3.3). Then x^{opt} is a point of absolute maximum in the elementary problem*

$$f(x) = \max f_0(x) - f_0(x^{opt}), \, f_1(x) - a_1, ..., f_m(x) - a_m + \chi(A \bigcap B) \to \inf,$$

where

$$\chi(A \bigcap B)$$

is the indicator function of the set $A \bigcap B$.

The subdifferential version of the classic Kuhn–Tucker theorem (Theorem 2.19) can now be stated as follows.

Theorem 3.17. *Assume that functions*

$$f_i,\ i = 0, 1, ..., m,$$

in (3.3) are continuous at an optimal point

$$x^{opt} \in A \bigcap B.$$

Then there are numbers

$$\hat{\lambda}_i \geq 0$$

such that

$$\sum_{i=1}^{m} \hat{\lambda}_i = 1,\ \hat{\lambda}_i (f_i(x) - a_i) = 0,\ i \geq 1,$$

and moreover, there is an element

$$\hat{x} \in \partial\chi(A \bigcap B)$$

for which

$$0 \in \sum_{i=1}^{m} \hat{\lambda}_i \partial f_i(x^{opt}) + \hat{x}.$$

We now consider the special case $A \equiv X$.

Theorem 3.18. *Let $A = X$. Assume all the conditions of Theorem 3.17 are fulfilled and the image of X under the mapping Λ is closed in Y. Then there are Lagrange multipliers*

$$\lambda_0^{opt},\ \lambda^{opt} \in \mathbb{R}^m,$$

and $p^{opt} \in Y^*$ such that

$$\lambda_i^{opt} \geq 0,\ i \geq 0,\ \lambda^{opt}(f_i(x^{opt}) - a_i) = 0,$$

and

$$\min_x L(x, p^{opt}, \lambda^{opt}, \lambda_0^{opt}) = L(x^{opt}, p^{opt}, \lambda^{opt}, \lambda_0^{opt}).$$

3.4 Duality

The duality concept constitutes an important part of the general convex programming. In fact one investigates under which assumptions is it possible to associate an equivalent maximization problem to a given convex minimization problem. This resulting maximization problem is called "dual" to the minimization problem. The main topic in duality is to investigate the relationships between both optimization problems (primal (2.1) and its dual). Let C_H be an ordering cone in the real Hilbert space H. Recall that every real Hilbert space can be partially ordered. Define the admissible set Q in (2.1) as follows:

$$Q := x \in \tilde{Q} \mid g(x) \in -C_H,$$

where

$$\tilde{Q} \subseteq H$$

(nonempty) and $g : \tilde{Q} \to H$ is a constraint mapping. We next call the resulting problem (2.1) a "primal problem."

Definition 3.5. *Let $\tilde{Q}$ be a nonempty subset of a real Hilbert space H with an ordering cone C_H. A mapping $g : \tilde{Q} \to H$ is called convex-like, if the set*

$$g(\tilde{Q}) + C_H$$

is convex.

Let us firstly consider an interesting result

Theorem 3.19. *Let the ordering cone C_H be closed and the joint mapping*

$$(J, g) : \tilde{Q} \to \mathbb{R} \times H$$

be convex-like. Then x^{opt} is a minimal solution of the problem (2.1) if and only if x^{opt} is a minimal solution of the following problem:

$$\min_{x \in \tilde{Q}} \sup_{v(\cdot) \in C_H^*} \{J(x) + v(g(x))\},$$

where C_H^* is a dual cone of C_H. In this case the extremal values of both problems are equal.

We refer to [276] for the corresponding proof of Theorem 3.19.

Now consider the "dual problem" associated with the originally given primal problem (2.1). We have

$$\max_{v \in C_H^*} \inf_{x \in \tilde{Q}} (J(x) + v(g(x))). \tag{3.4}$$

This problem is evidently equivalent to the following:

$$\begin{aligned} &\text{maximize } \lambda \\ &\text{subject to } J(x) + v(g(x)) \\ &x \in \tilde{Q},\ \lambda \in \mathbb{R},\ v(\cdot) \in C_H^*. \end{aligned} \tag{3.5}$$

If $v^{opt} \in C_H^*$ is a maximal solution of the dual problem (3.4) with the maximal value λ^{opt}, then

$$(v^{opt}, \lambda^{opt})$$

is a maximal solution of the problem (3.5). Conversely, for every maximal solution

$$(v^{opt}, \lambda^{opt})$$

of (3.5) v^{opt} is a maximal solution of the dual problem with the maximal value λ^{opt}.

We next study the weak and strong duality theorems. These classic results establish the sense in which the primal and dual problems (2.1) and (3.4) are equivalent.

Theorem 3.20. *Let all the above assumptions be satisfied. For every $\hat{x} \in \tilde{Q}$ and for every $v(\cdot) \in C_H^*$ the following inequality is satisfied:*

$$\inf_{x \in \tilde{Q}} \{J(x) + v(g(x))\} \leq J(\hat{x}).$$

It follows immediately from the above weak duality theorem that the maximal value of the dual problem is bounded from above by the minimal value of the primal problem (if these values exist). We obtain a lower bound of the minimal value of the primal problem, if one determines the value of the objective functional of the dual problem at an arbitrary element of the constraint set of the dual problem. If the primal and dual problem are solvable, then it is not guaranteed in general that the extremal values of these two problems are equal. If these two problems are solvable and the extremal values are not equal, then one speaks of a so-called "duality gap."

We now formulate the strong duality result.

Theorem 3.21. *Assume that all the above conditions are satisfied and, moreover, the ordering cone C_H has a nonempty interior* $\mathrm{int}(C_H)$. *If the primal problem (2.1) is solvable and the generalized Slater condition is satisfied, i.e., there is a vector*

$$\hat{x} \in \tilde{Q}$$

with

$$g(\hat{x}) \in \text{int}(C_H),$$

then the dual problem (3.4) is also solvable and the extremal values of the two problems are equal.

Relationships between the primal and the dual problem can also be described by a saddle point behavior of the Lagrange functional. These relationships will be investigated in this section. First, we define the notion of the Lagrange functional which has already been mentioned in the context of the generalized Lagrange multiplier rule (see Section 3.3). Consider the Lagrange functional

$$L : \tilde{Q} \times C_H^* \to \mathbb{R}$$

with

$$L(x, v(\cdot)) = J(x) + v(g(x)).$$

Definition 3.6. *A point*

$$(\hat{x}, \hat{v}(\cdot)) \in \tilde{Q} \times C_H^*$$

is called a saddle point of the Lagrange functional $L(\cdot, \cdot)$ if

$$L(\hat{x}, v(\cdot)) \leq L(\hat{x}, \hat{v}(\cdot)) \leq L(x, \hat{v}(\cdot))$$

for all $x \in \tilde{Q}$ and all $v(\cdot) \in C_H^*$.

The next result is a celebrated John von Neumann theorem.

Theorem 3.22. *Let all the above assumptions be satisfied. A point*

$$(\hat{x}, \hat{v}(\cdot)) \in \tilde{Q} \times C_H^*$$

is a saddle point of the Lagrange functional $L(\cdot, \cdot)$ if and only if

$$L(\hat{x}, \hat{v}(\cdot)) = \min_{x \in \tilde{Q}} \sup_{v(\cdot) \in C_H^*} L(x, v) = \max_{v(\cdot) \in C_H^*} \inf_{x \in \tilde{Q}} L(x, v).$$

Theorem 3.22 makes it possible to establish a relationship between a saddle point of the Lagrange functional and the solutions of the primal and dual problems.

Theorem 3.23. *Let all the above assumption be satisfied and, in addition, let the ordering cone C_H be closed. A point*

$$(\hat{x}, \hat{v}(\cdot)) \in \tilde{Q} \times C_H^*$$

is a saddle point of the Lagrange functional $L(\cdot, \cdot)$ if and only if $\hat{x}$ is a solution of the primal problem (2.1), $\hat{v}(\cdot)$ is a solution of the dual problem (3.4), and the extremal values of the two problems are equal.

We refer to [178] for a formal proof of Theorem 3.23. Let us close this section by presenting the specific sufficient condition for the existence of a saddle point of the Lagrange functional.

Theorem 3.24. *Assume that all the technical conditions given above are satisfied. Let the ordering cone C_H be closed and have an interior $\mathrm{int}(C_H)$. If x^{opt} is an optimal solution of the primal problem (2.1) and the generalized Slater condition is satisfied, i.e., there is one*

$$\hat{x} \in \tilde{Q}$$

with

$$g(\hat{x}) \in \mathrm{int}(C_H),$$

then there is $\hat{v}(\cdot) \in C_H^*$ so that

$$(x^{opt}, \hat{v}(\cdot))$$

is a saddle point of the Lagrange functional $L(\cdot, \cdot)$.

Theorem 3.24 can be proved by a direct application of the above generic results (Theorem 3.23). Note that it can also be derived without the assumption that the ordering cone C_H is closed.

3.5 Well-Posedness and Regularization

The notations of well-posedness in the optimization theory are significant as far as the numerical solution is involved. Ill-posed problems in the sense of Tikhonov or Hadamard should be handled with special care, since numerical methods will fail in general, and regularization techniques will be required.

Let us consider a proper extended real-valued functional

$$\Upsilon : \mathcal{W} \to (-\infty, \infty]$$

on a convergence space $\mathcal{W}$. The (global) minimization problem

$$\min_{w \in \mathcal{W}} \Upsilon(w) \tag{3.6}$$

is called "Tikhonov well-posed" if and only if there exists exactly one global minimizer w^{opt} and every minimizing sequence for (3.6) converges to w^{opt}. Otherwise, (3.6) is said to be "Tikhonov ill-posed." Note that uniqueness of the global minimizer to (3.6) does not imply well-posedness. The first special concept for ill-posed problems, suggested by Tikhonov, was proposed in [303]. In [304] Tikhonov pointed out that many optimal control problems (OCPs) (involving ordinary differential equations) are ill-posed with respect to the convergence of the minimizing sequence. Essential progress in the development of solution methods for ill-posed optimization problems was initiated by the work of Mosco [250]. A regularization method using the stabilizing properties of the proximal mapping (see [275,153]) was introduced by Martinet in [237].

We give here easy examples of well- and ill-posed OCPs governed by ordinary differential equations (see [144,308,336] and [26,28]).

Example 3.3. *Consider the following ill-posed OCP:*

$$\begin{aligned} &\text{minimize } J(u(\cdot)) := \int_0^1 x^2(t)dt \\ &\text{subject to } \dot{x}(t) = u(t), \ \text{ a.e. on } [0, 1], \\ &x(0) = 0, \\ &u(\cdot) \in \mathbb{L}^2([0, 1]), \ |u(t)| \leq 1. \end{aligned}$$

By $\mathbb{L}^2([0, 1])$ *we denote the standard Lebesgue space of all square integrable functions* $u : [0, 1] \to \mathbb{R}$. *The unique optimal control is*

$$u^{opt}(t) = 0$$

a.e., however the following minimizing sequence

$$u_r(t) = \sin(2\pi rt)$$

does not converge in the sense of $|| \cdot ||_{\mathbb{L}^2([0,1])}$. *Evidently,*

$$x_r(t) = \frac{1}{2\pi r}(1 - \cos(2\pi rt)), \ x_r(1) = 0,$$

and

$$\lim_{r \to \infty} J(u_r(\cdot)) = \lim_{r \to \infty} \frac{3}{8\pi^2 r^2} = J(u^{opt}(\cdot)) = 0.$$

In [29] we extend the ill-posed OCP from Example 3.3 by the additional target condition $x(1) \leq 0$. The extended OCP is also ill-posed. It is evident that well- or ill-posedness properties of an optimization problem depend strongly on the norm under consideration.

Example 3.4. *The minimization of*

$$\int_0^1 (x^2(t) + u^2(t))dt$$

subject to

$$\begin{aligned} \dot{x}(t) &= u(t), \\ x(0) &= 0, \\ |u(t)| &\leq 1 \text{ a.e.} \end{aligned}$$

has the unique optimal control

$$u^{opt}(t) = 0$$

a.e. on $[0, 1]$. *This OCP is well-posed with respect to* $|| \cdot ||_{\mathbb{L}^2([0,1])}$. *If one strengthens the convergence in* $\mathbb{L}^\infty(0, 1)$, *then the problem becomes ill-posed. For example, the minimizing sequence*

$$u_r(t) = \begin{cases} 0 & \text{if } t > 1/r, \\ s & 0 \leq t \leq 1/r, \end{cases}$$

where

$$s = \text{const},$$

does not converge in the sense of the following norm:

$$||u_r(\cdot) - u^{opt}||_{\mathbb{L}^\infty([0,1])} = s \quad \forall r \in \mathbb{N}.$$

One of the most prominent OCPs, namely, the Fuller problem, is also an ill-posed OCP (see, e.g., [166]). The ill-posed OCPs governed by partial differential equations are of frequent occurrence. For ill-posed OCPs with partial differential equations see, e.g., [202] and the references therein. For the further examples of well- and ill-posed OCPs see [336]. In our work we deal with OCPs in the presence of additional constraints. Generally the proof of the well-posedness for a constrained OCP is a very sophisticated theoretical problem. Therefore, an effective stable computational method for OCPs with additional constraints contains, as a rule, a numerical procedure for regularizing possible ill-posed problems.

An OCP with different additional constraints does not need to have only one optimal solution. On the other hand, our prime interest is to study minimizing sequences generated by a concrete numerical method. In practice, we focus our attention on the convergence properties and on the consistence of a numerical method. A general theoretic investigation (in the sense of Tikhonov) of the convergence properties for every minimizing sequence is usually not necessary. Therefore, in our works [27–29] we modify the standard Tikhonov approach and use the "stability" concept (see [28], Definition 2.1, p. 395). Needless to say that the concept of the numerical stability is weaker in comparison to the Tikhonov definition of well-posedness and is closely related to a chosen computational method. In [27,29] we propose some modifications of the classic proximal point approach and construct numerically stable approximations for constrained OCPs. With the aid of the presented proximal-based theory, usable numerical methods and optimization methods (see, e.g., [291]) can also be applied to ill-posed OCPs.

It should be mentioned that in the literature alternative definitions of well-posedness in optimization occur that are based on other viewpoints. Note that well-posedness and ill-posedness concepts for variational and optimization problems are comprehensively discussed in the books of Dontchev and Zolezzi [144], Kaplan and Tichatschke [202], and Vasil'ev [308]. For the theory of "locally ill-posed problems" see [191] and [33]. An optimization problem of the type (1.1) is called Hadamard well-posed if there exists exactly one global minimizer w^{opt} and, roughly speaking, w^{opt} depends continuously upon the parameters of the given problem. The notation of Hadamard well-posedness in optimization reminds us of the analogous concept for boundary value problems in mathematical physics [182]. More important than the mere similarity, there are significant results, showing that many linear operator equations, or variational inequalities, are well-posed in the classical sense of Hadamard if and only if an associated minimization problem has a unique solution, which depends continuously on the parameters of the problem. There are many links between Tikhonov and Hadamard definitions of well-posedness in the optimization theory [144]. In the book [33] we consider the linkages between both well-posedness concepts in the framework of stable operators in Banach spaces (see Chapter 2 of [33]). This investigation has culminated in the proof of main theorems for linear and nonlinear differentiable stable operators.

We now present our results in connection with the convergence analysis of a proximal-like method. Moreover, we discuss some relations to the Tikhonov regularization technique. As mentioned above the proximal point (regularization) method, suggested by Martinet in [237] and developed by Rockafellar [275], is one of the most popular stable methods for solving nonlinear equations, convex ([23,24]) and nonconvex ([164,203]) optimization problems, and variational inequalities. In parallel with Tikhonov's regularization [303] the proximal point algorithm is the main method for treating ill-posed problems of mathematical programming (see, e.g., [237,202,203]). The proximal point method is rich in applications. The first application of this method for solving the problems of determining an element $z \in X$ such that

$$0 \in T(z),$$

where X is a real Hilbert space and a multifunction $T : X \to X$ is a maximal monotone operator, was suggested by Rockafellar [275]. The proximal point method can be used for example for solving the variational inequalities [203] and for studying the asymptotic behavior at infinity solutions of evolution equations [8]. The idea of approximation by application of the proximal-like methods was extended by Benker, Hamel, and Tammer [82] to OCPs. A great number of works is devoted to the classical variant of the proximal point method and its various modifications (see, e.g., [276,24,180,181]). One can find a fairly complete review of the main results in [203] and in [27,28].

Let Z be a real Hilbert space. We examine the problem of convex minimization (similar to problem (2.1)), i.e.,

$$\begin{aligned} &\text{minimize } \psi(z) \\ &\text{subject to } z \in Q, \end{aligned} \tag{3.7}$$

where

$$\psi : Z \to \bar{\mathbb{R}}$$

is a proper convex lower semicontinuous functional and Q is a bounded, convex, closed subset of Z. By $\bar{\mathbb{R}}$ we denote here the extended real axis, i.e.,

$$\bar{\mathbb{R}} := \mathbb{R} \bigcup \{\infty\}.$$

It is evident that the abovementioned problem of finding a zero of a maximal monotone (multivalued) operator $T \equiv \partial\psi$ is equivalent to the following problem:

$$\psi(z) \to \min, \ z \in Z.$$

Here, $\partial\psi$ is the subdifferential of ψ. Let $F \subseteq Q$ be the set of optimal solutions of problem (3.7). Note that F is a convex and closed set [153]. We now introduce the *proximal mapping* [275,203], i.e.,

$$\begin{aligned} &\mathcal{P}_{\psi,Q,\chi} : \alpha \to \operatorname{argmin}_{z \in Q}[\psi(z) + \frac{\chi}{2}||z - \alpha||^2], \\ &\chi > 0, \ \alpha \in Z. \end{aligned} \tag{3.8}$$

The proximal mapping in (3.8) possesses the following properties:

- for all $z_1, z_2 \in Z$

$$||\mathcal{P}_{\psi,Q,\chi}(z_1) - \mathcal{P}_{\psi,Q,\chi}(z_2)||^2 \leq ||z_1 - z_2||^2 - \\ - ||z_1 - z_2 + \mathcal{P}_{\psi,Q,\chi}(z_2) - \mathcal{P}_{\psi,Q,\chi}(z_1)||^2;$$

- $\mathcal{P}_{\psi,Q,\chi}(z) = z$ if and only if $z \in F$;
- the functional

$$\eta(\alpha) := \min_{z \in Z}[\psi(z) + \frac{\chi}{2}||z - \alpha||^2]$$

is convex and continuously Fréchet differentiable on Z and

$$\nabla\eta(\alpha) = \chi(\alpha - \mathcal{P}_{\psi,Q,\chi}(\alpha))$$

(differentiability of ψ is not supposed).

Note that we use all these properties consistently in proofs of our main results [27–29]. Following Rockafellar [275] and Kaplan [203], we define the iterations of the classical proximal point method,

$$z_{cl}^0 \in Q, \\ z_{cl}^{i+1} \approx \mathcal{P}_{\psi,Q,\chi_i}(z_{cl}^i), \ i = 0, 1, \dots, \tag{3.9}$$

where $\{\chi_i\}$ is a given sequence with

$$0 < \chi_i \leq C < \infty$$

and

$$||z_{cl}^{i+1} - \mathcal{P}_{\psi,Q,\chi_i}(z_{cl}^i)|| \leq \epsilon^i, \ i = 0, 1, \dots, \ \sum_{i=0}^{\infty} \frac{\epsilon^i}{\chi_i} < \infty. \tag{3.10}$$

For the special case

$$\epsilon^i = 0, \ i = 0, 1, \dots,$$

the method (3.9)–(3.10) reduces to the exact scheme

$$\tilde{z}_{cl}^0 \in Q, \ \tilde{z}_{cl}^{i+1} = \mathcal{P}_{\psi,Q,\chi_i}(\tilde{z}_{cl}^i), \ i = 0, 1, \dots.$$

Under the condition (3.10), the sequence $\{z_{cl}^i\}$ converges in the weak topology to some element

$$z_{cl}^{opt} \in F.$$

Some alternative convergence conditions are discussed in [275,181]. Besides, (2.4) implies the convergence of the objective values

$$\psi(z_{cl}^i)$$

to $\psi(z_{cl}^{opt})$. In other words

$$\{z_{cl}^i\},\ i=0,1,...,$$

is a minimizing sequence. The weak convergence of $\{z_{cl}^i\}$ may fail if instead of

$$\sum_{i=0}^{\infty}\frac{\epsilon_i}{\chi_i}<\infty\ (\text{or}\ \sum_{i=0}^{\infty}\epsilon_i<\infty)$$

in (3.10), one has only $\epsilon_i \to 0$ (see [275]). Moreover, in the specific cases the corresponding sequence of the classic proximal point method converges weakly, but not strongly to a minimizing point of ψ [181]. Under some restrictive additional assumptions, Rockafellar proves the strong convergence of $\{z_{cl}^i\}$ to a unique solution of (3.7) (see [275], Theorem 2). Note that the condition

$$\sigma_i \to \infty,$$

where

$$\sigma_i := \sum_{p=0}^{i}\frac{1}{\chi_p},$$

is the weakest condition in order to ensure that

$$\psi(z_{cl}^i) \downarrow \inf_{z\in Q}\psi(z).$$

If

$$\sigma_i \to \sigma < \infty,$$

then $\{z_{cl}^i\}$ always converges strongly, i.e.,

$$||z_{cl}^{i+r}-z_{cl}^i|| \le \sum_{p=i+1}^{i+r}||z_{cl}^{p-1}-z_{cl}^p|| = \sum_{p=i+1}^{i+r}\frac{1}{\chi_p}||y_p|| \le$$
$$\le (\sum_{p=i+1}^{i+r}\frac{1}{\chi_p})||y_{i+1}||,$$

where

$$y_p := \chi_i(z_{cl}^{p-1} - z_{cl}^p).$$

Since

$$\sigma_i \to \sigma,$$

we see that $\{z_{cl}^i\}$ is a Cauchy sequence, and therefore converges strongly to some point z^∞, even if ψ does not have a minimizer (see [181])! If

$$F \neq \emptyset,$$

we have

$$||z - z^\infty|| \leq \sum_{p=1}^{\infty} ||z_{cl}^{p-1} - z_{cl}^p|| = \sum_{p=1}^{\infty} \frac{1}{\chi_p} ||y_p|| \leq \sigma ||y_1||,$$

and

$$\text{dist}(z^\infty, F) \geq \text{dist}(z, F) - ||z - z^\infty|| \geq \text{dist}(z, F) - \sigma ||y_1||,$$

where $\text{dist}(\cdot, \cdot)$ is a distance function in Z. If σ is small, then

$$\text{dist}(z^\infty, F) > 0,$$

and $z^\infty \notin F$. Finally, note that in [181] Güler introduced a variant of the proximal point method for (3.7). This method converges under the condition

$$\sum_{i=0}^{\infty} \frac{1}{\sqrt{\chi_i}} < \infty.$$

In [27] we applied a modification of this method to OCPs with constraints.

From the viewpoint of numerical analysis and constructive computational methods for optimization problems (for example, for OCPs) the strong convergence of a minimizing sequence is of practical significance. Numerical methods for OCPs based on the minimizing sequence of controls [155,269] are especially attractive when we have the strongly convergent minimizing sequence of control functions. We understand the "consistence" and "numerical stability" of an approximating technique for (2.1) in the sense of the following definition (see [28], Definition 2.1, p. 395).

Definition 3.7. *A method for solving the problem* (3.7) *is called stable if the associated minimizing sequence* z^k, $k = 0, 1, \dots$, *converges strongly to some element*

$$z^{opt} \in F,$$

i.e., if from

$$z^k \in Q, \ \psi(z^k) \to \psi(z^{opt}) \equiv \min_{z \in Q} \psi(z)$$

follows that $z^k \to z_{opt}$ *strongly for all* $z^0 \in Q$.

Some stability criteria in the sense of Definition 3.7 are also considered by Polyak [259]. Note that the approach of E. Polak uses the technique of the classic Lyapunov functions. Now we suggest the normality definition in the sense of Vasil'ev [308] (see also [28], Definition 2.2, p. 395)

Definition 3.8. *Let* $Q_\Omega \subseteq Q$ *and*

$$F_\Omega := Q_\Omega \cap F \neq \emptyset.$$

An optimal solution $z^{opt} \in F_\Omega$ *of* (3.7) *is called normal with respect to some function*

$$\Omega : Q_\Omega \to \bar{\mathbb{R}}$$

if

$$\min_{F_\Omega} \Omega(z) = \Omega(z^{opt}).$$

The classical proximal mapping can be used for creating a strong convergent minimizing sequence for problem (3.7). The questions of the *strong convergence* of proximal-like methods were considered for example by Bakushinskii [71], and Azhmyakov and Schmidt [28]. In our paper [28] we examined the strong convergence of a new proximal-based method for the problem (3.7). Using the classical proximal mapping $\mathcal{P}_{\psi,Q,\chi}$, we can construct the iterative procedure

$$\begin{aligned} z^i &= a^i z^0 + (1 - a^i)\mathcal{P}_{\psi,Q,\chi_i}(z^i), \\ i &= 0, 1, \dots, \end{aligned} \tag{3.11}$$

where $\{\chi_i\}$ is a given sequence with

$$0 < \chi_i \leq C < \infty$$

and

$$a^0 = 1,\ 0 < a^i < 1,\ \forall i = 1, 2, \dots,\ a^i \downarrow 0.$$

The sophisticated fixed point problem (2.5) can be approximated by the following iterative scheme ([26], pp. 399–400):

$$\begin{aligned} &z^i_{r+1} = a^i z^0 + (1 - a^i)\mathcal{P}_{\psi,Q,\chi_i}(z^i_r), \\ &z^i_0 = z^{i-1}_{N(i-1)},\ z^0_0 = z^0 \in Q, \\ &i = 0, 1, \dots,\ r = 0, 1, \dots N(i) - 1, \end{aligned} \tag{3.12}$$

where

$$N(i) \in \mathbb{N},\ N(0) = 1$$

and $\{a^i\}$ is the sequence given above. Let us now formulate the main results of [28] (see [28], Theorem 3.1, p. 398; Theorem 4.1, p. 401; Theorem 4.3, p. 404).

Theorem 3.25. *The sequence $\{z^i\}$ of the solutions of* (3.11) *converges strongly to some solution $z^{opt} \in F$ of problem* (3.7). *Let*

$$\Omega(z) := ||z - z^0||,\ z \in F,\ z^0 \in Q.$$

Then the element $z^{opt} \in F$ is a unique normal solution of problem (3.7) *with respect to the function $\Omega(\cdot)$.*

Theorem 3.26. *Suppose the sequences $\{N(i)\}$, $\{a^i\}$, $\{\chi_i\}$ in* (3.12) *satisfy the following conditions:*

$$\sum_{i=0}^{\infty} \frac{a^i N(i)}{\chi_i} < \infty,\ 0 < \chi_i \le C < \infty.$$

Then the sequence

$$\{z^i_{r+1}\},\ i = 0, 1, \dots,\ r = 0, \dots, N(i) - 1,$$

generated by (3.12), *is a minimizing sequence. This sequence converges weakly to an element of the set F.*

Theorem 3.27. *Suppose the sequence $\{N(i)\}$ satisfies the condition*

$$\begin{aligned} &\log_{b_i}\left(\frac{a^i}{qa^{i+1}}\right) + N(i)\log_{b_i}\left(\frac{1}{1 - a^i}\right) \le N(i + 1), \\ &i = 1, 2, \dots, \\ &0 < q < 1,\ N(1) \in \mathbb{N}, \end{aligned}$$

where

$$b_i := \frac{1}{1 - a^{i+1}}, \; i = 1, 2, ...$$

Assume that $\{z^i\}$ *generated by Eq.* (3.11) *converges strongly to some element* $z^{opt} \in F$. *Then the minimizing sequence* $\{z^i_{N(i)}\}$ *generated by the method* (3.12) *converges strongly to the same element* $z^{opt} \in F$.

The presented theorems provide a possible answer to Rockafellar's question: "the question of whether the weak convergence established by Martinet can be improved to strong convergence thus remains open" (see [275], p. 879). Note that the answer is known to be affirmative in the case of a quadratic functional ψ. This follows from the Krasnoselskii theorem [214]. The proofs of the formulated theorems are based on the Browder theorem [109] and on the analytic facts for *sunny retracts* developed by Shioji and Takahashi [297] (see also [216]). Moreover, we use some standard techniques of nonlinear analysis. It is well known that a minimizing sequence generated by the Tikhonov method converges to the set of normal solutions of problem (2.1) (in the sense of Definition 3.7) [303,308]. In the general case the sequence $\{z^i_{cl}\}$ does not possess this property. Note that the concept of normal solutions plays an important role in operations research [308].

In our case we deal with a proximal-like method (3.11)–(3.12) that converges to a normal solution,

$$z^{opt} \in F,$$

of problem (3.7). Moreover, the proposed iterative procedures are "dissipative." Let

$$z^0, \mathcal{P}_{\psi,Q,\chi_i}(z^i) \in \mathcal{W}^i_F,$$

where $\mathcal{W}^i_F$ is a convex neighborhood of the bounded, convex, closed set F. One can prove that

$$z^i \in \mathcal{W}^i_F.$$

Therefore, we call the method (3.11) "dissipative." On the other hand, the iterative schemes described in [71] do not possess (in general) these properties. In [28] (Theorem 3.2, pp. 398–399) we also extend the strongly convergent proximal point approach to the convex optimization problems in uniformly convex and uniformly smooth real Banach spaces (see, e.g., [211]).

Theorem 3.28. *Let $\{y^i\}$ be the sequence of solutions of the equations*

$$y^i = \tilde{a}^i y^0 + (1-\tilde{a}^i)\frac{1}{i+1}\sum_{j=0}^{i} \mathcal{P}^j_{\tilde{\psi},K,\chi_i}(y^i),\ y^0 \in \tilde{Q},$$

where $i = 0, 1, \ldots$ and $\tilde{Q}$ is a bounded, convex, closed subset of a uniformly convex, uniformly smooth real Banach space $\tilde{Z}$,

$$\tilde{\psi} : \tilde{Z} \to \bar{\mathbb{R}}$$

is a proper convex lower semicontinuous functional, and $\{\tilde{a}^i\}$ is a real sequence such that

$$\tilde{a}^0 = 1,\ 0 < \tilde{a}^i \leq 1,$$

and $\tilde{a}^i \to 0$. Let

$$\mathcal{P}_{\tilde{\psi},K,\chi_i}(y^i) := \operatorname{argmin}_{y\in\tilde{Q}}[\tilde{\psi}(y) + \frac{\chi_i}{2}||y - y^i||^2].$$

Then $\{y^i\}$ converges strongly to some solution of the following problem:

$$\begin{aligned}&\text{minimize } \tilde{\psi}(y)\\&\text{subject to } y \in \tilde{Q}.\end{aligned}$$

The classical proximal point method (3.8)–(3.10) is a variant of the general proximal-like methods using so-called Bregman functions and Bregman distances in a real Hilbert space ([115]) or in a reflexive real Banach space ([119,195,114–117]). The proximal-like method in Banach spaces was analyzed for the case of a quadratic Bregman function [4] and for the case of a general Bregman function ([119,75,76]). The approach proposed in [28] can also be used for the construction of the corresponding strongly convergent variant of the general proximal-like method with Bregman distances considered by Burachik and Iusem [115]. However, even in the case of real Hilbert space, the square of the norm leads always to simpler numerical algorithms. The scheme with a nonquadratic Bregman function has been proposed mainly with penalization purposes. We refer to [126,195] for details.

In [28] we also derive the useful estimates for the presented method (2.6) ([28], Lemma 4.1, pp. 400–401), i.e.,

$$||z_0^i - \mathcal{P}_{\psi,Q,\chi_i}(z_0^i)|| \leq \frac{2}{\sqrt{3\chi_i}}\sqrt{\psi(z_0^i) - \psi(\mathcal{P}_{\psi,Q,\chi_i}(z_0^i))}. \tag{3.13}$$

Note that the following estimate is also correct for an arbitrary point $z \in Q$:

$$\sqrt{\frac{2}{\chi}(\psi(z) - \psi(\mathcal{P}_{\psi,Q,\chi}(z)))} \geq ||z - \mathcal{P}_{\psi,Q,\chi}(z)||.$$

In the case of a Lipschitz continuous functional ψ one can use (3.13) for creating the general estimate ([28], Theorem 4.2, pp. 402–403).

Theorem 3.29. *Let $\{z^i\}$ be the sequence generated by Eq.* (3.11). *Assume that $\{z^i\}$ converges strongly to some element*

$$z^{opt} \in F$$

and that the functional ψ is Lipschitz *continuous with* Lipschitz *constant L. Let*

$$\{z^i_{r+1}\}$$

be the sequence generated by (2.6). *Then the estimation*

$$||z^{opt} - z^i_{N(i)}|| \leq ||z^{opt} - z^i|| + (1 - a^i)^{N(i)} \mathrm{diam} Q +$$
$$+ \frac{2(1 - a^i)^{N(i)}}{a^i \sqrt{3\chi_i}} \sqrt{|\psi(z^i_0) - \psi(z^i_1)| + a^i L \mathrm{diam} Q},$$

where

$$||z^{opt} - z^i|| \to 0,$$

holds.

Example 3.5. *We now apply our method* (3.11) *to* the OCP *from* Example 3.3 *and compute the approximate optimal control $\tilde{u}(\cdot)$. Clearly, we have*

$$Z = \mathbb{L}^2(0, 1)$$

and

$$J(u(\cdot)) = \int_0^1 (\int_0^t u(\tau) d\tau)^2 dt.$$

It is easy to see that the functional J here is lower semicontinuous and proper convex. The set of all admissible control functions is a bounded, convex, and closed subset of the Hilbert *space $\mathbb{L}^2(0, 1)$. The computed optimal control $\tilde{u}(\cdot)$ satisfies the inequality*

$$||u^* - \tilde{u}||_{\mathbb{L}^2} \leq \delta = 10^{-5}.$$

We have considered seven i-steps. Therefore, we have

$$\sum_{i=0}^{6} N(i) = 60$$

iterations of the algorithm. The approximation $\tilde{u}(\cdot)$ of the optimal control has the following bang-bang *type:*

$$\tilde{u}(t) = \begin{cases} h & \text{if } t \in [t_{2j-1}, t_{2j}), \\ -h & \text{otherwise}, \end{cases}$$

where h is a positive constant $h < \delta$ and

$$[t_{2j-1}, t_{2j}) \subset [0, 1], \; j = 1, \ldots, M \; (M \in \mathbb{N}),$$

are some equidistant half-open intervals. The OCP *from* Example 3.3 *has been solved (with the preassigned exactness $\delta = 10^{-5}$) by application of the* Bakushinskii *method* [71]. *In this case one has used more iterations.*

3.6 Numerical Methods in Convex Programming

In this section we study first-order numerical schemes for the main convex optimization problem (2.1). Additionally to the basic assumptions from Section 2.2 suppose that the objective functional $J(\cdot)$ is differentiable and consider a general variant of the projected gradient method for the concrete numerical treatment of (2.1). It can be expressed as follows:

$$x_{(l+1)}(\cdot) = \gamma_l P_Q\left[x_{(l)}(\cdot) - \alpha_l \nabla J(x_{(l)}(\cdot))\right] + (1 - \gamma_l)x_{(l)}(\cdot), \; l \in \mathbb{N}, \tag{3.14}$$

where P_Q is the operator of projection on the convex set Q (see Definition 3.3) and $\{\alpha_l\}$ and $\{\gamma_l\}$ are sequences of some suitable step sizes associated with the method. By ∇ we denote here the Fréchet derivative of $J(\cdot)$. Note that we consider here the space $X = H$ (a real Hilbert space). Therefore we identify here the abstract real Hilbert space H with a suitable space of functions, i.e.,

$$x : R \to \mathbb{R}^m.$$

In that case we use the natural notation $x(\cdot)$ for the elements of this space of functions.

Note that several choices are possible for the step sizes α_l and γ_l in (3.14). Let us describe the established strategies for the constructive step size selection.

- The constant step size: $\gamma_l = 1$ and

$$\alpha_l = \alpha > 0$$

  for all $l \in \mathbb{N}$.

- Armijo line search along the boundary of Q: $\gamma_l = 1$ for all $l \in \mathbb{N}$ and α_l is determined by

$$\alpha_l := \bar{\alpha}\theta^{\chi(l)}$$

for some

$$\bar{\alpha} > 0,$$

$\theta,\ \delta \in (0, 1)$, where

$$\chi(l) := \min\{\chi \in \mathbb{N} \mid J(P_Q\left[x_{(l,s)}(\cdot)\right]) \leq J(x_{(l)}(\cdot)) - \delta\langle\nabla J(x_{(l)}(\cdot)), x_{(l)}(\cdot) - P_Q[x_{(l,s)}(\cdot)]\rangle\}$$

and

$$x_{(l,s)}(\cdot) := x_{(l)}(\cdot) - \bar{\alpha}\theta^{\chi}\nabla J(x_{(l)}(\cdot)).$$

- Armijo line search along the feasible direction:

$$\{\alpha_l\} \subset [\bar{\alpha}, \hat{\alpha}]$$

for some

$$\bar{\alpha} < \hat{\alpha} < \infty$$

and γ_l is determined by the following Armijo rule:

$$\gamma_l := \theta^{\chi(l)},$$

for some $\theta,\ \delta \in (0, 1)$, where

$$\chi(l) := \min\{\chi \in \mathbb{N} \mid J(x_{(l,s)}(\cdot)) \leq J(x_{(l)}(\cdot)) - \theta^{\chi}\delta\langle\nabla J(x_{(l)}(\cdot)), x_{(l)}(\cdot) - P_Q[w_l(\cdot)]\rangle\}$$

and

$$x_{(l,s)}(\cdot) := \theta^{\chi} P_Q[w_l(\cdot)] + (1 - \theta^{\chi})x_{(l)}(\cdot).$$

- Exogenous step size before projecting: $\gamma_l = 1$ for all $l \in \mathbb{N}$ and α_l given by

$$\alpha_l := \frac{\delta_l}{||\nabla J(x_{(l)}(\cdot))||_H}, \quad \sum_{l=0}^{\infty}\delta_l = \infty, \ \sum_{l=0}^{\infty}\delta_l^2 < \infty.$$

Recall that under some nonrestrictive assumptions the projected gradient iteration (3.14) generates a minimizing sequence for the main problem (2.1). Many useful and mathematically exact convergence theorems for the gradient iterations (3.14) can be found in [175,178,259]. A comprehensive discussion of the weakly and strongly convergent variants of the basic gradient method can be found in [79,87,88]. We also refer to [27,29,30,150,269,300,317] for some specific convergence results obtained for the gradient-based schemes applied to hybrid and switched OCPs.

The first strategy from the ones presented above was analyzed in [178] and its weak convergence was proved under Lipschitz continuity of $\nabla J(\cdot)$. The main difficulty here is the necessity of taking

$$\alpha \in (0, 2/L),$$

where L is the Lipschitz constant for $\nabla J(\cdot)$ (see also [87]).

Note that the second gradient-based strategy requires one projection onto Q for each step of the inner loop resulting from the Armijo line search. Therefore, many projections might be performed for each iteration l, making the second strategy inefficient when the projection onto the set Q cannot be computed explicitly. On the other hand, the third strategy demands only one projection for each outer step, i.e., for each iteration l. The second and third optimization strategies from the variants presented above are the constrained versions of the line search proposed in [16] for solving unconstrained optimization problems. Under existence of minimizers and some convexity assumptions for the minimization problem under consideration, it is possible to prove, for the second and third strategies, convergence of the whole sequence to a solution in finite-dimensional spaces (see [115]).

The last strategy from the approaches presented above, as its counterpart in the unconstrained case, fails to be a descent method. Furthermore, it is easy to show that this approach implies

$$||x_{(l+1)}(\cdot) - x_{(l)}(\cdot)|| \leq \delta_l$$

for all l, with δ_l given above. This reveals that convergence of the sequence of points generated by this exogenous approach can be very slow (step sizes are small). Note that the second and third strategies allow for occasionally large step sizes because both strategies employ all information available at each l-iteration. Moreover, the last strategy does not take into account the values of the objective functional for determining the "best" step sizes. These characteristics, in general, entail poor computational performance. The basic convergence results for the obtained approximating sequence

$$\{x_{(l)}(\cdot)\}$$

can be stated as follows.

Theorem 3.30. *Assume that all hypotheses from* Section 3.1 *are satisfied. Consider a sequence* $\{x_{(l)}(\cdot)\}$ *generated by method* (3.14) *with a constant step size* α. *Then for an admissible initial point*

$$x_{(0)}(\cdot) \in Q$$

the resulting sequence $\left\{x_{(l)}(\cdot)\right\}$ *is a minimizing sequence for problem* (2.1), *i.e.,*

$$\lim_{l\to\infty} J(x_{(l)}(\cdot)) = J(x^*(\cdot)).$$

Additionally assume that $\nabla J(\cdot)$ *is Lipschitz continuous and*

$$\alpha \in (0, 2/L),$$

where $\{x_{(l)}(\cdot)\}$ *converges* H *weakly to a solution*

$$x^{opt} \in Q$$

of (2.1).

Let us now discuss shortly the main relaxation idea following the example set by the generic optimization problem (2.1). In (2.1) we have assumed the convexity of the admissible set $Q \subset H$. If this is not the case we follow the main relaxation concept and replace a nonconvex Q by the convexification (convex combination) conv(Q).

The second crucial assumption in (2.1) is the convexity of the objective functional $J(\cdot)$ on the set Q. In the case of a violation of this basic hypothesis we have to consider a suitable convexification of this functional. For example, one can study the closed convex hull

$$\bar{co}\{J(x(\cdot))\}$$

of the objective $J(x(\cdot))$. We refer to [190] for the exact concept of

$$\bar{co}\{J(x(\cdot))\}.$$

Note that the biconjugate $J^{**}(x(\cdot))$ of $J(x(\cdot))$ (determined on a real Hilbert space H) is equal to $\bar{co}\{J(x(\cdot))\}$ (see [190] for the formal proof). For a given objective functional $J(\cdot)$ getting its closed convex hull is a complicated, but at the same time fascinating, operation. Note that the objective functional $J(u(\cdot))$ in OCPs is usually a composite functional. This situation is a typical scenario of the general optimal control processes governed by ordinary differential equations, by hybrid and switched systems. The numerical calculation of $\bar{co}\{J(x(\cdot))\}$ is not broached in this book. A novel computational method for a practical evaluation of

$$\bar{co}\{J(u(\cdot))\}$$

is proposed in [243,287,288]. This method constitutes a generalization of the celebrated McCormic relaxation scheme in the specific case of a composite functional in an OCP. The closed convexification of such a functional is realized by solving an "auxiliary control system" (see [287] for details).

Using the above convex constructions of the objective functional and of the convexified set of admissible control functions, we can now formulate the following relaxed (convex) optimization problem:

$$\begin{aligned} &\text{minimize } \bar{co}\{J(x(\cdot))\} \\ &\text{subject to } x(\cdot) \in \text{conv}(Q). \end{aligned} \tag{3.15}$$

The problem (3.15) is in fact a classical relaxation of the possible nonconvex problem (2.1). The optimization procedure in (3.15) is determined over a convex set of admissible control inputs $\text{conv}(Q)$. The approximability property of the fully relaxed problem (3.15) can be expressed as follows ([190]):

$$\inf_{x(\cdot)\in H} J(x(\cdot)) = \inf_{x(\cdot)\in H} \bar{co}\{J(x(\cdot))\}.$$

This is simply a consequence of the following simple facts:

$$\begin{aligned} &\inf_{x(\cdot)\in H} J(x(\cdot)) = -J^*(0), \\ &J^*(x(\cdot)) = \bar{co}\{J(x(\cdot))\}, \end{aligned}$$

where $J^*(x(\cdot))$ is a conjugate of $J(x(\cdot))$. Finally note that Theorem 3.30 is also true for the relaxed (convexified) nonlinear program (3.15) with the replacement of the nonconvex elements $J(\cdot)$ and Q (in the nonconvex variant of problem (2.1)) by

$$\bar{co}\{J(x(\cdot))\}$$

and $\text{conv}(Q)$, respectively.

Let us note that there are many alternative numerical schemes for the algorithmic treatment of the convex program (2.1). The celebrated sequential programming method (SQP) is comprehensively discussed in [112,113]. One of the oldest methods in convex programming, namely, the ellipsoid method, is presented in [267]. We also refer to [260,263] for further numerical approaches.

We next present a specific numerical concept, namely, the "separate convex programming" (or nested) approach we use in relaxation of hybrid OCPs (HOCPs) and switched OCPs (SOCPs). Let

$$H = H_1 \otimes H_2$$

be a Cartesian product of two real Hilbert spaces. Consider the following minimization/maximization problem:

$$\begin{aligned} &\text{extremize } J(v_1, v_2) \\ &\text{subject to } (v_1, v_2) \in \mathcal{V}_1 \otimes \mathcal{V}_2 \subset H, \end{aligned} \tag{3.16}$$

where

$$\mathcal{V}_1 \subset H_1, \ \mathcal{V}_2 \subset H_2$$

are bounded, convex sets and

$$J(\cdot, v_2) : H_1 \to [-\infty, \infty], \ J(v_1, \cdot) : H_2 \to [-\infty, \infty]$$

are proper convex or concave (not mandatorily simultaneous) functionals for

$$v_1 \in H_1, \ v_2 \in H_2,$$

respectively. We next use the following natural notation:

$$v := (v_1, v_2), \ \mathcal{V} := \mathcal{V}_1 \otimes \mathcal{V}_2.$$

Recall that a convex functional $J(\cdot, v_2)$ is called "proper" if

$$J(\cdot, v_2) \neq -\infty$$

and its effective domain

$$\operatorname{dom}\{J(\cdot, v_2)\} := \{v_1 \in H_1 \mid J(v_1, v_2) < \infty\}$$

is a nonempty set. Evidently, the same concept can also be applied to $J(v_1, \cdot)$. We next omit the word "proper," because only such convex functionals will be considered in this book. Let us note that the optimization problem (2.1) belongs to the "separate convex programming" and constitutes a useful abstract framework for many practically oriented optimization problems (see, e.g., [23]). The "separate" functionals $J(v_1, \cdot)$ and $J(\cdot, v_2)$ can be convex or concave. Note that the "partial" convexity (or concavity) of $J(\cdot, v_2)$ and $J(v_1, \cdot)$ does not imply the "global" convexity (concavity) property of $J(\cdot)$. For example, the bilinear functional

$$J(v) = \langle v_1, v_2 \rangle_H,$$

where $\langle \cdot, \cdot \rangle_H$ denotes a scalar product in H, is a nonconvex functional. In the case of a convex $J(\cdot, v_2)$ and concave $J(\cdot, v_2)$ we call (2.1) a convex–concave problem. In addition to the above formal conditions, we next suppose that $J(\cdot)$ is bounded on a set

$$\mathcal{V} + \epsilon \mathcal{B} \subset \operatorname{int}\{\operatorname{dom}\{J(\cdot, \cdot)\}\},$$

where $\epsilon > 0$ and $\mathcal{B}$ is the open unit ball of H. We call the basic problem (3.16) a "minimization problem" if "extremize" in (3.16) is replaced by "minimize." Consequently, the "maximization version" of (2.1) corresponds to "maximize" instead of "extremize."

Since $J(\cdot)$ is assumed to be bounded on $\mathcal{V} + \epsilon\mathcal{B}$, we conclude that this objective functional is continuous on this set (see [153,192,275]). Consequently, it is also Lipschitz continuous on the set $\mathcal{V}$ (see [275], Theorem 10.4) and the existence of an optimal solution

$$v^{opt} := (v_1^{opt}, v_2^{opt}) \in \mathcal{V}$$

to problem (3.16) is guaranteed by application of the Weierstrass theorem (see, e.g., [131, 192]). We now consider the "minimization" version of the basic optimization problem (3.16) and introduce the following function:

$$\begin{aligned} &F(\hat{v}_2) := \min_{v_1 \in \mathcal{V}_1} J(v_1, \hat{v}_2), \\ &\hat{v}_2 \in \mathcal{V}_2. \end{aligned} \tag{3.17}$$

Here $\hat{v}_2 \in \mathcal{V}_2$ is a fixed element. Consider now the auxiliary minimizing problem

$$\begin{aligned} &\text{minimize } F(\hat{v}_2) \\ &\text{subject to } \hat{v}_2 \in \mathcal{V}_2. \end{aligned} \tag{3.18}$$

Since (3.17) and (3.16) constitute the conventional convex (or concave) programs, the auxiliary function $F(\cdot)$ in (2.2) is well defined and the existence of an optimal solution

$$v_1^0,\ v_2^o \in \mathcal{V}_2$$

to the above problems is guaranteed (see [275]). Evidently, v_1^o depends on a concrete selection

$$\hat{v}_2 \in \mathcal{V}_2$$

in (3.16). The next fundamental result provides a theoretic basis for the proof of numerical consistence of some computational algorithms we will propose in the next sections.

Theorem 3.31. *The value*

$$\theta := \min_{\hat{v}_2 \in \mathcal{V}_2} F(\hat{v}_2)$$

in (3.18) is the overall minimal value for the initially given problem (3.16). A solution set

$$\text{Argmin}_{v \in \mathcal{V}} J(v)$$

of (3.16) is given as follows:

$$\text{Argmin}_{v \in \mathcal{V}} J(v) = J^{-1}(\theta).$$

In the case $(v_1^{opt}, v_2^{opt}) \in \text{Argmin}_{v\in\mathcal{V}} J(v)$ *we also have*

$$(v_1^{opt}, v_2^{o}),\ (v_1^{o}, v_2^{opt}) \in \text{Argmin}_{v\in\mathcal{V}} J(v).$$

Proof. Since the objective function $J(\cdot)$ is Lipschitz continuous on the set $\mathcal{V}$, the operation $J^{-1}(\theta)$ is well defined.

By definition of v^{opt} we evidently have

$$J(v_1^o, v_2^o) \geq J(v_1^{opt}, v_2^{opt}), \tag{3.19}$$

where

$$v_1^o \in \mathcal{V}_1$$

is an optimal solution of the minimization problem in (3.17). Moreover,

$$\begin{aligned} J(v_1^o, \hat{v}_2) = \min_{v_1\in\mathcal{V}_1} J(v_1, \hat{v}_2) \leq \\ J(v_1^{opt}, \hat{v}_2)\ \forall v_2 \in \mathcal{V}_2. \end{aligned}$$

Therefore we obtain

$$J(v_1^o, v_2^o) = \min_{\hat{v}_2\in\mathcal{V}_2} \min_{v_1\in\mathcal{V}_1} J(v_1, \hat{v}_2) \leq J(v_1^{opt}, v_2^{opt}). \tag{3.20}$$

From (3.19) and (3.20) it follows that

$$(v_1^o, v_2^2) \in \text{Argmin}_{v\in\mathcal{V}} J(v).$$

We next observe

$$J^{-1}(\theta) = \{(v_1, \hat{v}_2) \in \mathcal{V}_1 \times \mathcal{V}_2 \mid v_1 \in \chi_1(\hat{v_2}),\ \hat{v}_2 \in \chi_2\},$$

where

$$\begin{aligned} \chi_1(\hat{v_2}) &:= \{v_1 \in \mathcal{V}_1 \mid J(v_1, v_2) = F(v_2)\}, \\ \chi_2 &:= \{v_2 \in \mathcal{V}_2 \mid F(v_2) = \theta\}. \end{aligned}$$

Therefore

$$\text{Argmin}_{v\in\mathcal{V}} J(v) = J^{-1}(\theta).$$

By definition of v^{opt} we also have

$$J(v_1^{opt}, v_2^o) \geq J(v_1^{opt}, v_2^{opt}). \tag{3.21}$$

Moreover,

$$\min_{v_1 \in \mathcal{V}_1} J(v_1, \hat{v}_2) \leq J(v_1^{opt}, \hat{v}_2)$$

for all $v_2 \in \mathcal{V}_2$. This implies

$$J(v_1^{opt}, v_2^o) = \min_{\hat{v}_2 \in \mathcal{V}_2} \min_{v_1 \in \mathcal{V}_1} J(v_1, \hat{v}_2) \leq J(v_1^{opt}, v_2^{opt}). \tag{3.22}$$

From (3.21)–(3.22) we deduce that

$$(v_1^{opt}, v_2^o) \in \mathrm{Argmin}_{v \in \mathcal{V}} J(v)$$

and the symmetry argument also implies

$$(v_1^o, v_2^{opt}) \in \mathrm{Argmin}_{v \in \mathcal{V}} J(v).$$

The proof is completed. □

Finally note that Theorem 3.31 provides an analytic basis for various "splitting" methods in the general mathematical programming. The main difficulty of this "nested optimization" approach consists in a constructive determination of an expression $v_1^o(\hat{v}_2)$. For a convex (or concave, or convex–concave) case Theorem 3.31 constitutes a natural solution approach.

3.7 Notes

Let us complete here with useful references for the topics discussed in this chapter: [4,8,11, 13,16,23–25,27–33,71,75–77,79,82,83,87,88,91,114–117,119,126,137,153,158,162,164,165, 178,180–182,188,190,191,195,198,202–204,213,220,221,227,228,237,246–250,252,256, 259–263,275,276,278,291,303,304,308,327,336,337].

CHAPTER 4

Short Course in Continuous Time Dynamic Systems and Control

4.1 Carathéodory Differential Equations

Let us start with the fundamental concept of an abstract function of two variables that is measurable in one variable and continuous in another. We refer to [6] for the measurability (Borel measurability) concepts and for some future mathematical details.

Definition 4.1. *Let* (S, Σ) *be a measurable space, and let* X *and* Y *be topological spaces. A function* $f : S \times X \to Y$ *is a Carathéodory function if for each* $x \in X$ *the function*

$$f^x := f(\cdot, x) : S \to Y$$

is $(\Sigma, \mathcal{B}_Y)$ *measurable and for each* $s \in S$ *the function*

$$f^s := f(s, \cdot) : X \to Y$$

is continuous.

We next recall some important properties of the Carathéodory functions. These functions are jointly measurable.

Theorem 4.1. *Let* (S, Σ) *be a measurable space,* X *a separable metrizable space, and* Y *a metrizable space. Then every Carathéodory function*

$$f : S \times X \to Y$$

is jointly measurable.

Proof. Let d and ρ be compatible metrics on X and Y, respectively. Let $\{x_1, x_2, ...\}$ be a countable dense subset of X and observe that, since $f(s, \cdot)$ is continuous, $f(s, x)$ belongs to the closed set F if and only if for each n there is some x_m with

$$d(x, x_m) < 1$$

A Relaxation-Based Approach to Optimal Control of Hybrid and Switched Systems
https://doi.org/10.1016/B978-0-12-814788-7.00010-2

and

$$f^{-1}(F) = \bigcap_{n=1}^{\infty} \bigcup_{m=1}^{\infty} \{s \in S : f(s, x_m) \in N_{1/n}(F)\} \times B_{1/n}(x_m),$$

where $B_{1/n}(x_m)$ is a ball of the radius $1/n$ around the point x_m and

$$N_{1/n}(F) := \{y \in Y : \rho(y, F) < 1/n\}.$$

Since f is measurable in s and $N_{1/n}(F)$ is open (and hence Borel),

$$\{s \in S : f(s, x_m) \in N_{1/n}(F)\}$$

is measurable. Thus $f^{-1}(F)$ is measurable. The proof is completed. □

Recall now a useful abstract result from the general measurability theory [6].

Theorem 4.2. *Let (X, Σ), (X_1, Σ_1), and (X_2, Σ_2) be measurable spaces, and let*

$$f_1 : X \to X1$$

and $f_2 : X \to X_2$. Define $f : X \to X_1 \times X_2$ by

$$f(x) := (f_1(x), f_2(x)).$$

Then

$$f : (X, \Sigma) \to (X_1 \times X_2, \Sigma_1 \otimes \Sigma_2)$$

is measurable if and only if the two functions $f_1 : (X, \Sigma) \to (X_1, \Sigma_1)$ and $f_2 : (X, \Sigma) \to (X_2, \Sigma_2)$ are both measurable.

The measurability of a composition of the Carathéodory functions is established in the following theorem.

Theorem 4.3. *Let (S, Σ) be a measurable space, X, Y_1, and Y_2 be separable metrizable spaces, and Z be a topological space. If*

$$f_i : S \times X \to Y_i, \ i = 1, 2,$$

are Carathéodory functions and $g : Y_1 \times Y_2 \to Z$ is Borel measurable, then the composition $h : S \times X \to Z$ defined by

$$h(s, x) = g(f_1(s, x), f_2(s, x))$$

is jointly measurable.

Proof. By Theorem 4.1 each f_i is

$$(\Sigma \bigotimes \mathcal{B}_X, \mathcal{B}_{Y_i})$$

measurable from $S \times X$ into Y_i. Therefore, by Theorem 4.2, the function

$$(s, x) \to (f_1(s, x), f_2(s, x))$$

is $(\Sigma \otimes \mathcal{B}_X, \mathcal{B}_{Y_1} \otimes \mathcal{B}_{Y_2})$ measurable. Now g is

$$(\mathcal{B}_{Y_1} \times Y_2, \mathcal{B}_Z)$$

measurable, and since each Y_i is separable,

$$\mathcal{B}_{Y_1} \otimes \mathcal{B}_{Y_2} = \mathcal{B}_{Y_1 \times Y_2},$$

so the composition

$$h = g \circ (f_1, f_2)$$

is measurable. The proof is completed. □

We next present the last necessary measurability result related to the Carathéodory functions, namely, a theorem about the measurability of the Carathéodory functions into $\mathbb{C}(X, Y)$. Let us use a simplified notation for $f : S \to Y^X$ and write f_s for $f(s)$. Given a function

$$f : S \times X \to Y$$

define $\hat{f} : S \to Y^X$ by

$$\hat{f}_s(x) = f(s, x).$$

By the same manner we determine

$$\bar{g} : S \times X \to Y$$

by $g(s, x) = g_s(x)$.

Theorem 4.4. *Let (S, Σ) be a measurable space, X a compact metrizable space, (Y, d) a separable metric space, and $\mathbb{C}(X, Y)$ be endowed with the topology of d-uniform convergence.*

- *If $f : S \times X \to Y$ is a Carathéodory function, then $\hat{f}$ maps S into $\mathbb{C}(X, Y)$ and is Borel measurable.*
- *If $g : S \to \mathbb{C}(X, Y)$ is Borel measurable, then g is a Carathéodory function.*

Proof. Let $f : S \times X \to Y$ be a Carathéodory function. Let us show that

$$\hat{f}^{-1}(B) \in \Sigma$$

for each set B of the form

$$B = e_x^{-1}(F) := \{h \in \mathbb{C}(X, Y) : d(h(x), F) = 0\},$$

where F is an arbitrary closed subset of Y. Define

$$\theta : S \times X \to \mathbb{R}$$

by

$$\theta(s, x) = d(f(s, x), F).$$

By Theorem 4.3, θ is jointly measurable, so θ^x defined by

$$\theta^x(s) = \theta(s, x)$$

is measurable. Then

$$\hat{f}^{-1}(B) = \{s \in S : d(f(s, x), F) = 0\} \equiv (\theta^x)^{-1}(0),$$

which belongs to Σ, so $\hat{f}$ is Borel measurable.

Let $g : S \to \mathbb{C}(X, Y)$ be Borel measurable, and define

$$\bar{g} : S \times X \to Y$$

by $\bar{g}(s, x) = g_s(x)$. Clearly g is continuous in x for each s. To see that $\hat{g}(\cdot, x)$ is Borel measurable, let U be an open subset of Y. Now the pointwise open set

$$G := \{h \in \mathbb{C}(X, Y) : h(x) \in U\}$$

is an open subset of $\mathbb{C}(X, Y)$. But

$$g^{-1}(G) \in \Sigma,$$

since g is Borel measurable. Thus g is a Carathéodory function. The proof is completed. □

We now recall the fundamental definition of an absolutely continuous function and consider the Carathéodory-type initial value problem (i.v.p.).

Definition 4.2. *A function ξ on an interval $T \subset \mathbb{R}$ is called absolutely continuous iff*

$$\xi(t_1) - \xi(t_2) = \int_{t_1}^{t_2} \frac{d\xi(\tau)}{dt} d\tau$$

for all $t_1, t_2 \in T$.

We have

$$\begin{aligned} &\dot{x}(t) = f(t, x(t)) \text{ a.e. on } [t_0, t_1] \subset \mathbb{R}_+, \\ &x(t_0) = x_0, \end{aligned} \tag{4.1}$$

where $x \in \mathbb{R}^n$, $t \in \mathbb{R}_+$, and function $f(\cdot, \cdot)$ is determined on a rectangular domain

$$R := R_t \otimes R_x := \{(t, x) \in \mathbb{R}_+ \times \mathbb{R}^n : |t - t_0| \leq T, \ ||x - x_0|| \leq a\}.$$

Here $x : [t_0, t_1] \to R_x$ is assumed to be absolutely continuous function (a solution of (4.1)). When g is continuous, the local existence of solutions is provided by the Peano theorem. Several existence and uniqueness results are known also in the case of a discontinuous right-hand side. We next assume that $f(\cdot, \cdot)$ is a Carathéodory function, namely, it is measurable in t for each fixed x and continuous in x for each fixed t. The next theorem constitutes a fundamental result from the classic theory of ordinary differential equations (ODEs).

Theorem 4.5. *Assume that the above conditions associated with the i.v.p. (4.1) are satisfied. Let additionally*

$$f(t, x) \leq q(t),$$

for a measurable function $q(\cdot)$. Then problem (4.1) has at least one (absolute continuous) solution. We have

$$x(t) = x_0 + \int_{t_0}^{t} f(\tau, x(\tau)) d\tau, \ t \in [t_0, t_0 + r],$$

where $r > 0$. If additionally the map $x \to (t, x)$ is Lipschitz continuous for each t, with a uniform Lipschitz constant, then the Cauchy problem (4.1) has a unique solution, depending Lipschitz continuously on the initial condition x_0.

Note that one can consider the right-hand side in (4.1) also determined as $F : [t_0, t_1] \times \mathbb{R}^n \to \mathbb{R}^n$ and require the boundedness of this function $f(\cdot, \cdot)$.

We close this section with an interesting extension of the conventional result Theorem 4.5 for the autonomous i.v.p. We have

$$\begin{aligned} &\dot{x}(t) = f(x(t)) \text{ a.e. on } [t_0, t_1] \subset \mathbb{R}_+, \\ &x(t_0) = x_0 \end{aligned} \tag{4.2}$$

and refer to [6,282] for the additional analytical material.

Theorem 4.6. *Assume that f in (4.2) has the form*

$$f(x) = F[g_1(\tau_1(x), x), ..., g_N(\tau_N(x), x)],$$

where $F : \mathbb{R}^N \to \mathbb{R}^n$ is continuous and

- *each map*

$$\tau_i : \mathbb{R}^n \to \mathbb{R}$$

  is continuously differentiable;
- *each*

$$g_i : \mathbb{R} \times \mathbb{R}^n \to \mathbb{R}$$

  is a Carathéodory function;
- *for some compact set $K \subset \mathbb{R}^m$, at every point x we have*

$$f(x) \in K, \; \nabla \tau_i(x) \cdot z > 0, \; z \in K.$$

Then the Cauchy problem (4.2) has at least one solution.

The first assumption in Theorem 4.6 is in fact a transversality condition. From the second assumption of Theorem 4.6 we easily deduce that every trajectory of (4.2) satisfies the differential inclusion (DI) $\dot{x}(t) \in K$. We will discuss the DIs in detail in Section 4.5. The second assumption of Theorem 4.6 implies that the trajectory of (4.2) must cross transversally any hypersurface of the form

$$\tau_i(x) = \text{constant}.$$

According to the structure of $f(\cdot)$ in Theorem 4.6, these are the surfaces across which $f(\cdot)$ can jump.

To guarantee the existence of solutions, some kind of transversality condition is necessary. The next simple example lacks such a transversality (jump) condition.

Example 4.1. *Consider a specific right-hand side in (4.2) given by*

$$f(x) = 1(\text{if } x < 0) \text{ and } -1(\text{if } x \geq 0),$$

where $x(0) = 0$.

Let us finally refer to [185,192] for various additional useful facts from the theory of Carathéodory-type ODEs.

4.2 Absolute Continuity

We discuss here shortly a fundamental analytic concept, namely, absolute continuity of a function.

Definition 4.3. *Let $[a, b]$ be an interval in $\mathbb{R}$. A function $x : [a, b] \to \mathbb{R}$ is said to be absolutely continuous if for every $\epsilon > 0$ there exists $\delta > 0$ such that, for every finite collection $\{[a_i, b_i]\}$ of disjoint subintervals of $[a, b]$, we have*

$$\sum_i (b_i - a_i) < \delta \Rightarrow \sum_i |x(b_i) - x(a_i)| < \epsilon.$$

Evidently we now need to establish a relation between two definitions of an absolutely continuous function (Definition 4.2 and Definition 4.3). As shown in integration theory a continuous function $x(\cdot)$ is absolutely continuous if and only if it is an indefinite integral, that is, there exists a function $y(\cdot) \in \mathbb{L}_1(a, b)$ such that

$$x(t) = x(a) + \int_a^t y(\tau)d\tau, \ t \in [a, b]. \tag{4.3}$$

From the theory of integration it follows that $x(\cdot)$ is differentiable at almost every point in (a, b), with

$$\dot{x}(t) = y(t), \ t \in (a, b) \text{ a.e.}$$

In some sense the absolutely continuous functions are well behaved, namely, they coincide with the integral of their derivative. For this reason, they constitute the customary class in which the theory of ODEs is developed. These functions play a central role in the conventional and hybrid optimal control (OC).

The vector space of absolutely continuous functions on an interval $[a, b]$ is temporarily denoted $AC[a, b]$. It is endowed with the norm

$$||x(\cdot)||_{AC[a,b]} = |x(a)| + \int_a^b |\dot{x}(t)|dt.$$

For

$$1 \leq p \leq \infty$$

we next denote by $AC^p[a, b]$ the class of continuous functions $x(\cdot)$ which admit the specific representation (4.3) with $y(\cdot) \in \mathbb{L}_p(a, b)$. The norm on $AC^p[a, b]$ is given as follows:

$$||x(\cdot)||_{AC^p[a,b]} = |x(a)| + ||\dot{x}(\cdot) \cdot ||_{\mathbb{L}_p(a,b)}.$$

Recall that a function $x(\cdot)$ is called Lipschitz continuous if there exists a number $L > 0$ such that

$$|x(s) - x(t)| \leq L|s - t| \ \forall s, t \in [a, b].$$

A Lipschitz continuous function $x(\cdot)$ is easily seen to be absolutely continuous, with a derivative $\dot{x}(\cdot)$ (almost everywhere) that satisfies

$$|\dot{x}(t)| \leq L$$

a.e. Thus, a Lipschitz function belongs to $AC^{\infty}[a, b]$. On the other hand, if an element $x(\cdot) \in AC^{\infty}[a, b]$ satisfies the Lipschitz condition given above, the minimal Lipschitz constant L for this is $||\dot{x}(\cdot)||_{\mathbb{L}_{\infty}(a,b)}$. Note that there are absolutely continuous functions that do not possess the Lipschitz continuity property (for example

$$x(t) = \sqrt{t}$$

considered on the interval [0, 1]).

Recall that a sum, difference, product, and correctly defined quotient of absolutely continuous functions are absolute continuous. A composite function $x_1(x_2(\cdot))$, where $x_1(\cdot)$ is Lipschitz continuous and $x_2(\cdot)$ is an absolutely continuous function, is absolutely continuous [208,209].

We now consider a conventional control system in the context of the theory of absolute continuous functions. Recall that control system theory constitutes an important part of modern electrical engineering. We refer to some classic works devoted to the ODE-based control systems [192,185]. We have

$$\begin{aligned} &\dot{x}(t) = f(t, x(t), u(t)) \text{ a.e. on } [0, t_f], \\ &x(0) = x_0 \in \mathbb{R}^n, \end{aligned} \tag{4.4}$$

where $x(t) \in \mathbb{R}^n$ (the state variable) and the control variable (input)

$$u(t) \in U \subset \mathbb{R}^m.$$

We assume the set $\mathcal{U}$ of admissible controls to be a set of bounded measurable mappings of the type

$$u : [0, t_f] \to \mathbb{R}^m.$$

The map $u(\cdot)$ can be naturally extended on $[0, \infty[$ and the set of controls can be endowed with the $\mathbb{L}_{\infty}$-norm. Under some mild assumptions the parametric i.v.p. (4.4) has a solution. For example, if we suppose that $f(\cdot, \cdot, u)$ is a Carathéodory function and $f(t, x, \cdot)$ is Lipschitz, we deduce the Carathéodory property of the combined function, i.e.,

$$\tilde{f}(t, x) := f(t, x, u(t)),$$

and using theorems from Section 4.1 we easily establish the solvability of (4.4) for every

$$u(\cdot) \in \mathcal{U} \subset \mathbb{L}_\infty(0, t_f).$$

A solution of (4.4) generated by an admissible control $u(\cdot) \in \mathcal{U}$ is next denoted by $x^u(\cdot)$.

This book is mainly devoted to the relaxations methods and is oriented towards numerical applications. Therefore, we have decided to make some regularity assumptions which are twofold. First of all, the systems and the cost functions are taken relatively "smooth." Secondly, the class of admissible controls under consideration is the set of bounded measurable functions and hence the trajectories have bounded derivatives almost everywhere. Both assumptions are in our opinion reasonable in a book devoted to relaxed trajectories and do not avoid complexity met in the optimal control problem (OCP). On the other hand, a reader interested in dealing with general absolutely continuous curves associated to unbounded control problems and corresponding optimality principles under "minimal hypothesis" can consult [127] and [296].

4.3 Sobolev Spaces

This section is devoted to a necessary introduction to the theory of Sobolev spaces. We firstly need to introduce the concept of a weak derivative (see, e.g., [1]).

Definition 4.4. *Let $1 \leq p < \infty$. A function $w : \Omega \subset \mathbb{R}^d \to \mathbb{R}$ is said to be locally p-integrable,*

$$w(\cdot) \in \mathbb{L}_{p,loc}(\Omega),$$

if for every $x \in \Omega$ there is an open neighborhood $\mathcal{B}_x$ of x such that

$$\bar{\mathcal{B}_x} \subset \Omega$$

and $w(\cdot) \in \mathbb{L}_p(\mathcal{B}_x)$.

Notice that a locally integrable function can behave arbitrarily badly near the boundary $\partial\Omega$.

Definition 4.5. *Let Ω be a nonempty open set in*

$$\mathbb{R}^d, \ v(\cdot), w(\cdot) \in \mathbb{L}_{1,loc}(\Omega).$$

Then $w(\cdot)$ is called a weak αth derivative of $v(\cdot)$ if

$$\int_\Omega v(x)\partial^\alpha \phi(x)dx = (-1)^{|\alpha|} \int_\Omega w(x)\phi(x)dx$$

for all $\phi(\cdot) \in \mathbb{C}_0^\infty$. We also say $w(\cdot)$ is a weak derivative of $v(\cdot)$ of order $|\alpha|$.

Recall that a weak derivative, if it exists, is uniquely defined up to a set of measure zero [1,328]. Moreover, if $v(\cdot) \in \mathbb{C}^m(\Omega)$, then for each α with

$$|\alpha| \leq m,$$

the classical partial derivative of order α is also the weak αth partial derivative of $v(\cdot)$. Let us refer to [1] for the complete theory of weakly differentiable functions.

Let Ω be open and bounded in $\mathbb{R}^d$, and let V denote a function space on $\mathbb{R}^{d-1}$. We say $\partial\Omega$ is of class V if for each point

$$x_0 \in \partial\Omega$$

there exist an $r > 0$ and a function $g(\cdot) \in V$ such that upon a transformation of the coordinate system, if necessary, we have

$$\Omega \bigcap B(x0, r) = \{x \in B(x0, r) : x_d > g(x_1, ..., x_{d-1})\}.$$

Here, $B(x_0, r)$ denotes the d-dimensional ball centered at x_0 with radius r. In particular, when V consists of Lipschitz continuous functions, we say Ω is a Lipschitz domain. When V consists of $\mathbb{C}^k$ functions, we say it is a $\mathbb{C}^k$ domain.

Definition 4.6. *Let k be a nonnegative integer, $p \in [1, \infty]$. The Sobolev space $\mathbb{W}^{k,p}(\Omega)$ is the set of all the functions*

$$v(\cdot) \in \mathbb{L}_p(\Omega)$$

such that for each multiindex α with

$$|\alpha| \leq k,$$

the αth weak derivative $\partial^\alpha v(\cdot)$ exists and

$$\partial^\alpha v(\cdot) \in \mathbb{L}_p(\Omega).$$

The norm in the space $\mathbb{W}^{k,p}(\Omega)$ *is defined as*

$$||v(\cdot)||_{\mathbb{W}^{k,p}(\Omega)} := [\sum_{|\alpha|\leq k} ||\partial^{\alpha} v(\cdot)||^{p}_{\mathbb{L}_p(\Omega)}]^{1/p}$$

for

$$1 \leq p < \infty$$

and

$$||v(\cdot)||_{\mathbb{W}^{k,p}(\Omega)} := \max_{|\alpha|\leq k} ||\partial^{\alpha} v(\cdot)||_{\mathbb{L}_\infty(\Omega)}$$

for $p = \infty$*. When* $p = 2$*, we write*

$$\mathbb{H}^k(\Omega) \equiv \mathbb{W}^{k,2}(\Omega).$$

Let us now recall the principal characterizations of a Sobolev space $\mathbb{W}^{k,p}(\Omega)$ [1,21].

Theorem 4.7. *The Sobolev space is a Banach space.*

Theorem 4.8. *The Sobolev space* $\mathbb{H}^k(\Omega)$ *is a Hilbert space with the inner product*

$$\langle v(\cdot), w(\cdot)\rangle_k = \int_{\Omega} \sum_{|\alpha|\leq k} \partial^{\alpha} v(\cdot) \partial^{\alpha} w(\cdot),$$

where $v(\cdot), w(\cdot) \in \mathbb{H}^k(\Omega)$*.*

Since Sobolev spaces are defined through Lebesgue spaces, an element of a Sobolev space is an equivalence class of measurable functions that are equal a.e. When we say a function from a Sobolev space is continuous, we mean that from the equivalence class of the function, we can find one function which is continuous.

Various estimations and inequalities involving Sobolev functions (functions from a Sobolev space) are usually proved for smooth functions first, followed by a density argument. A theoretical basis for this technique is density results of smooth functions in Sobolev spaces. We next present a collection of some approximability results associated with the Sobolev spaces.

Theorem 4.9. *Assume*

$$v(\cdot) \in \mathbb{W}^{k,p}(\Omega),\ 1 \leq p < \infty.$$

Then there exists a sequence

$$\{v(\cdot)_l\} \subset \mathbb{C}^{\infty}(\Omega) \bigcap \mathbb{W}^{k,p}(\Omega)$$

such that

$$||v(\cdot) - v(\cdot)_l||_{\mathbb{W}^{k,p}(\Omega)} \to 0 \text{ as } n \to \infty.$$

Theorem 4.10. *Assume Ω is a Lipschitz domain. Assume*

$$v(\cdot) \in \mathbb{W}^{k,p}(\Omega), \ 1 \leq p < \infty.$$

Then there exists a sequence

$$\{v(\cdot)_l\} \subset \mathbb{C}^{\infty}(\bar{\Omega}) \bigcap \mathbb{W}^{k,p}(\Omega)$$

such that

$$||v(\cdot) - v(\cdot)_l||_{\mathbb{W}^{k,p}(\Omega)} \to 0 \text{ as } n \to \infty.$$

Theorem 4.11. *Assume*

$$k \geq 0, \ p \in [1, \infty[.$$

Then the space $\mathbb{C}_0^{\infty}(\mathbb{R}^d)$ is dense in $\mathbb{W}^{k,p}(\mathbb{R}^d)$.

We finally present an important Sobolev theorem on compact embedding (see, e.g., [1]). Recall that a Banach space X is compactly embedded in Y (denoted as $X \hookrightarrow\hookrightarrow Y$), if

$$||v(\cdot)||_Y \leq c||v(\cdot)||_X$$

for all $v(\cdot) \in X$ and each bounded sequence in X has a subsequence converging in Y.

Theorem 4.12. *Let $\Omega \subset \mathbb{R}^d$ be a Lipschitz domain. Then the following statements are valid.*

- *If $k < d/p$, then*

$$\mathbb{W}^{k,p}(\Omega) \hookrightarrow\hookrightarrow \mathbb{L}_p(\Omega)$$

  for any $q < p^*$, where p^* is defined by $1/p^* = 1/p - k/d$.
- *If $k = d/p$, then*

$$\mathbb{W}^{k,p}(\Omega) \hookrightarrow\hookrightarrow \mathbb{L}_q(\Omega)$$

  for any $q < \infty$.
- *If $k > d/p$, then*

$$\mathbb{W}^{k,p}(\Omega) \hookrightarrow\hookrightarrow \mathbb{C}^{k-[d/p]-1,\beta}(\Omega),$$

  where $\beta \in [0, [d/p] + 1 - d/p)$.

We now observe that in the one-dimensional case, with $\Omega = (a, b)$ being a bounded interval, we have

$$\mathbb{W}^{k,p}(a, b) \hookrightarrow \mathbb{C}[a, b]$$

for any $k \geq 1, \ p \geq 1$.

For the control system (4.4) introduced in Section 4.2 and taking into consideration the main assumptions for (4.4), we have $x^u(\cdot) \in AC^p[0, t_f]$. We now can identify

$$AC^p[0, t_f] \equiv \mathbb{W}^{1,p}(0, t_f)$$

and consider the control system (4.4) over this space such that an admissible pair for (4.4) is defined as

$$(u(\cdot), x^u(\cdot)) \in \mathcal{U} \bigotimes \mathbb{W}^{1,p}(0, t_f) \subseteq \mathbb{L}_\infty(0, t_f) \bigotimes \mathbb{W}^{1,p}(0, t_f),$$

where $x^u(\cdot)$ is a solution of the i.v.p. (4.4) generated by an admissible control $u(\cdot) \in \mathcal{U}$.

4.4 Impulsive Control Systems

In parallel with the conventional control systems formalized by the i.v.p. (4.4) we also consider the impulsive dynamic models. These models are given by the ODEs with some "impulsive components" on the right-hand sides. Let us note that impulsive dynamic systems constitute a challenging and very important modeling approach to many real-world time-dependent systems. In this section we study a class of impulsive hybrid systems (IHSs) and consider the following system.

Definition 4.7. *An IHS is a 7-tuple*

$$\mathcal{IHS} := \{Q; X; U; \mathcal{U}; F; \Theta; \mathcal{S}\},$$

where

- *Q is a finite set of locations;*
- *$X = \{X_q\}, \ q \in Q$, is a collection of state spaces such that $X_q \subseteq \mathbb{R}^n$;*
- *$U \subseteq \mathbb{R}^m$ is a control set;*
- *$\mathcal{U}$ is a set of admissible control functions;*
- *$F = \{f_q\}, \ g \in Q$, is a family of vector fields*

$$f_q : [0, t_f] \times X_q \times U!\mathbb{R}^n;$$

- *$\Theta = \{\Theta_q\}_{q \in Q}$ is a collection of maximal (constant) jump amplitudes;*

- $\mathcal{S} = \{\mathcal{S}_{q_1,q_2}\}_{q_1,q_2 \in Q}$ *is a family of switching sets such that*

$$\mathcal{S} \subseteq \Xi_{q_1,q_2} := \{(q_1, x_1, q_2, x_2) : q_1, q_2 \in Q, \ x_1 \in X_{q_1}, \ x_2 \in X_{q_2}\}.$$

We next consider IHSs that satisfy the following assumptions:

- the functions $f_q(t, \cdot, \cdot)$, where $q \in Q$, are continuously differentiable and $f_q(\cdot, x, u)$ are measurable;
- there exists a constant $K < \infty$ such that

$$||\frac{\partial f(t, x, u)}{\partial x}|| < K$$

  for all $(t, x; u) \in [0, t_f] \times X_q \times U$ and all $q \in Q$;
- the control set U is a compact subset of $\mathbb{R}^m$.

Moreover, we assume that smooth functions

$$m_{q_1,q_2} : \mathbb{R}^n \to \mathbb{R}, \ q_1, \ q_2 \in Q,$$

with nonzero gradients are given, such that the hypersurfaces

$$M_{q_1,q_2} := \{x \in \mathbb{R}^n : m_{q_1,q_2}(x) = 0\}$$

are pairwise disjoint. Note that in this case a hypersurface M_{q_1,q_2} characterizes the set $\mathcal{S}$ at which a switch from location q_1 to location q_2 can take place. Evidently, the projection of $\mathcal{S}$ on the product space

$$X_{q_1} \bigotimes X_{q_2}$$

is a subset of M_{q_1,q_1}. A pair $(q, x(t))$ represents the generalized hybrid state of the IHS determined above.

Let us also note that we come back to the conventional (nonhybrid) control system (4.4) for the case of a one-point set of locations Q. Evidently, every function $f_q \in F$ is a Carathéodory function.

Let us now introduce some standard spaces, namely, the space $\mathbb{C}_0^\infty(0, t_f)$ of all vector functions from $\mathbb{C}^\infty(0, t_f)$ that vanish outside a compact subset of $(0, t_f)$ and the space $\mathcal{D}(0, t_f)$ of generalized functions (the Schwartz distributions). Recall that $\mathcal{D}(0, t_f)$ can be considered as a space of linear, sequentially continuous functionals, with respect to the convergence on the space $\mathbb{C}_0^\infty(0, t_f)$. We next introduce the following concept of a hybrid trajectory of an IHS from Definition 4.7.

Definition 4.8. *A hybrid trajectory of an IHS is a triple*

$$\{x(\cdot), \{q_i\}_{i \in Q}, \tau\},$$

where $x(\cdot) \in \mathcal{D}(0, t_f)$ *is a discontinuous trajectory,* $\{q_i\}_{i \in Q}$ *is a finite sequence of the concretely realized locations, and* τ *is the corresponding sequence of realized switching times*

$$0 = t_0 < t_1 < \ldots < t_i < \ldots < t_r = t_f$$

such that for each $i = 1, \ldots, r$ *there exists* $u(\cdot) \in \mathcal{U}$ *and*

- $x(0) = x_0 \in \mathbb{R}^n$, *and*

  $$x_i(\cdot) = x(\cdot)|_{t \in (t_{i-1}, t_i)}$$

  is an absolutely continuous function on (t_{i-1}, t_i)*;*

- $$\dot{x}_i(t) = f_{q_i}(t, x(t), u(t)) + \theta_{q_i}\delta(t - t_i) \text{ a.e.on } [\mathrm{t}_{\mathrm{i}-1}, \mathrm{t}_{\mathrm{i}}],$$

  where $\delta(\cdot)$ *is the Dirac function and*

  $$||\theta_{q_i}|| \le \Theta_{q_i}.$$

Let us note that the derivative $\dot{x}_i(\cdot)$ in Definition 4.8 is considered as a weak derivative of the generalized function $x_i(\cdot)$ defined on the full time interval $[t_{i-1}, t_i]$ (see Definition 4.5). It is also evident that a function $x(\cdot)$ from Definition 4.8 consists of an absolutely continuous part defined on the open intervals (t_{i-1}, t_i) and involves jumps of magnitude θ_{q_i} at the switching times t_i. Evidently, a uniform representation of the evolution equation for the complete trajectory $x(\cdot)$ of the given IHS can be represented as follows:

$$\begin{aligned} \dot{x}(t) &= \sum_{i=1}^{r} \beta_{[t_{i-1}, t_i)}(t) f_{q_i}(t, x(t), u(t)) + \theta_{q_i}\delta(t - t_i), \\ x(0) &= x_0, \end{aligned} \tag{4.5}$$

where $\beta_{[t_{i-1}, t_i)}(\cdot)$ is a characteristic function of the interval $[t_{i-1}, t_i)$, $i = 1, \ldots, r$. Note that the i.v.p. (4.5) is also considered in the sense of weak derivatives. Under the assumptions presented above, for each $u(\cdot) \in \mathcal{U}$ and all

$$||\theta_{q_i}|| \le \Theta_{q_i}$$

the i.v.p. (4.5) has a unique solution in $\mathcal{D}(0, t_f)$ [1,185].

We now consider the following auxiliary i.v.p.:

$$\dot{y}(t) = \sum_{i=1}^{r} \beta_{[t_{i-1},t_i)}(t) f_{q_i}(t, \tilde{y}(t), u(t)),$$
$$y(0) = x_0, \tag{4.6}$$

where

$$\tilde{y}(t) = y(t) + \sum_{i=1}^{r} \theta_{q_i} \eta(t - t_i)$$

for $i = 1, ..., r$, and $\eta(\cdot)$ is the Heaviside step function. Note that $\eta(\cdot)$ can also be considered as an element of the space $\mathcal{D}(0, t_f)$. Under the assumptions stated in Section 4.3, the i.v.p. (4.6) has a unique absolutely continuous solution for each $u(\cdot) \in \mathcal{U}$. Next we consider the function $y(\cdot)$ in (4.6) as an element of the Sobolev space $\mathbb{W}_n^{1,\infty}(0, t_f)$, i.e., the space of absolutely continuous (n-dimensional) functions with essentially bounded derivatives. We are now able to formulate the equivalence result.

Theorem 4.13. *Under the basic assumptions presented above, the unique solution $x(\cdot) \in \mathcal{D}(0, t_f)$ of the initially given i.v.p. (4.5) can be represented in the following form:*

$$x(t) = y(t) + \sum_{i=1}^{r} \theta_{q_i} \eta(t - t_i), \tag{4.7}$$

where

$$y(\cdot) \in \mathbb{W}_n^{1,\infty}(0, t_f)$$

is a (unique) solution to the auxiliary i.v.p. (4.6).

Proof. Since the weak derivative of the Heaviside step function $\eta(t - t_i)$ is equal to the Dirac function $\delta(t - t_i)$, the weak derivative of the right-hand side of (4.7) is

$$\dot{x}(t) = \dot{y}(t) + \sum_{i=1}^{r} \theta_{q_i} \delta(t - t_i).$$

For an absolutely continuous function $y(\cdot)$ its weak derivative coincides with the classical derivative. Using (4.7), the i.v.p. (4.5) can be written in the following form:

$$\dot{x}(t) = \dot{y}(t) + \sum_{i=1}^{r} \theta_{q_i} \delta(t - t_i) =$$
$$\sum_{i=1}^{r} \beta_{[t_{i-1},t_i)}(t) f_{q_i}(t, y(t) + \sum_{i=1}^{r} \theta_{q_i} \eta(t - t_i), u(t)) + \sum_{i=1}^{r} \theta_{q_i} \delta(t - t_i) =$$
$$\sum_{i=1}^{r} \beta_{[t_{i-1},t_i)}(t) f_{q_i}(t, x(t), u(t)) + \theta_{q_i} \delta(t - t_i).$$

Moreover, we have

$$x(0) = y(0).$$

The uniqueness arguments for solutions of (4.5) and (4.6) complete the proof. □

The proposed representation (4.6) constitutes in fact a simple transformation of values. The transformed control system has no state jumps (in comparison to the original system (4.5)). The affine structure of (4.7) makes it possible to deduce the following natural characterization of the above transformation.

Theorem 4.14. *The transformation (4.7) from Theorem 4.13 is a bijective mapping*

$$\mathcal{D}(0, t_f) \to \mathbb{W}_n^{1,\infty}(0, t_f),$$

and the solutions $x(\cdot)$ and $y(\cdot)$ of problems (4.5) and (4.6) are related by relation (4.7).

We refer to [1,39,159] for the corresponding proof.

Note that after application of the transformation (4.7) we get a nonimpulsive HS from Definition 1.1. The concepts of HS mentioned above exclude the possibility of sliding mode dynamics with respect to the given switching hypersurfaces

$$M_{q_i, q_{i+1}}, \; i = 1, ..., r-1.$$

The condition that guarantees the absence of the sliding mode regimes along the manifolds $M_{q_i,q_{i+1}}$ for the auxiliary HS (4.6) can also be formulated geometrically. Let $\mathcal{T}M_{q_i,q_{i+1}}(z)$ be the tangent (vector) space to $M_{q_i,q_{i+1}}$ at a point

$$z \in M_{q_i,q_{i+1}}.$$

Then in conformity with the well-known concept from [6], we obtain the following condition:

$$(\lambda \dot{y}_i(t_i) + (1-\lambda)\dot{y}_i(t_{i-1})) \bigcap \mathcal{T}M_{q_i,q_{i+1}}(y_i(t_i)) = \emptyset.$$

Evidently, this relation expresses the condition for the sliding mode free dynamics of the auxiliary HS (4.6). Using Theorem 4.14 we also conclude that the originally given IHS (4.5) does not include a sliding mode behavior.

4.5 Set-Valued Functions and Differential Inclusions

Consider a subset of $\Omega \subset \mathbb{R}^m$. A set-valued function $F(\cdot)$ from Ω to $\mathbb{R}^n$ is a mapping from Ω to the subsets of $\mathbb{R}^n$. This concept means that we associate with each $\omega \in \Omega$ a subset

$$F(\omega) \subset \mathbb{R}^n.$$

Note that $F(\omega)$ can also be empty. Such mappings arise rather frequently in abstract control theory (CT).

Definition 4.9. *A measurable function $\gamma : \Omega \to \mathbb{R}^n$ such that*

$$\gamma(\omega) \in F(\omega)$$

for almost all $\omega \in \Omega$ is called a measurable selection associated with the set-valued function $F(\cdot)$.

To find a measurable selector for a given set-valued function is a challenging topic of the set-valued analysis (see [6,159]). In this and the following sections we study some necessary facts from this part of the general analysis and apply these facts in relaxation theory. A set-valued function $F(\cdot)$ is next also denoted as follows: $F : \Omega \rightrightarrows \mathbb{R}^n$.

Definition 4.10. *The set-valued function $F : \Omega \rightrightarrows \mathbb{R}^n$ is measurable provided that Ω is measurable and provided that the set*

$$F^{-1}(W) = \{\omega \in \Omega : F(\omega) \bigcap W \neq \emptyset\}$$

is Lebesgue measurable for every closed $W \subset \mathbb{R}^n$.

Note that Definition 4.10 can be equivalently rewritten taking a compact set W in the definition. This fact is an immediate consequence of the representation of a closed set W as a union of countably many compact sets W_j. In the case of a singleton $F(\cdot)$ this set-valued function is measurable if and only if the selector $\gamma(\cdot)$ is measurable (as a conventional function). Similar to the functions considered in Chapter 3 the effective domain

$$\mathrm{dom}(F(\cdot))$$

of $F(\cdot)$ can be defined as follows:

$$\mathrm{dom}(F(\cdot)) := \{\omega \in \Omega : F(\omega) \neq \emptyset\}.$$

Naturally $F(\cdot)$ is said to be closed-valued when $F(\omega)$ is a closed set for each $\omega \in \Omega$.

Let us consider an abstract construction of a set-valued function that is typical for CT. Let $\{\gamma_j(\cdot)\}$ be a sequence of measurable functions. The following "synthetic" set-valued function (determined by using the above sequence)

$$F(\omega) := \{\gamma_j(\omega) : j \geq 1\}$$

is measurable. The corresponding (stronger) result can be formulated as follows.

Theorem 4.15. *Let $F : \Omega \rightrightarrows \mathbb{R}^n$ be a closed-valued and measurable set-valued function. Then there exists a countable family*

$$\{\gamma_j : \mathrm{dom}(F(\cdot)) \to \mathbb{R}^n\}$$

of measurable functions such that

$$F(\omega) = \mathrm{cl}\{\gamma_j(\omega) : j \geq 1\} \ \omega \in \mathrm{dom}(F(\cdot)) \text{ a.e.}$$

Proof. Let $y \in \mathbb{R}^n$. The function $x \to \mathrm{dist}_{\mathbb{R}^n}(F(x), (y))$ restricted to $\mathrm{dom}(F(\cdot))$ is measurable. Here $\mathrm{dist}_{\mathbb{R}^n}(\cdot, \cdot)$ denotes a usual (Euclidean) distance. The above fact follows from the identity (for $0 \leq r < R$), i.e.,

$$\mathrm{dist}_{\mathbb{R}^n}(F(\cdot), (y))^{-1}(r, R) = \{s \in \mathrm{dom}(F(\cdot)) : F(x) \bigcap B(y, r) = \emptyset\} \bigcap F^{-1}(B_0(y, R)).$$

By $B(y, r)$ and $B_0(y, R)$ we denote the balls (closed and open) centered in y and with the corresponding radius, respectively. Let

$$\{y_j\}_{j \geq 1}$$

be a sequence that is dense in $\mathbb{R}^n$. Now we define an auxiliary function $f_0 : \mathrm{dom}(F(\cdot)) \to \mathbb{R}^n$. We have

$$f_0(x) = \text{the first } y_j \text{ with } \mathrm{dist}_{\mathbb{R}^n}(F(x), (y))(y_j) \leq 1.$$

The above functions are measurable on $\mathrm{dom}(F(\cdot))$. The proof is completed. □

The main (classic) result about the measurable selections is summarized in Theorem 4.16. The corresponding proof can be found in [6].

Theorem 4.16. *Let $F : \Omega \rightrightarrows \mathbb{R}^n$ be closed-valued and measurable. Then there exists a measurable function*

$$\gamma : \mathrm{dom}(F(\cdot)) \to \mathbb{R}^n$$

such that $\gamma(\omega) \in F(\omega)$, where $\omega \in \mathrm{dom}(F(\cdot))$ a.e.

Note that a sum and an intersection of two measurable set-valued functions are measurable. The next result is of crucial importance for the theory of convexified DIs [6,86] and is used for the measurability proof of the right-hand sides of relaxed control systems. Moreover, this result establishes a typical structure of the set-valued mappings used in modern CT.

Theorem 4.17. *Let $\Omega \subset \mathbb{R}^m$ be measurable and $\phi : \Omega \times \mathbb{R}^n \times \mathbb{R}^l \to \mathbb{R}$ a function with the following properties:*

- *the mapping* $x \to \phi(x, p, q)$ *is measurable on* Ω *for each*

$$(p, q) \in \mathbb{R}^n \times \mathbb{R}^l;$$

- *the mapping* $(p, q) \to \phi(\omega, p, q)$ *is continuous for each* $\omega \in \Omega$.

Let $P, Q : \Omega \rightrightarrows \mathbb{R}^n$ *be measurable closed-valued set-valued functions and* $c, d : \omega \to \mathbb{R}$ *measurable functions. Then* $F : \Omega \rightrightarrows \mathbb{R}^n$ *defined by*

$$F(\omega) = \{p \in P(x) : c(x) \leq \phi(\omega, p, q) \leq d(x) \text{ for some } q \in Q(x)\}$$

is measurable.

As an immediate consequence of Theorem 4.17 we obtain the main measurability result for a convex combination of the given set-valued function. Consider a measurable set-valued function $G(\cdot)$ defined on Ω and recall that for an $\omega \in \Omega$ the notation $\text{co}\{G(\omega)\}$ denotes the convex hull of the set $G(\omega)$. We next call the mapping $\omega \to \text{co}\{G(\omega)\}$ a convexified set-valued function.

Theorem 4.18. *Let* $G : \Omega \rightrightarrows \mathbb{R}^n$ *be a measurable and closed-valued set-valued function. Then the set-valued function* $F(\cdot)$ *defined by*

$$F(\omega) = \text{co}\{G(x)\}$$

is also measurable.

Proof. Let Υ denote the set of all nonnegative vectors

$$\lambda = (\lambda_0, \lambda_1, \ldots, \lambda_n)^T \in \mathbb{R}^{n+1}$$

whose coordinates sum is equal to 1. Then the set-valued function

$$Q(\omega) := \Upsilon \times G(\omega) \times G(\omega) \times \cdots \times G(\omega)$$

is measurable. Let f be defined as follows:

$$f(\lambda, g_0, g_1, \ldots, g_n) = \sum_{i=0}^{n} \lambda_j g_i,$$

where each g_i lies in $\mathbb{R}^n$. Then, the set $F(\omega)$ is described by

$$\{v \in \mathbb{R}^n : |v - f(\lambda, g_0, g_1, \ldots, g_n)| = 0\}$$

for some $(\lambda, g_0, g_1, \ldots, g_n)^T \in Q(x)$. Applying now Theorem 4.17 we deduce the statement of the theorem. □

The above facts from the set-valued analysis are widely used in the theory of DIs of the following type:

$$\dot{x}(t) \in F(t, x(t)), \text{ a.e. on } t \in [0, t_f], \tag{4.8}$$

where $F(\cdot)$ is a set-valued function. For example, a simple IHS from Section 4.4 with $\theta_{q_i} \equiv 0$ and a singleton Q constitutes a conventional control system. An adequate mathematical model for this dynamic system is the DI (4.8) with

$$F(t, x) := f(t, x(t), U).$$

The corresponding convexified set-valued function, namely,

$$\tilde{F}(t, x) := \text{co}\{f(t, x(t), U)\},$$

describes the relaxed control system that corresponds to the initially given model. We will discuss these questions in details in Section 4.11.

4.6 Lipschitz Set-Valued Functions

This section contains a short remark about a specific class of set-valued functions, namely, Lipschitz set-valued functions.

Definition 4.11. *The set-valued function $F(\cdot)$ is said to be Lipschitz on a set Γ if $F(\omega)$ is defined and nonempty for every $\omega \in \Gamma$ and, for some constant L, we have*

$$\omega_1, \omega_2 \in \Gamma \Rightarrow F(\omega_1) \in F(\omega_2) + K|\omega_1 - \omega_2|B,$$

where B is a unit ball. A set-valued function $F(\cdot)$ is said to be Lipschitz near ω if there is a neighborhood B_ω of ω such that $F(\cdot)$ is Lipschitz on B_ω.

Note that in the case we require $\omega_1, \omega_2 \in B\omega_0, r$, where $r > 0$, we get a natural concept of a locally Lipschitz set-valued function $F(\cdot)$ in Definition 4.11. We next present an important technical result for a class of Lipschitz set-valued functions. This theorem will be used in some proofs of necessary optimality conditions for the relaxed OCPs (see Chapter 8).

Theorem 4.19. *Let $F(\cdot)$ be a set-valued function from $\mathbb{R}^n$ to $\mathbb{R}^m$ with a closed graph Gr, and let dist_G denote the Euclidean distance function of Gr. Suppose that $F(\cdot)$ satisfies the local Lipschitz condition*

$$\Gamma(y) \in \Gamma(z) + B(0, k|y - z|) \; \forall y, z \in B(\omega_0, r), \tag{4.9}$$

where

$$\omega_0 \in \mathbb{R}^n, \ r > 0.$$

Let $v_0 \in F(\omega_0)$. *Then*

$$(a, b) \in N_G^L(\omega_0, v_0) \Rightarrow |a| \leq k|b|.$$

If the Lipschitz condition in (4.9) holds globally (that is, for all $y, z \in \mathbb{R}^n$*), then, for any point*

$$(\omega, v) \in \mathbb{R}^n \times \mathbb{R}^m,$$

we have

$$(a, b) \in \partial_L d_G(\omega, v) \Rightarrow |a| \leq k|b|$$

and

$$d_G(\omega, v) > 0, \ (a, b) \in \partial_L d_G(\omega, v) \Rightarrow |b| \geq (1 + k^2)^{-1/2} - 1/2.$$

The proof of Theorem 4.19 can be found in [6]. In general the global and local Lipschitz set-valued functions for the DIs play a similar role as the corresponding (global and local) Lipschitz right-hand sides of ODEs.

Consider a set-valued function $F(\omega)$, $\omega = (t, x) \in \Omega$, and determine the following function (see [130,131]):

$$\rho(t, x, v) := \{|v - y| : y \in F(t, x)\}.$$

In [131] it was shown that $v \in F(t, x)$ iff $\rho(t, x, v) = 0$. The next result clarifies the formal relation between $F(\cdot)$ and $\rho(\cdot, \cdot, \cdot)$.

Theorem 4.20. *If a set-valued function* $F(\cdot)$ *is measurable and Lipschitz, then*

- *for each* $x, v \in \mathbb{R}^n$, *the function* $t \to \rho(t, x, v)$ *is measurable;*
- *for any* (t, x_1) *and* (t, x_2) *in* Ω *and for any* $v_l, v2 \in \mathbb{R}^n$, *one has*

$$|\rho(t, x_1, v_1) - \rho(t, x_2, v_2)| \leq k(t)|x_1 - x_2| + |v_1 - v_2|.$$

The proof of the above result can be found in [130,131].

4.7 Measurable Selections

The material of this section generalizes the finite-dimensional concepts and results of Section 4.6. We study here the measurability of the general correspondences. Results presented in this section make it easier to understand some necessary aspects of the theory of DIs. Then again the DIs represent the useful mathematical formalism of the ODEs involved in control systems and specifically the relaxed systems. The main methodological interest here is whether a correspondence admits a selector that is measurable.

Given two abstract sets X and Y, we consider the concept of a correspondence.

Definition 4.12. *A correspondence Φ from X to Y is a mapping from X to the subsets $\Phi(x)$ of Y, and we write $\Phi : X \twoheadrightarrow Y$.*

The graph $Gr\{\Phi\}$ of Φ can be defined as follows:

$$\{(x, y) \in X \times Y : y \in \Phi(x)\}.$$

A correspondence $\Phi : X \twoheadrightarrow Y$ has two natural inverses. Let us consider only the so-called lower inverse Φ^l (also called the weak inverse) defined by

$$\Phi^l(A) := \{x \in X : \Phi(x) \bigcap A \neq \emptyset\}.$$

We next need some classic results related to the measurability of correspondences. By Σ we next denote the σ-algebra of measurable subsets of a measurable space S.

Definition 4.13. *Let (S, Σ) be a measurable space and X a topological space. We say that a correspondence $\Phi : S \twoheadrightarrow X$ is:*

- *weakly measurable, if*

$$\Phi^l(G) \in \Sigma$$

  for each open subset $G \subset X$;
- *measurable, if*

$$\Phi^l(G) \in \Sigma$$

  for each closed subset $G \subset X$;
- *Borel measurable, if*

$$\Phi^l(G) \in \Sigma$$

  for each Borel subset $B \subset X$.

In the case of a singleton Φ the concepts of measurability, weak measurability, and Borel measurability of Φ all coincide with Borel measurability of Φ as a function.

Theorem 4.21. *Suppose $f : S \times X \to Y$ is a Carathéodory function, where (S, Σ) is a measurable space, X is a separable metrizable space, and Y a topological space. For each subset G of Y define the correspondence $\varphi_G : S \twoheadrightarrow X$ by*

$$\phi_G(s) := \{x \in X : f(s, x) \in G\}.$$

If G is open, then ϕ_G is a measurable correspondence.

Theorem 4.22. *For a correspondence $\phi : (S, \Sigma) \twoheadrightarrow X$ from a measurable space into a metrizable space we have the following.*

- *If ϕ is measurable, then it is also weakly measurable.*
- *If ϕ is compact-valued and weakly measurable, then it is measurable.*

We refer to [6] for a formal proof of Theorem 4.21 and Theorem 4.22. The two next results are in fact simple consequences of the structure of open sets of a metrizable space (see [6]).

Theorem 4.23. *A correspondence $\phi : S \twoheadrightarrow X$ from a measurable space into a topological space is weakly measurable if and only if its closure correspondence $\bar{\phi}$ is weakly measurable.*

We finally need the last analytic fact related to the measurability of correspondences (see [6]).

Theorem 4.24. *For a sequence $\{\phi_n\}$ of correspondences from a measurable space (S, Σ) into a topological space X we have the following.*

- *The union correspondence $\phi : S \twoheadrightarrow X$, defined by*

$$\phi(s) = \bigcup_{n=1}^{\infty} \phi_n(s),$$

  is (a) weakly measurable, if each ϕ_n is weakly measurable; (b) measurable, if each ϕ_n is measurable; and (c) Borel measurable, if each ϕ_n is Borel measurable.
- *If X is a separable metrizable space and each ϕ_n is weakly measurable, then the product correspondence $\psi : S \twoheadrightarrow X^N$, defined by*

$$\psi(s) = \prod_{n=1}^{\infty} \phi_n(s),$$

  is weakly measurable.

- *If X is a separable metrizable space, each ϕ_n is weakly measurable with closed values, and for each s there is some k such that $\phi_k(s)$ is compact, then the intersection correspondence $\theta : S \twoheadrightarrow X$, defined by*

$$\theta(s) = \bigcap_{n=1}^{\infty} \phi_n(s),$$

is measurable (and hence weakly measurable).

As mentioned above we are mainly interested to establish the existence of a measurable selector. The celebrated axiom of choice (see [208,282]) guarantees that nonempty-valued correspondences always admit selectors, but they may have no additional useful properties. Let us give the fundamental (but natural) concepts of selectors from a given correspondence.

Definition 4.14. *A selector from a correspondence $\Phi : X \rightrightarrows Y$ is a function $f : X \to Y$ that satisfies $f(x) \in \Phi(x)$ for each $x \in X$. In the case X and Y are topological spaces, we say that $f(\cdot)$ is a continuous selector if it is a selector and continuous.*

Definition 4.15. *A measurable selector from a correspondence $\Phi : S \rightrightarrows X$ between measurable spaces is a measurable function $f : S \to X$ satisfying $f(s) \in \Phi(s)$ for each $s \in S$.*

Evidently, Definition 4.15 generalizes the finite-dimensional concept, namely, Definition 4.9. Let us formulate here the first necessary result that establishes the existence of a measurable selector for weakly measurable correspondences.

Theorem 4.25. *A correspondence $\phi : (S, \Sigma) \twoheadrightarrow X$ with nonempty compact values from a measurable space into a separable metrizable space is weakly measurable if and only if there exists a sequence $\{f_n\}$ of measurable selectors from ϕ satisfying*

$$\phi(s) = \{f_1(s), \bar{f}_2(s), ...\}$$

for each s. In particular, every weakly measurable correspondence with nonempty compact values from a measurable space into a separable metrizable space admits a measurable selector.

Note that the above result is a variant of the celebrated Kuratowski–Ryll-Nardzewski selection theorem [6,330]. This theorem states that a weakly measurable correspondence with nonempty closed values from a measurable space into a Polish space admits a measurable selector.

4.8 The Filippov–Himmelberg Implicit Functions Theorem

The Kuratowski–Ryll-Nardzewski Theorem 4.25 from Section 4.7 provides a theoretical basis for the formal proof of the next selection theorem, known as Filippov's implicit function theorem (Filippov's lemma). This result constitutes a specific generalized version of the fundamental implicit function theorem and is very useful for the abstract theory of DIs.

Theorem 4.26. *Consider separable metric spaces X and Y and a measurable space $\{S, \Sigma\}$. Suppose that $f : S \times X \to Y$ is a Carathéodory function and that $\phi : S \twoheadrightarrow X$ is weakly measurable with nonempty compact values. Assume also that $\pi : S \to Y$ is a measurable selector from the range of f on ϕ in the sense that g is measurable and for each s there exists $x \in \phi(s)$ with*

$$\pi(s) = f(s, x).$$

Then the correspondence $\gamma : S \twoheadrightarrow X$, defined by

$$\gamma(s) := \{x \in \phi(s) : f(s, x) = \pi(s)\},$$

is measurable and admits a measurable selector. That is, in addition to γ being measurable, there exists a measurable function $\xi : S \to X$ with $\xi(s) \in \phi(s)$ and

$$\pi(s) = f(s, \xi(s))$$

for each $s \in S$.

Proof. Fix a compatible metric d on Y and consider the continuous function $g : Y \times Y \to R$ defined by

$$g(y1, y2) = d(y1, y2).$$

Also, consider the function $h : S \times X \to R$ defined by

$$h(s, x) = g(f(s, x), \pi(s)) = d(f(s, x), \pi(s)).$$

Since f is continuous in x, it follows immediately that h is continuous in x. Now since the functions

$$(s, x) \to f(s, x)$$

and $(s, x) \to \pi(s)$ (both from $S \times X$ to Y) are clearly Carathéodory functions, it follows that h is jointly measurable. In particular, h is measurable in s and so h is itself a Carathéodory function.

Next for each n define the correspondence $\psi_n : S \twoheadrightarrow X$ by

$$\psi_n := \{x \in X : d(f(s,x), \pi(s) < 1/n)\} = \{x \in X : h(s,x) \in (-\infty, 1/n)\}.$$

By Theorem 4.21 each ψ_n is measurable. Thus, by Theorem 4.22 and 4.23 the correspondence ψ_n is weakly measurable. Observe that

$$\gamma(s) = \phi(s) \bigcap \psi_1(s) \bigcap \psi_2(s) \bigcap \cdots$$

and that

$$\{\psi, \psi_1, \psi_2, \ldots\}$$

satisfies the hypotheses of Theorem 4.24. Therefore γ is measurable (and hence weakly measurable) and has compact values. By hypothesis, γ has also nonempty values. By Theorem 4.25, the correspondence γ has measurable selectors, any of which will do for ξ. The proof is finished. □

Theorem 4.26 constitutes an abstract result from modern set-valued analysis. The original Filippov lemma can be formulated as follows (see, e.g., [86]).

Theorem 4.27. *Let T be a measure space, let Z be a Hausdorff space, and let D be a topological space that is the countable union of compact metric spaces. Let Γ be a measurable map from T to Z and let ϕ be a continuous map from D to Z such that*

$$\Gamma(T) \in \phi(D).$$

Then there exists a measurable map μ from T to D such that the composite map $\phi \circ \mu$ from T to Z is equal to Γ.

We refer to [86] for a formal proof of Theorem 4.27. In the control theoretical applications we evidently have the concrete spaces $T,\ Z, D$, i.e.,

$$T = \mathbb{R}_+,\ Z = \mathbb{R}^n,\ D = \mathbb{R}^m.$$

The next result constitutes a finite-dimensional version of Theorem 4.26 and Theorem 4.27.

Theorem 4.28. *Let $[a,b] \subset \mathbb{R}_+$ be an interval, U an interval in $\mathbb{R}^m$, and h a map from $[a,b] \times U$ into $\mathbb{R}^n$ that is continuous on U for a.e. $t \in [a,b]$ and is measurable on $[a,b]$ for each $u \in U$. Let W be a measurable function on $[a,b]$ with range in $\mathbb{R}^n$, and let $\tilde{V}$ be a function from $[a,b]$ to U such that*

$$W(t) = h(t, eV(t)) \text{ a.e.}$$

Then there exists a measurable function $V : [a,b] \to U$ such that $W(t) = h(t, V(t))$ a.e.

Let us finally note that the theory of measurable selectors associated with the general correspondences is a well-developed part of the set-valued analysis. One also can study the existence question for the selectors that are measurable in the sense of the weak star topology. Let us mention the classic results of Strassen, the results of T. Kim, K. Prikry, and N.C. Yannelis related to the Carathéodory-type selectors from a correspondence [6], the celebrated Aumann selection theorem, the Jankov–von Neumann selection theorem [22], the Jacobs selection theorem, and the Castaing selection theorem (see [6]).

4.9 Continuous Selections of the Differential Inclusions and the Michael Theorem

In this section we continue the study of the selectors associated with the correspondences. We also apply some of the presented analytic facts to the abstract formal models of the general control systems, namely, to DIs. The next classic result is due to F.E. Browder [109], and while it is one of the more straightforward selection theorems, it is also very useful. Let us firstly recall some basic concepts. Let $AX \times Y$. Then for each

$$x \in X, \; y \in Y$$

the x- and y-sections of A are defined as follows:

$$A_x = \{y \in Y : (x, y) \in A\}, \; A_y = \{x \in X : (x, y) \in A\}.$$

Consider now two measurable spaces (X, Σ_X) and (Y, Σ_Y). A subset $A \subset X \times Y$ has measurable sections if $A_x \in \Sigma_Y$ for each $x \in X$ and $A_y \in \Sigma_X$ for each $y \in Y$. Let us also recall that a Hausdorff space is called paracompact if every open cover of the space has an open locally finite refinement cover.

Theorem 4.29. *Assume a correspondence has nonempty convex values and open lower sections from a paracompact topological space to a Hausdorff topological vector space. Then this correspondence admits a continuous selector.*

Proof. Let $\phi : X \twoheadrightarrow Y$ satisfy the above assumptions. Let F denote the image of X under ϕ. Since ϕ has nonempty values, the family

$$\{\phi^{-1}(y) : y \in F\}$$

of lower sections is an open cover of X. From the paracompactness of X, it follows that there exists a locally finite continuous partition of unity $\{g_y\}_{y \in F}$ such that

$$g_y(x) = 0$$

for each x that does not belong to $\phi^{-1}(y)$. In particular, notice that

$$g_y(x) > 0$$

implies $y \in \phi(x)$.

For each $x \in X$ now determine $f(x) = \sum_{y \in F} g_y(x)y$ and note that the local finiteness of $\{g_y\}_{y \in F}$ in conjunction with the convexity of $\phi(x)$ guarantees that

$$f(x) \in \phi(x)$$

for all $x \in X$. Now let us observe that the formula $f(x)$ defines in fact a function $f : X \to F \subset Y$ that is continuous and hence it is a continuous selector from the correspondence $\phi(\cdot)$. □

The above Browder selection theorem applies to any topological vector space but requires a strong assumption of the existence of open lower sections. This fact is hard to establish for the right-hand sides of DIs. Therefore, we next present a (simplified) version of the celebrated Michael selection theorem. Let us recall that a Fréchet space is a completely metrizable locally convex space.

Theorem 4.30. *A lower hemicontinuous correspondence from a paracompact space into a Fréchet space with nonempty closed convex values admits a continuous selector.*

The main proof idea for Theorem 4.30 includes a construction of some approximating sequences of continuous selectors. These sequences are cleverly defined so that one finally gets a uniformly Cauchy property. Since Fréchet spaces are complete, it converges to a continuous function, which turns out to be the desired selector from the correspondence under consideration.

Recall that a correspondence $\phi : X \twoheadrightarrow Y$ between topological spaces is lower hemicontinuous at x if for every open set W that meets $\phi(x)$ the lower inverse image $\phi^l(W)$ is a neighborhood of x. We now close our consideration of the abstract correspondences and the corresponding selectors by the following useful result [6].

Theorem 4.31. *Let Z be a locally compact Polish space, and let X be a compact metrizable space. Let μ be a finite Borel measure on X. Then every lower hemicontinuous correspondence $\phi : Z \twoheadrightarrow \mathbb{L}_1(\mu, \mathbb{R}^n)$ with decomposable values admits a continuous selector.*

Finally note that if Y is a topological space, then we say that the correspondence $\phi : X \twoheadrightarrow Y$ (as well as the set-valued function) is "closed-valued" or has closed values if $\phi(\cdot)$ is a closed set for each x. The terms "open-valued," "compact-valued," "convex-valued," etc., are defined similarly.

We now consider some elements of the abstract theory of DIs. Assume $F : [0, T] \times \mathbb{R}^n \rightrightarrows R^n$ is a set-valued function, and consider the DI

$$\dot{x}(t) \in F(t, x).$$

In order to construct solutions of this DI, one can consider a naive approach: to find a selection $f(t, x) \in F(t, x)$ and solve the ODE $\dot{x} = f(t, x)$.

As follows from the important Theorem 4.30 for the lower semicontinuous set-valued function $F(\cdot, \cdot)$ with closed convex values, we obtain an existence of a continuous selector $f(\cdot, \cdot)$. The local existence of solutions to the above ODE with continuous right-hand side is then a direct consequence of the classic Peano theorem. Dropping the assumption that the values of $F(\cdot, \cdot)$ are convex, the above approach runs into difficulties. Indeed, in this case no continuous selection of the right-hand side of the above DI may exist; hence the first step in the above naive approach cannot be accomplished. Evidently, a measurability property of the selection $f(\cdot, \cdot)$ can imply the discontinuity of the right-hand side of the above ODE. This property is insufficient for the guaranteed existence of a solution. In order to follow the "naive approach" (considered above) in the nonconvex case, we thus need to isolate a property of functions $f : [0, t_f] \times \mathbb{R}^n \to \mathbb{R}^n$ which is stronger than measurability but weaker than continuity; it must be strong enough so that the ODE under consideration always has solutions and the corresponding selector problem can be solved.

Taking into consideration the above motivation we next present a very useful approach based on the concept of directionally continuous selections.

Definition 4.16. *A mapping $f : \mathbb{R} \times \mathbb{R}^n \to Y$ is Γ^M-continuous if, for every (t, x), one has*

$$\lim_{s \to \infty} f(t_s, x_s) = f(t, x)$$

for every sequence $(t_s, x_s) \to (t, x)$ such that

$$(t_s - t, x_s - x) \in \Gamma^M,$$

where

$$\Gamma^M := \{(t, x) \in \mathbb{R}^{n+1} : ||x|| \leq Mt\}.$$

Note that a (possibly discontinuous) function $f(\cdot, \cdot)$ from Definition 4.16 satisfying the above conditions for some $M > 0$ is in fact "directionally continuous." Observe that, in the case $n = 0$, the above definition simply means that $f : \mathbb{R} \to Y$ is right-continuous. The importance of the above Γ^M-continuity concept is illustrated by the following result.

Theorem 4.32. *Let $f : [0, t_f] \times \mathbb{R}^n \to \mathbb{R}^n$ be Γ^M-continuous and satisfy*

$$||f(t, x)|| \leq L$$

for a positive constant $L < M$ and all (t, x). Then, for every $x_0 \in \mathbb{R}^n$, the i.v.p.

$$\dot{x}(t) = f(t, x(t)),$$
$$x(O) = x_0$$

has a Carathéodory solution on $[0, t_f]$.

The main result for the directionally continuous correspondences can be stated as follows.

Theorem 4.33. *For any constant $M > 0$, every lower semicontinuous set-valued function (correspondence) $F : [0, t_f] \times \mathbb{R}^n \rightrightarrows \mathbb{R}^n$ with closed values admits a Γ^M-continuous selector.*

We now combine Theorem 4.32 and Theorem 4.33 and obtain a general existence result for the generic i.v.p. associated with a DI.

Theorem 4.34. *Let*

$$F : [0, t_f] \times \mathbb{R}^n \rightrightarrows \mathbb{R}^n$$

be a bounded lower semicontinuous set-valued function with compact values. Then the i.v.p.

$$\begin{aligned} \dot{x}(t) &\in F(t, x(t)), \\ x(0) &= x_0 \end{aligned} \tag{4.10}$$

has a solution on $[0, t_f]$.

Finally note that the celebrated Michael selection theorem (Theorem 4.30) provides a theoretical basis for the existence of a solution of a typical i.v.p. for a convexified DI, namely, for the generalized dynamic system in a classic relaxed OCP.

4.10 Trajectories of Differential Inclusions

Let us generalize to a n-dimensional case the concept of an absolutely continuous function.

Definition 4.17. *A vector-valued function $x : [0, t_f] \to \mathbb{R}^n$ whose components are absolutely continuous functions is called an arc.*

The next result constitutes an abstract sequential compactness property for inclusions.

Theorem 4.35. *Consider system (4.10) and assume that the set-valued function $F(\cdot,\cdot)$ is restricted to $[0,t_f]\times R$, where R is a closed subset of $\mathbb{R}^n$. Assume that the image of $F(t,x)$ is a closed subset of $\mathbb{R}^n$. We next assume that*

- *for each $t\in[0,t_f]$, the set*

$$G(t)=\{(y,z):(t,y,z)\in[0,t_f]\times R\times\mathbb{R}^n,\ z\in F(t,y)\}$$

  is closed and nonempty;
- *for every measurable function y on $[0,t_f]$ satisfying*

$$(t,y(t))\in[0.t_f]\times R$$

  a.e. and every $p\in\mathbb{R}^n$, the support function map

$$t\to\{(p,v):v\in F(t,y(t))\}$$

  is measurable;
- *for an integrable function $\kappa(\cdot)$, we have*

$$F(t,y)\subset B(0,\kappa(t))\ \forall(t,y)\in[0,t_f]\times R.$$

Let now y_i be a sequence of measurable functions on $[0,t_f]$ having $(t,y_i(t))\in[0,t_f]\times R$ a.e. and converging almost everywhere to y^*, and let $z_i:[0,t_f]\to\mathbb{R}^n$ be a sequence of functions satisfying $||z_i(t)||\le\kappa(t)$ a.e. whose components converge weakly in $\mathbb{L}_1(0,t_f)$ to those of *. Suppose that, for certain measurable subsets π_i of $[0t_f]$ satisfying

$$lim_{i\to\infty}\text{meas}(\pi_i)=t_f,$$

we have

$$z_i(t)\in F(t,y_i(t))+B(0,r_i(t)),\ t\in\pi_i\ \text{a.e.},$$

where $r_i(\cdot)$ is a sequence of nonnegative functions converging in $\mathbb{L}_1(0,t_f)$ to 0. Then

$$z^*(t)\in F(t,y^*(t)),t\in[0,t_f]\ \text{a.e.}$$

We refer to [6,86] for a formal proof of the presented result.

Consider now the i.v.p. with a bounded (possibly discontinuous) right-hand side

$$\dot{x}(t)=f(t,x(t)),$$
$$x(0)=x_0.$$

Let us now define the convex-valued, upper semicontinuous regularization of the right-hand side $f(\cdot,\cdot)$ of the above problem, i.e.,

$$G(t,x) := \bigcap_{\epsilon>0} \bar{co}\{f(t_1,x_1);\ |t_1 - t| < c,\ |x_{-}x| < c\},$$

for $\epsilon > O$. Introduce the following i.v.p. associated with the DI

$$\begin{aligned} &\dot{x}(t) \in G(t, x(t)) \text{ a.e. } [0, t_f], \\ &x(0) = x_0. \end{aligned} \tag{4.11}$$

We next use the above introduced concept of the upper semicontinuous regularization to a discontinuous right-hand side and establish an interesting equivalence result.

Theorem 4.36. *Let $f : \mathbb{R} \times \mathbb{R}^n \to \mathbb{R}^n$ be Γ^M-continuous and satisfy $||f(t,x)|| \leq L$ for some $L < M$ and all (t,x). Call G defined in (4.11) the upper semicontinuous, compact convex-valued regularization of $f(\cdot,\cdot)$. Then every solution of*

$$\dot{x}(t) \in G(t, x(t))$$

is also a Carathéodory solution of the original i.v.p., i.e.,

$$\begin{aligned} &\dot{x}(t) = f(t, x(t)), \\ &x(0) = x_0. \end{aligned}$$

One can also prove an interesting fact about the trajectories of the DI (4.10) with a bounded lower semicontinuous set-valued function $F(\cdot)$ that is in some sense "inverse" to the result of the above Theorem 4.36. For the original DI (4.10) there exists a bounded upper semicontinuous, compact convex-valued multifunction $G(\cdot)$ such that every solution of the i.v.p. (4.11) is also a solution of (4.10).

Let us now mention two fundamental results on the continuous dependency of trajectories of the DIs on the initial data.

Theorem 4.37. *Let $f : \mathbb{R} \times \mathbb{R}^n \to \mathbb{R}^n$ be any map which satisfies $||f(t,x)|| \leq L$ for all (t,x). Assume that, for some $M > L$, as $v(\cdot)$ varies over all Lipschitz continuous functions with constant M, the total variation of $t \to f(t, v(t))$ is uniformly bounded. Then each i.v.p.*

$$\begin{aligned} &\dot{x}(t) = f(t, x(t)), \\ &x(0) = x_0 \end{aligned}$$

has a unique solution, depending continuously on x_0.

Theorem 4.38. *Let $F : [0, t_f] \times \mathbb{R}^n \rightrightarrows \mathbb{R}^n$ be a bounded, Lipschitz continuous set-valued function with compact values. Then there exists a directionally continuous selector $f(\cdot, \cdot)$ of $F(\cdot, \cdot)$ such that, for every $x_0 \in \mathbb{R}^n$, the i.v.p.*

$$\dot{x}(t) = f(t, x(t)),$$
$$x(0) = x_0$$

has a unique solution, depending continuously on the initial data.

We refer to [159] for the formal proofs of the above fundamental continuity properties.

4.11 Differential Inclusions in Control Theory

Let us now present a theoretic result that in fact implies an "equivalence" between a control system of the type (4.4) (in fact the HS (4.6) with Q singleton) and a DI. Note that the equivalence mentioned above is understood here in the sense of the set of trajectories of DIs. This concept, namely, the coincidence of trajectories of two dynamic systems, constitutes a fundamental approach for an adequate "comparison" of these two systems. Note that the proof of the following result makes it in fact possible to establish the above equivalence. Recall that for each $u(\cdot) \in \mathcal{U}$ the generated solution $x^u(\cdot)$ is an element of $\mathbb{W}_n^{1,\infty}(0, t_f)$ (see Section 4.2).

Theorem 4.39. *Consider the control system (4.4) from Section 4.2 and assume*

- *$f(t, u, x)$ is continuous in (u, x) and measurable in t and U is a compact subset of $\mathbb{R}^m$;*
- *$f(t, \cdot, \cdot)$ is Lipschitz;*
- *the set (sometimes called an orientor field) $f(t, U, x)$ is convex for each (t, x).*

Let $(u_i(\cdot), x_i(\cdot))$ be a sequence of admissible pairs for (4.4) such that the set $\{x_i(t) : i \geq 1\}$ is bounded. Then there exists a subsequence of $x_i(\cdot)$ converging uniformly to a state trajectory of the control system (4.4).

Proof. Define

$$F(t, x) := f(t, U, x),$$

which is a mapping from $[0, t_f] \times \mathbb{R}^n$ to the subsets of $\mathbb{R}^n$. Then $x_i(\cdot)$ satisfies the DI

$$x_i(t) \in F(t, x_i(t)) \text{ a.e.}$$

The Lipschitz hypothesis implies that a subsequence of the $x_i(\cdot)$ converges uniformly to an arc (in the sense of Definition 4.17) $x^*(\cdot)$ and therefore the corresponding sequence of derivatives $\{\dot{x}_i(\cdot)\}$ converges weakly in $\mathbb{L}_1(0, t_f)$ (see [192,193]).

From the hypotheses of Theorem 4.39 it follows that $F(\cdot, \cdot)$ is convex-valued and that, for each t, the graph of $F(t, \cdot)$ is closed. Let S be a compact subset of $[0, t_f] \times \mathbb{R}^n$ containing the graphs of all the functions $x_i(\cdot)$. Then, for a certain constant k, for all $(t, x) \in S$, the set $F(t, x)$ is bounded (by the Lipschitz condition).

Since $f(t, \cdot, x)$ is continuous, the support function map is measurable. By Theorem 4.35, the limit arc $x^*(\cdot)$ satisfies the DI (4.10) with the specific set-valued function $F(\cdot, \cdot)$ defined above. Applying the finite-dimensional version (Theorem 4.28) of the Filippov–Himmelberg implicit functions theorem we deduce the existence of a measurable (admissible) $u(\cdot)$ (in fact $u(\cdot) \in \mathcal{U}$). The proof is complete. □

Note that the Lipschitz condition from Theorem 4.39 can be replaced by the condition of the linear growth. Then we have

$$(t, x) \in [0, t_f] \times \mathbb{R}^n, u\ U \Rightarrow ||f(t, x, u)|| \leq \Delta(t)(1 + ||x||)$$

for an integrable function $\Delta(\cdot)$. Let us also note that Theorem 4.39 does not assume any convergence condition for the control variable $u(\cdot)$. Theorem 4.39 means that the limit $x^*(\cdot)$ is an arc such that there exists a control function

$$u^*(\cdot) \in \mathcal{U}$$

satisfying

$$\dot{x}(t) = f(t, u^*(t), x^*(t)),\ t \in [0, t_f] \text{ a.e.}$$

The analytic result of Theorem 4.39 constitutes in fact a sequential compactness of the trajectories of the DI (4.10) with $F(t, x) := f(t, U, x)$ and for the corresponding control system (4.4).

Recall that a function of two variables $\chi(y, x) \in \mathbb{R},\ (y, x) \in \mathbb{R}^m \times \mathbb{R}^n$ is measurable when it is measurable with respect to the σ-algebra generated by products of the Lebesgue measurable subsets of $\mathbb{R}^m$ (for y) and Borel measurable subsets of $\mathbb{R}^n$ (for x). Given an arc $x(\cdot)$ that solves (4.10) for

$$F(t, x) := f(t, U, x)$$

we can deduce that (4.4) holds. The main question here is whether there is a measurable function $u(\cdot)$ doing this. The direct application of the Filippov–Himmelberg implicit functions theorem (Theorem 4.26 from Section 4.8) (see also the finite-dimensional version, namely, Theorem 4.28) to the specific DI (4.10) with the set-valued function $F(t, x) := f(t, U, x)$ (see the proof of Theorem 4.39) implies the following fundamental result.

Theorem 4.40. *Consider the control system (4.4) with a Carathéodory function $f(t,u,x)$ (continuous in x for each (t,u) and measurable in (t,u) for each x) and assume that U is compact. Then the control system (4.4) and the DI (4.10), where $F(t,x) := f(t,U,x)$, have the same trajectories.*

The proof of the above result is based on the general Theorem 4.26. As in the case of Theorem 4.39 the above result states that for a solution $x(\cdot)$ of the specific DI (4.10) with $F(t,x) := f(t,U,x)$ there exists an admissible control $u(\cdot) \in \mathcal{U}$ such that these two trajectories coincide. We can finally write

$$\{x(\cdot)\} \equiv \{x^u(\cdot)\},$$

where $u(\cdot) \in \mathcal{U}$, for the corresponding sets of solutions.

In parallel with (4.10) for

$$F(t,x) := f(t,U,x),$$

where $f(\cdot,\cdot,\cdot)$ is the right-hand side of the control system (4.4), we also consider the relaxed DI. This DI (and the corresponding i.v.p.) can also be determined as (4.10) for the convexified orientor field

$$F(t,x) := \bar{co} f(t,U,x).$$

Evidently, the right-hand sides of the control systems of the type (4.4) not always satisfy the hypotheses of Theorem 4.39. The most critical property from the assumptions of Theorem 4.39 is the convexity assumption. If $f(t,U,x)$ fails to have the required convexity property (the convexity of the velocity sets), the sequential compactness of trajectories and the established equivalence of (4.4) and (4.10) can also fail. This observation is the main motivation for the concept of a relaxed DI, i.e.,

$$\begin{aligned} \dot{x}(t) &\in \bar{co}(f(t,U,x)), \\ x(0) &= x_0. \end{aligned} \tag{4.12}$$

Evidently, the orientor field in (4.12) satisfies the convexity condition from Theorem 4.39 and the sequential compactness of the corresponding trajectories is now reestablished.

Definition 4.18. *The solutions $x^r(\cdot)$ of the i.v.p. (4.12) are called the relaxed trajectories associated with the initially given control system (4.4).*

Note that under the generic assumptions related to the right-hand side $f(\cdot,\cdot,\cdot)$ of (4.4) (see, e.g., Section 4.2 or [159,173]) the relaxed DI (4.12) has an absolutely continuous solution $x^r(\cdot)$.

4.12 Constructive Approximations of Differential Inclusions

Similar to the numerical analysis of ODEs and the corresponding computational methods the theory of DIs also includes some results related to the discretization schemes and constructive approaches to the i.v.p. of the type (4.10). This part of the abstract theory is of crucial importance for the engineering applications (including OC applications). In this section we give a short introduction to the discrete approximations of DIs and also cite the most important works from this area.

Consider the i.v.p. (4.10) under the assumption of an absolutely continuous solution $x(\cdot)$ to this problem. Let $\mathcal{F}$ be the set of solutions to (4.10). Usually this set consists of more than one element, that is, we have a bundle of trajectories. Moreover, from the point of view of numerical analysis one can include into an extended bundle various consistent approximating solutions $y(\cdot) \in \mathcal{F}$. In this book we only deal with a simple Euler-type discretization scheme for DIs.

Consider a number $N \in \mathbb{N}$ and introduce a time grid and the corresponding step size, i.e.,

$$0 < ... < t_i < \cdots < t_N = t_f,$$
$$h := t_j - t_{j_1} = t_f / N, \ j = 1, \cdots, N.$$

Put now $\xi_0 \equiv x_0$, and calculate the $(j+1)$ iteration from

$$\xi_{j+1} \in \xi_j + hF(t_j, \xi_j), \ j = 1, ..., N-1. \tag{4.13}$$

We will call the discrete scheme (4.13) a difference inclusion (DifI) associated with the initially given DI (4.10). We plan to consider the abstraction (4.13) as an adequate approximating scheme for (4.10).

Definition 4.19. *A solution of the approximating DifI (4.13) for the given parameters N, h is any continuous and piecewise linear function*

$$\xi^N : [0, t_f] \to \mathbb{R}^n$$

such that

$$\xi^N(t) = \xi_j + h(t - t_j)(\xi_{j+1} - \xi_j), \ t \in [t_{j+1}, t_j], \ j = 1, ..., N-1.$$

Let us now present a classic convergence result associated with the Euler method (4.13) for DIs.

Theorem 4.41. *Suppose that the set-valued function $F(\cdot,\cdot)$ is nonempty, compact- and convex-valued, and, moreover, upper semicontinuous in $[0,t_f]\times\mathbb{R}^n$. Assume that there exist constants k and a, such that*

$$||z|| < k||x|| + a,$$

whenever $z \in F(t,x)$, $x \in \mathbb{R}^n$, and $t \in [0,t_f]$. Then every sequence $\{\xi^N(\cdot)\}$ generated by the method (4.13) has a subsequence which converges as $N \to \infty$, uniformly in $[0,t_f]$, to a solution of the original i.v.p. (4.10).

The corresponding proof uses the celebrated Arzela theorem and in fact is based on the idea of the classical Peano theorem to prove existence of solutions to differential equations. The next result (obtained under stronger assumptions than Theorem 4.41) establishes the approximability property of the sequence $\{\xi^N(\cdot)\}$ generated by the Euler method (4.13).

Theorem 4.42. *Suppose that the conditions of Theorem 4.41 hold and additionally assume that $F(\cdot,\cdot)$ is continuous on $[0,t_f]\times\mathbb{R}^n$ and Lipschitz continuous in x on some bounded sets in $\mathbb{R}^n$. Then for every $\epsilon > 0$ there exists $\tilde{N} \in \mathbb{N}$ such that for every number*

$$N > \tilde{N}$$

and for every solution $\xi^N(\cdot)$ of the DifI (4.13) there exists a solution $x(\cdot)$ of the original i.v.p. (4.10) such that

$$||\xi^N(\cdot) - x(\cdot)||_{\mathbb{C}(0,t_f)} < \epsilon.$$

As usually the norm $||\cdot||_{\mathbb{C}(0,t_f)}$ denotes the $\mathbb{C}(0,t_f)$-norm. Note that the technical proof of the above result is based on a very interesting extension of the classic Gronwall lemma, namely, on the following Gronwall–Filippov–Waiewski theorem.

Theorem 4.43. *Let $y : [0,t_f] \to \mathbb{R}^n$ be an absolutely continuous function with $y(0) = x_0$, γ being a positive constant. Define*

$$\Theta := \{(t,x) \in [0,t_f]\times\mathbb{R}^n : ||x - y(t)|| < \gamma\}.$$

Let $F(\cdot,\cdot)$ be nonempty closed-valued and continuous and satisfy

$$d_H(F(t,x), F(t,z)) < k(t)||x - z||$$

for all (t,x) and (t,z) from Θ with a measurable $k(\cdot)$, where

$$d_H(\cdot,\cdot)$$

is the Hausdorff distance.

Assume, moreover, that

$$d(y(t), F(t, y(t))) < p(t)$$

for almost all $t \in [0, t_f]$ *for a measurable* $p(\cdot)$ *such that*

$$\zeta(t) = \int_0^t \exp[\int_\tau^t k(s)ds]p(\tau)d\tau \leq \gamma.$$

Then there exists a solution $x(\cdot)$ *to the i.v.p. (4.10) such that*

$$||x(t) - y(t)|| < \zeta(t)$$

for all $t \in [0, t_f]$.

As usual in numerical analysis one is interested in estimating the distance between the sets of solutions of the initially given i.v.p. (4.10) for the DI and the auxiliary DifI (4.13). This error estimation associated with the proposed numerical approximation, namely, with the discrete scheme (4.13), has to be realized by a reasonable manner, by the step size h (parameter of the method). The classic result in this sense is due to B.N. Pshenichny [28], who proved the following fact.

Theorem 4.44. *Let* $F(\cdot)$ *be a (stationary) compact- and convex-valued set-valued function. Let* $x(\cdot)$ *be a solution of*

$$\dot{x}(t) \in F(x(t)) \text{ for a.e. } t \in [0, t - f],$$
$$x(0) = x_0.$$

Suppose that there exists $\epsilon > 0$ *such that* $F(\cdot)$ *is Lipschitz continuous on the set*

$$\{\xi \in \mathbb{R}^n : ||\xi - x(t)|| < \epsilon, \; t \in [0, t_f]\}.$$

Then there exist constants $c > 0$ *and* $\tilde{N} \in \mathbb{N}$ *such that for all* $N > \tilde{N}$ *there exists a solution* $\xi^N(\cdot)$ *of the Euler-type DifI (4.13) such that*

$$\max_{0 \leq j \leq N} ||\xi^N(t_j) - x(t_j)|| \leq ch.$$

We refer to [28] for a formal proof of Theorem 4.44. Let us also note that there are many advanced numerical approximations of the i.v.p. of the type (4.10). A fairly complete overview can be found in [144–146]. We also refer to [147] for some recent results from the numerical analysis of DIs.

4.13 Notes

Let us here give some references related to the abstract theory of DIs and to the concrete applications of this abstraction to OCPs [6,20,22,27,29–32,74,86,92–94,96,102,127,142,143, 130–132,135,144–147,154,161,167,173,192,198,226,234,235,246–248,254,264,269,279,280, 293,309,314,321,326].

CHAPTER 5

Relaxation Schemes in Conventional Optimal Control and Optimization Theory

5.1 Young Measures

Extensions in problems of variational calculus, beginning with the idea of Hilbert, were realized many times for various purposes. Let us consider the regular variational problem

$$\int_{\Upsilon} \mathcal{L}(s, \omega(s), \dot{\omega}(s))ds \to \min$$
$$\omega = \varsigma \text{ on } \partial\Upsilon.$$

In the 20th problem of his Paris lecture, Hilbert asked the following question: "Is it true that the presented variational problem has a solution $\omega(\cdot)$ in case we generalize the notion of solution in an appropriate sense?" Today we know the answer is positive. Under some natural assumption (see, e.g., [328]) for a bounded region Υ in $\mathbb{R}^n$ and for the smooth functions

$$\mathcal{L}(\cdot, \cdot, \cdot), \ \varsigma(\cdot)$$

there exists a solution $\omega(\cdot)$ from the Sobolev space

$$\mathbb{W}_l^1(\Upsilon), \ 1 < l < \infty.$$

The historically first investigation on this subject was a work of N.N. Bogoljubov [94]. The concept of relaxed controls was introduced by L.C. Young in 1937 under the name of generalized curves and surfaces [326]. It has been used extensively in the literature for the study of diverse optimal control problems (OCPs) [321,127,173,192,154,279]. As Clarke points out in [130], a relaxed problem is, in general, the only one for which existence theorems can be proved and, for this reason, there are many who deem it the only reasonable problem to consider in practice. However, though relaxation of a problem is important for the existence theory, we are also interested in the development of the relaxation theory (RT)-based numerical solution algorithms.

Recall the following famous example of L.C. Young.

A Relaxation-Based Approach to Optimal Control of Hybrid and Switched Systems
https://doi.org/10.1016/B978-0-12-814788-7.00011-4

Example 5.1.

$$\text{minimize } J(v(\cdot)) := \int (v_x(x)^2 - 1)^2 + v^2(x)dx$$
$$\text{subject to } v(0) = v(1) = 0,$$

where $v(\cdot)$ is a differentiable function.

The minimum of the objective functional is zero, but it cannot be attained since there is no function that satisfies the above bound conditions almost everywhere. A suitable minimizing sequence is an oscillating sequence. It converges in the weak topology to zero.

Let us make some remarks about the necessary notation. By $\mathbb{C}_0(\mathbb{R}^n)$ we denote a closure of the space of continuous functions on $\mathbb{R}^n$ with a compact support. The dual space $\mathbb{C}_0(\mathbb{R}^n)^*$ is the space of signed Radon measures with finite mass (see, e.g., [282]). The duality pairing is given by

$$\langle \mu(\cdot), \lambda(\cdot) \rangle = \int_{\mathbb{R}^n} \lambda(x) d\mu(x).$$

We also need the celebrated concept of the weak* measurability (see, e.g., [279,280]).

Definition 5.1. *A map $\mu : \Omega \to \mathcal{M}(\mathbb{R}^n)$ is called weakly* measurable if the functions $x \to \langle \mu(x), \lambda \rangle$ are measurable for all $\lambda \in \mathbb{C}_0(\mathbb{R}^n)$.*

The fundamental theorem of the Young measures can now be formulated as follows.

Theorem 5.1. *Let $\Omega \subset \mathbb{R}^n$ and let $z_j : \Omega \to \mathbb{R}^n$ be a sequence of measurable functions. Then there exists a subsequence $\{z_{j_k}\}$ and a weakly* measurable map $\mu : \Omega \to \mathcal{M}(\mathbb{R}^n)$ such that:*

- $$\mu_x \geq 0, \ ||\mu_x|| := \int_{\mathbb{R}^n} d\mu_x \leq 1$$

  for $x \in \Omega$;
- *for all $\lambda(\cdot) \in \mathbb{C}_0(\mathbb{R}^n)$ we have*

  $$\lambda(z_{j_k}) \rightharpoonup \langle \mu_x, \lambda \rangle \text{ in } \mathbb{L}^\infty(\Omega);$$

- *in the case $||\mu_x|| = 1$ for a.a. $x \in \Omega$ we have*

  $$\lim M \to \infty \sup_k ||z_{j_k}|| \geq M = 0.$$

The map μ from Theorem 5.1 is called the Young measure generated by the sequence $\{z_{j_k}\}$. An interesting analytic result for the Young measures is a natural generalization of the classic Jensen inequality (see, e.g., [213]).

Theorem 5.2. *Let $g : \mathbb{R}^n \to \mathbb{R}^n$ be strictly convex and μ be a probability measure on $\mathbb{R}^n$ with compact support. Then*

$$\langle \mu, \lambda \rangle \geq g(\langle \mu, \mathrm{id} \rangle),$$

with equality occurring iff μ is a Dirac measure.

Let us consider a simple example.

Example 5.2. *We have*

$$\text{minimize}_{v(\cdot \in X \subset \mathbb{C}_0(\mathbb{R}^n))} J(v(\cdot)) = \int_\Omega f(\nabla v(x))dx.$$

Assume that $\{v_l\}_{l \in \mathbb{N}} \subset \mathbb{C}_0(\mathbb{R}^n)$ is a minimizing sequence for the above minimization problem. Also assume that this problem has an optimal solution $v^{opt}(\cdot)$. Then using the above Jensen inequality we deduce

$$\liminf_{l \to \infty} J(v_l) = \int_\Omega f(\nabla v^{opt}(x))dx = J(v^{opt}(\cdot)).$$

Moreover, from Theorem 5.2 we obtain that μ (a generated Young measure) is a Dirac measure.

More generally, the following result is true.

Theorem 5.3. *Assume $z_j : \Omega \to \mathbb{R}^n$ generates the Young measure $\mu : \Omega \to \mathcal{M}(\mathbb{R}^n)$. Then $z_j \to z$ in measure implies $\mu = \delta_{z(x)}$.*

We refer to [280,321] for the corresponding proof.

5.2 The Gamkrelidze–Tikhomirov Generalization

Consider the following OCP:

$$\begin{aligned}
&\text{minimize } J(x(\cdot), u(\cdot)) = \int_0^{t_f} f_0(t, x(t), u(t))dt \\
&\text{subject to } \dot{x}(t) = f(t, x(t), u(t)) \text{ a.e. on } [0, t_f], \\
&x(0) = x_0, \\
&u(t) \in U \text{ a.e. on } [0, t_f], \\
&h_j(x(t_f)) \leq 0 \; \forall j \in I, \\
&q(t, x(t)) \leq 0 \; \forall t \in [0, t_f],
\end{aligned} \tag{5.1}$$

where $f_0 : [0, t_f] \times \mathbb{R}^n \times \mathbb{R}^m \to \mathbb{R}$ is a continuously differentiable function,

$$f : [0, t_f] \times \mathbb{R}^n \times \mathbb{R}^m \to \mathbb{R}^n,$$
$$h_j : \mathbb{R}^n \to \mathbb{R} \text{ for } j \in I,$$
$$q : [0, t_f] \times \mathbb{R}^n \to \mathbb{R},$$

and $x_0 \in \mathbb{R}^n$ is a fixed initial state. By I we denote a finite set of index values associated with the corresponding constraints in (5.1). We next assume that the functions f,

$$h_j(\cdot), \ j \in I,$$

and $q(t, \cdot)$, $t \in [0, t_f]$ are continuously differentiable and f_0 is an integrable function. The set U of admissible controls in (5.1) is assumed to be a compact and convex subset of $\mathbb{R}^m$. We will restrict here our consideration to the simple "box-type" admissible control sets

$$U := \{u \in \mathbb{R}^m \ : \ b_-^i \leq u_i \leq b_+^i, \ i = 1, ..., m\},$$

where $b_-^i, b_+^i, i = 1, ..., m$, are some constants. The admissible control functions $u : [0, t_f] \to \mathbb{R}^m$ are square integrable functions of time. Let

$$\mathcal{U} := \{v(\cdot) \in \mathbb{L}_m^2([0, t_f]) \ : \ v(t) \in U \text{ a.e. on } [0, t_f]\}$$

be the set of admissible control functions. Without loss of generality we suppose only one state condition; it is possible to extend our approach to OCPs with several state conditions. Recall that by $\mathbb{L}_m^2([0, t_f])$ we denote the standard Lebesgue space of all square integrable functions $[0, t_f] \to \mathbb{R}^m$. In addition, we assume that for each $u(\cdot) \in \mathcal{U}$ the initial value problem

$$\dot{x}(t) = f(t, x(t), u(t)) \text{ a.e. on } [0, t_f], \ x(0) = x_0 \tag{5.2}$$

has a unique absolutely continuous solution $x^u(\cdot)$. For some constructive uniqueness conditions related to (5.2) see, e.g., [86,154,192,269]. Given an admissible control function $u(\cdot)$ the solution to the initial value problem (5.2) is an absolutely continuous function $x : [0, t_f] \to \mathbb{R}^n$. It is denoted by $x^u(\cdot)$. We assume that the problem (5.1) has an optimal solution. It is well known that the class of OCPs of the type (5.1) is broadly representative (see [86,5,192,154]).

We next also consider the basic OCP (5.1) with a terminal functional $\mathcal{J}(u(\cdot)) := \phi(x^u(1))$ and with an additional integral constraint

$$\int_0^{t_f} s(t, x(t), u(t))dt \leq 0,$$

where $s : [0, t_f] \times \mathbb{R}^n \times \mathbb{R}^m \to \mathbb{R}$ is a continuous function. Recall that an initially given OCP (5.1) with an integral costs functional $J(\cdot, \cdot)$ can be reformulated as an equivalent OCP with an associated terminal functional $\mathcal{J}(\cdot)$.

We next incorporate in our dynamic optimization framework an additional restriction, namely, the control inequality

$$\tilde{s}(u(\cdot)) \leq 0,$$

where

$$\tilde{s}(u(\cdot)) := \int_0^{t_f} s(t, x^u(t), u(t))dt.$$

Moreover, we use the following additional notation:

$$\tilde{h}_j(u(\cdot)) := h_j(x^u(t_f)) \; \forall j \in I,$$
$$\tilde{q}(u(\cdot))(t) := q(t, x^u(t)) \; \forall t \in [0, t_f].$$
$$\aleph(n+1) := \Big\{(\alpha^1(\cdot), ..., \alpha^{n+1}(\cdot))^T \; : \; \alpha^j(\cdot) \in \mathbb{L}_1^1([0, t_f]), \; \alpha^j(t) \geq 0,$$
$$\sum_{j=1}^{n+1} \alpha^j(t) = 1 \; \forall t \in [0, t_f]\Big\}, \; \alpha(\cdot) := (\alpha^1(\cdot), ..., \alpha^{n+1}(\cdot))^T.$$

In parallel with the given control system (5.2) we next consider the relaxed control system in the form of a Gamkrelidze system (sometimes also called the Gamkrelidze chattering) [173, 154], i.e.,

$$\begin{aligned} \dot{y}(t) &= \sum_{j=1}^{n+1} \alpha^j(t) f(t, y(t), u^j(t)) \text{ a.e. on } [0, t_f], \\ y(0) &= x_0, \end{aligned} \tag{5.3}$$

where $\alpha(\cdot) \in \aleph(n+1)$ and

$$u^j(\cdot) \in \mathcal{U}, \; j = 1, ..., n+1.$$

Under some generic assumptions for the initially given control system (5.2) there exists an absolutely continuous solution $y^\nu(\cdot)$ of the initial value problem (5.3) generated by the (admissible) generalized control

$$\nu(\cdot) \in \aleph(n+1) \times \mathcal{U}^{n+1},$$

where $\nu(t) := (\alpha^1(t), ..., \alpha^{n+1}(t), u^1(t), ..., u^{n+1}(t))^T$.

Recall that the Radon probability measure ς on the Borel sets of U is a regular positive measure ς such that $\varsigma(U) = 1$. Let

$$\mathcal{M}_{+}^{1}(U)$$

be the space of all probability measures on the Borel sets of U. A relaxed control $\mu(\cdot)$ is a measurable function

$$\mu : [0, t_f] \to \mathcal{M}_{+}^{1}(U),$$

where "measurability" is defined in [321]. Following [154] we denote by $R_c([0, t_f], U)$ the set of relaxed controls $\mu(\cdot)$. These controls give rise to the relaxed dynamics

$$\begin{aligned} \dot{\eta}(t) &= \int_U f(t, \eta(t), u)\mu(t)(du) \text{ a.e on } [0, t_f], \\ \eta(0) &= x_0. \end{aligned} \tag{5.4}$$

Note that the generalized control system (5.4) with the "measure-type controls" $\mu(\cdot) \in R_c([0, t_f], U)$ is sometimes called the Young relaxations of the dynamic system from (5.1). The initially given control system can also be rewritten as a control problem for the differential inclusion, i.e.,

$$\begin{aligned} \dot{x}(t) &\in f(t, x(t), U) \text{ a.e. on } [0, t_f], \\ x(0) &= x_0, \end{aligned}$$

where

$$f(t, x(t), U) := \{f(t, x(t), u) \; : \; u \in U\}.$$

Note that the differential inclusions of the type (5.4) constitute specific dynamic constraints in the OCP under consideration. Taking into account the abovementioned standard assumptions, a function $x^u(\cdot)$ is a solution of the initial control system (5.1) with $u(\cdot) \in \mathcal{U}$ if and only if it is a solution of the initial differential inclusion given above. For details, see [154] (and also Chapter 4). Related to the initially given differential inclusion we next consider the relaxed (convexified) differential inclusion (RDI)

$$\begin{aligned} \dot{x}(t) &\in \operatorname{conv} f(t, x(t), U) \text{ a.e. on } [0, t_f], \\ x(0) &= x_0. \end{aligned} \tag{5.5}$$

We are now interested to establish some expected relations between the originally given dynamic system (5.2) and the relaxed control systems (5.3), (5.4), and (5.5) given above. In that sense let us present the fundamental equivalence theorem (see [127,321,154] for the formulation and the corresponding formal proof).

Theorem 5.4. *Let the initially given control system* (5.2) *from OCP (5.1) satisfy the basic assumptions. A function $y(\cdot)$ is a solution of the* Gamkrelidze *system* (5.3) *with $\nu(\cdot) \in \aleph(n+1) \times \mathcal{U}^{n+1}$ if and only if it is a solution of the* RDI (5.5). *Moreover, a function $x(\cdot)$ is a solution of the* RDI (5.5) *if and only if it is a solution of the Young relaxed control system* (5.4) *with a relaxed control $\mu(\cdot)(\cdot) \in R_c([0, t_f], U)$.*

Let

$$\nu(\cdot) = (\alpha^1(\cdot), ..., \alpha^{n+1}(\cdot), u^1(\cdot), ..., u^{n+1}(\cdot))^T,$$
$$\bar{\nu}(\cdot) = (\bar{\alpha}^1(\cdot), ..., \bar{\alpha}^{n+1}(\cdot), \bar{u}^1(\cdot), ..., \bar{u}^{n+1}(\cdot))^T.$$

As a consequence of the basic existence and uniqueness theorems for differential equations (see, e.g., [86,269]) one can obtain the following useful estimates [269].

Theorem 5.5. *Let the initial control system* (5.2) *satisfy the basic assumptions. There exist finite constants c_1, c_2, c_3, and c_4 such that*

$$\begin{aligned}
&||x^u(\cdot)||_{\mathbb{L}^\infty_n([0,t_f])} \le c_1, \\
&||x^u(\cdot) - x^{\tilde{u}}(\cdot)||_{\mathbb{L}^\infty_n([0,t_f])} \le \\
&\le c_2 ||u(\cdot) - \tilde{u}(\cdot)||_{\mathbb{L}^2_m([0,t_f])}, \\
&||y^\nu(\cdot)||_{\mathbb{L}^\infty_n([0,t_f])} \le c_3, \\
&||y^\nu(\cdot) - y^{\bar{\nu}}(\cdot)||_{\mathbb{L}^\infty_n([0,t_f])} \le \\
&\le c_4 \sum_{j=1}^{n+1} ||u^j(\cdot) - \bar{u}^j(\cdot)||_{\mathbb{L}^2_m([0,t_f])},
\end{aligned}$$

for all $u(\cdot), \tilde{u}(\cdot) \in \mathcal{U}$ and for all

$$\nu(\cdot), \bar{\nu}(\cdot) \in \aleph(n+1) \times \mathcal{U}^{n+1},$$

where $x^u(\cdot), x^{\tilde{u}}(\cdot)$ are the solutions of (5.2) *and $y^\nu(\cdot), y^{\bar{\nu}}(\cdot)$ are the solutions of* (5.3) *associated with the admissible controls $u(\cdot), \tilde{u}(\cdot)$ and $\nu(\cdot), \bar{\nu}(\cdot)$, respectively.*

Using the solution $y^\nu(\cdot)$ of (5.3), we can define the Gamkrelidze relaxation of the OCP (5.1). We have

$$\begin{aligned}
&\text{minimize } \bar{\mathcal{J}}(\nu(\cdot)) \\
&\text{subject to (5.3) } \nu(\cdot) \in \aleph(n+1) \times \mathcal{U}^{n+1} \\
&\bar{h}(\nu(\cdot)) \le 0, \; \bar{q}(\nu(\cdot))(t) \le 0 \; \forall t \in [0, t_f], \; \bar{s}(\nu(\cdot)) \le 0,
\end{aligned} \tag{5.6}$$

where

$$\bar{\mathcal{J}}(\nu(\cdot)) := \phi(y^{\nu}(t_f)),$$
$$\bar{h}(\nu(\cdot)) := h(y^{\nu}(t_f)),$$
$$\bar{q}(\nu(\cdot))(t) := q(t, y^{\nu}(t)) \; \forall t \in [0, t_f],$$
$$\bar{s}(\nu(\cdot)) := \int_0^{t_f} \sum_{j=1}^{n+1} \alpha^j(t) s(t, y^{\nu}(t), u^j(t)) dt.$$

Note that under some mild assumptions the relaxed OCP (5.6) has an optimal solution $\nu^{opt} \in \aleph(n+1) \times \mathcal{U}^{n+1}$ (see [167]). We denote an optimal solution by $\nu^{opt}(\cdot)$. The following result uses the equivalence theorem (Theorem 5.4) and establishes the relation between the OCP (5.1) and the Gamkrelidze relaxation (5.6). First let us note that in the case of the originally given objective functional $J(\cdot, \cdot)$ from OCP (5.1) the corresponding relaxed functional can be written as follows:

$$\int_0^{t_f} \int_U f_0(t, x(t), u) \mu(t)(du) dt.$$

The next result represents a relation between the relaxed and original OCPs and can be proved as in [92].

Theorem 5.6. *Let $\nu(\cdot) \in \aleph(n+1) \times \mathcal{U}^{n+1}$ be an optimal solution of the relaxed OCP (5.6) and $u^{opt}(\cdot) \in \mathcal{U}$ be an optimal solution of the initially given OCP (5.1). Then*

$$\bar{\mathcal{J}}(\nu^{opt}(\cdot)) \leq \mathcal{J}(u^{opt}(\cdot)).$$

Let us note that the natural result of Theorem 5.6 indicates that (under natural conditions) positive gaps

$$Gap := \mathcal{J}(u^{opt}(\cdot)) - \bar{\mathcal{J}}(\nu^{opt}(\cdot))$$

may occur.

Let $\mathbb{L}^1([0, t_f], \mathbb{C}(U))$ be the space of absolutely integrable functions from $[0, t_f]$ to $\mathbb{C}(U)$ (the space of all continuous function on U). Note that the topology imposed on $R_c([0, t_f], U)$ is the weakest topology such that the mapping

$$\mu \rightarrow \int_0^{t_f} \int_U \psi(t, u) \mu(t)(du) dt$$

is continuous for all $\psi(\cdot, \cdot) \in \mathbb{L}^1([0, t_f], \mathbb{C}(U))$. Finally we recall ([321], p. 287) that the space of the Young measures $R_c([0, t_f], U)$ is a compact and convex space of the dual space to $\mathbb{L}^1([0, t_f], \mathbb{C}(U))^*$. The topology of the "weak" norm associated with the space

$$\mathbb{L}^1([0, t_f], \mathbb{C}(U))^*$$

and restricted to the space $R_c([0, t_f], U)$ coincides with the weak star topology [321].

5.3 Chattering Lemma and Relaxed Trajectories

The chattering lemma constitutes an important analytic tool of the classic theory of differential inclusions (see [10,15]). Moreover, it provides a necessary theoretic fundament for the analytic proof of the Pontryagin maximum principle (see, e.g., [10]). In Chapter 7 we will discuss an approach to the RT of hybrid systems (HSs) based on the classic chattering lemma. We call this approach a "fully relaxed" relaxation scheme. Note that a concrete practical application of this methodology to the numerical treatment of the classic and hybrid OCPs can cause significant relaxation gaps (see, e.g., [2,6,26,32]). Let us give here a generic formulation of the celebrated chattering lemma.

Theorem 5.7. *Let $\mathcal{Y}$ be a finite closed interval and let $\mathcal{X}$ be a compact set in $\mathbb{R}^n$. Let*

$$g_1, \ldots, g_K$$

be functions defined on $\mathcal{Y} \times \mathcal{X}$ with range in $\mathbb{R}^n$ and possessing the following properties:

- *each $g_k(\cdot, x)$ is a measurable function on $\mathcal{Y}$ for each $x \in \mathcal{X}$;*
- *each $g_k(t, \cdot)$ is continuous on $\mathcal{X}$ for each $t \in \mathcal{Y}$;*
- *there exists an integrable function $\mu(\cdot)$ defined on $\mathcal{Y}$ such that for all pairs*

$$(t, x), \;\; (t, x') \in \mathcal{Y} \times \mathcal{X}$$

and $k = 1, \ldots, K$

$$||g_k(t, x)|| \leq \mu(t), \;\; ||g_k(t, x) - g_k(t, x')|| \leq \mu(t)||x - x'||.$$

Let

$$\varsigma_k(\cdot), \;\; k = 1, \ldots, K$$

be real-valued nonnegative measurable functions defined on $\mathcal{Y}$ and satisfying

$$\sum_{k=1}^{K} \varsigma_k(t) = 1 \text{ (a.e.)}.$$

Then for every $\epsilon > 0$ there exists a subdivision of $\mathcal{Y}$ into a finite collection of nonoverlapping intervals

$$\mathcal{Y}_r, r = 1, \ldots, R$$

and an assignment of one of the functions $g_1, \ldots, g_K$ to each $\mathcal{Y}_r$ such that the following holds: if $g_{\mathcal{Y}_r}$ denotes the function assigned to $\mathcal{Y}_r$ and if $g(\cdot, \cdot)$ is a function that agrees with $g_{\mathcal{Y}_r}$ on the interior

$$\text{int}(\mathcal{Y}_r)$$

of each $\mathcal{Y}_r$, *i.e.,*

$$g(t,x) = g_{\mathcal{Y}_r}(t,x)$$

for $t \in \mathrm{int}(\mathcal{Y}_r)$, $r = 1, ..., R$, *then for every*

$$t', t'' \in \mathcal{Y}, \ x \in \mathcal{X}$$

we have

$$|| \int_{t'}^{t''} (\sum_{k=1}^{K} \varsigma_k(t) g_k(t,x) - g(t,x)) dt || \leq \epsilon.$$

We refer to [86] for complete proofs of Theorem 5.7 (the chattering lemma).

Let us now give here a useful qualitative interpretation of Theorem 5.7. In the classic control applications the functions $g_1, ..., g_K$ are obtained as in the Gamkrelidze system (5.3) (the Gamkrelidze chattering). The function $g(\cdot)$ from Theorem 5.7 is obtained in the same fashion as the right-hand side of (5.3). That is, the basic time interval is divided up into a large number of small intervals and on each subinterval we choose one of the controls $u_1, ..., u_K$ to build the control $\mu(\cdot)$. In a physical system, the relaxed control $\mu(\cdot)$ corresponds to a rapid switching back and forth among the constant controls $u_1, ..., u_K$. In the engineering praxis, the resulting control system is said to "chatter." From the proof of Theorem 5.7 one can also deduce that a relaxed trajectory (generated by a relaxed control $\mu(\cdot)$) can be approximated as close as we please by an ordinary trajectory (generated by an admissible control in the initial system).

The above result can also be applied to the relaxation procedure of some classes of HSs. The first application of the chattering lemma to the hybrid RT was discussed in [81]. The main theoretic results related to relaxations of the OCP with HSs are presented in [9,37,38]. These results are direct consequences of the above chattering lemma.

5.4 The Fattorini Approach

Relaxed trajectories of the abstract differential inclusions in a separable Banach Z space are related to trajectories of the integral evolution equations

$$\begin{aligned} &\dot{x}_\lambda(t) \in A(t)x_\lambda(t) + \int_{\Gamma(t)} g(t, x_\lambda(t), u) d\lambda_t(u) \text{ a.e. on } [0, t_f], \\ &x_\lambda(0) = x_0 \in \mathrm{dom}(A(\cdot)), \end{aligned} \tag{5.7}$$

where $A(\cdot)$ is a maximal monotone operator and $\lambda(\cdot)$ is an admissible relaxed control. Note that $\lambda(\cdot)$ is the measurable selection of the set-valued function

$$F(t) := \{\nu(\cdot) \in \mathcal{M}_1^+(\mathcal{U}),\ \nu(\Gamma(t)) = 1\},\ t \in [0, t_f].$$

Here $\mathcal{U}$ is the space of all admissible (measurable) control functions $u(\cdot)$ with $u(t) \in \Gamma(t)$. Since

$$\int_{\Gamma(t)} g(t, x_\lambda(t), u) d\lambda_t(u) \in \bar{\text{co}}\{g(t, x_\lambda(t), \Gamma(t))\}$$

for almost all $t \in [0, t_f]$, the trajectory $\dot{x}_\lambda$ of the differential inclusion (5.7) is an absolutely continuous function that satisfies the differential inclusion

$$x(t) \in A(t) x_\lambda(t) + \text{co}\tilde{F}(t, x(t)), \tag{5.8}$$

where

$$\tilde{F}(t, x) := g(t, x(t), \Gamma(t)).$$

Note that the relaxed trajectories of (5.7) coincide with the trajectories of the differential inclusion (5.8) (see [22,159]). The consideration of these trajectories leads to some integral representation theorem involving Young measures (see [280] for further details). The existence theorem for the above differential inclusion can be formulated as follows.

Theorem 5.8. *Suppose that $\Gamma : [0, t_f] \rightrightarrows K(\mathcal{U})$ is a compact-valued measurable set-valued function, $g(\cdot, \cdot)$ is a Carathéodory function such that*

$$t \to \sup\{||g(t, u)||,\ u \in \Gamma(t)\}$$

is integrable on $[0, t_f]$, and $v : [0, t_f] \to Z$ is an integrable function such that

$$v(t) \in \bar{\text{co}} g(t, \Gamma(t))$$

for a.e. $t \in [0, t_f]$. Then there exists a Young measure such that

$$v(t) = \int_{\Gamma(t)} g(t, u) d\lambda_t(u) \text{ a.e.}$$

5.5 Some Further Generalizations of the Young Measures

In this section we discuss (very briefly) some generalizations of the classic Young measures introduced in Section 5.1. While Young measures successfully capture a possible oscillatory behavior of sequences, they completely miss concentration effects. On the other side various concentrations appear in various problems of the calculus of variations and optimal control. These concentration effects may be dealt with by an appropriate generalization of the classic Young measures [280].

Let Ω be a bounded, open, measurable subset of $\mathbb{R}^n$ and $\{u_j\}$, $j \in \mathbb{N}$, be a sequence of measurable functions from Ω into $\mathbb{R}^m$. Recall that Young measures on Ω are weakly* measurable mappings $x \to \nu_x$, $x \in \Omega$, with

$$\nu_x : \Omega \to \mathcal{M}_1^+(\mathbb{R}^{M \times N}).$$

By $\mathcal{M}(X)$ we denote the set of Radon measures on a Borel set X and by

$$\mathcal{M}_1^+(X)$$

its subset of probability measures. It is known that the set of Young measures $Y(\Omega, \mathbb{R}^{M \times N})$ is a convex subset of the Lebesgue space

$$\mathbb{L}_*^\infty(\Omega, \mathcal{M}(\mathbb{R}^{M \times N})).$$

From the results of [22] it follows that for every bounded sequence

$$\{y(\cdot)_k\} \subset \mathbb{L}_1(\Omega, \mathbb{R}^{M \times N}), \quad k \in \mathbb{N},$$

there exists a subsequence (not relabeled) and a Young measure

$$\nu = \{\nu_x\}_{x \in \Omega} \in Y(\Omega, \mathbb{R}^{M \times N})$$

such that for all $v(\cdot) \in \mathbb{C}^1(\mathbb{R}^{M \times N})$ and all $g(\cdot) \in \mathbb{L}_\infty(\Omega)$

$$\lim_{k \to \infty} \int_\Omega g(x) v(y(x)) dx = \int_\Omega g(x) [\int_{\mathbb{R}^{M \times N}} v(A) d\nu_x(A)] dx. \tag{5.9}$$

By

$$Yo(\Omega, \mathbb{R}^{M \times N}) \subset Y(\Omega, \mathbb{R}^{M \times N})$$

we next define the set of all Young measures which are created using all bounded sequences in $\mathbb{L}_1(\Omega, \mathbb{R}^{M \times N})$.

In the case the function $v(\cdot)$ has linear growth at infinity the limit (5.9) generally does not hold due to the concentration effects mentioned above. These concentrations are caused by a not uniformly integrable sequence $\{y_k(\cdot)\}$. For this reason, various generalizations of the classic Young measures have been proposed to guarantee the existence of the limits in (5.9).

5.6 On the Rubio Relaxation Theory

Let us now consider a specific relaxation procedure proposed in [281] and extended to the OCPs with constraints in [92,93].

Consider the classic OCP

$$\begin{aligned}
&\text{minimize } \int_0^{t_f} f_0(t, x(t), u(t))dt \\
&\text{subject to} \\
&\dot{x}(t) = f(t, x(t), u(t)) \text{ a.e. on } [0, t_f], \\
&x(0) = x_0, \ x(t_f) = x_f, \\
&x(t) \in X, \ u(t) \in U, \text{ f.a.a. } t \in [0, t_f], \\
&g(t, x(t), u(t)) \geq 0 \text{ f.a.a. } t \in [0, t_f].
\end{aligned} \tag{5.10}$$

Note that, different from the basic OCP (5.1), the above problem contains a boundary value problem for the corresponding dynamic constraints (for ordinary differential equations). Moreover, it also includes a general state constraint $x(t) \in X$, where $X \subseteq \mathbb{R}^n$ is assumed to be compact. The above OCP is considered for the Carathéodory-type right-hand sides $f(\cdot, \cdot, \cdot)$ and for the continuous constraint functions $g(\cdot, \cdot, \cdot)$ and objective functions $f_0(\cdot, \cdot, \cdot)$. The control set $U \subset \mathbb{R}^m$ is also assumed to be compact.

The Rubio relaxation of the constrained OCP (5.10) can be stated as follows: find a positive linear bounded functional

$$\Lambda \in \mathbb{C}^*([0, t_f] \times X \times U)$$

which solves the abstract optimization problem

$$\begin{aligned}
&\text{minimize } \Lambda(f_0(\cdot, \cdot, \cdot)) \\
&\text{subject to} \\
&\Lambda(\varphi^f) = \Delta\varphi \text{ f.a. } \varphi \in \mathbb{C}^1(B_{[0,t_f]\times X}) \\
&\Lambda(\chi_{[0,t]})(g) \geq 0 \text{ f.a } [0, t] \subseteq [0, t_f].
\end{aligned} \tag{5.11}$$

Here φ^f denotes the formal derivative $\varphi_t + \varphi_x f$, and

$$\Delta\varphi := \varphi(t_f, x_f) - \varphi(0, x_0).$$

Additionally $B_{[0,t_f]\times X}$ in (5.11) is an open neighborhood of the space $[0, t_f] \times X$. By $\chi_{[0,t]}(\cdot)$ we denote here the characteristic function of the time interval $[0, t_f]$. Note that the existence of optimal solution

$$\Lambda^{opt} \in \mathbb{C}^*([0, t_f] \times X \times U)$$

to the abstract OCP (5.11) is a simple consequence of the linear program result in the dual space

$$\mathbb{C}^*([0, t_f] \times X \times U).$$

Moreover, the minimum in (5.11) coincides with the infimum (see [92,93,281] for details)

$$\Lambda^{opt}(f_0(\cdot, \cdot, \cdot)) \equiv \inf_{Problem(5.11)} \Lambda(f_0(\cdot, \cdot, \cdot)). \tag{5.12}$$

The Rubio relaxation (5.11) makes it possible to establish the lower bound for the Gamkrelidze (5.3) and Young (5.4) relaxations discussed in Section 5.2. Let $u^{opt}(\cdot)$ be an optimal solution to the original OCP (5.10). We present here a natural extension of the "gap" Theorem 5.6.

Theorem 5.9. *Assume that the conditions of Theorem 5.6 are satisfied. Then*

$$\inf_{Problem(5.11)} \Lambda(f_0(\cdot, \cdot, \cdot)) \leq \bar{\mathcal{J}}(v^{opt}(\cdot)) \leq \mathcal{J}(u^{opt}(\cdot)).$$

We refer to [281] for a formal proof of Theorem 5.9. Let us also note that the above result seems to be very natural due to the natural consideration of the powers of sets

$$\mathbb{C}^*([0, t_f] \times X \times U)$$

and $R_c([0, t_f], U)$; recall that $v(\cdot)$ in the relaxed OCP (5.3) is an element of the set of measures

$$R_c([0, t_f], U) \subset \mathcal{M}^1_+(U).$$

Taking into consideration (5.11) we now can write

$$Gap := \mathcal{J}(u^{opt}(\cdot)) - \Lambda^{opt}(f_0(\cdot, \cdot, \cdot))$$

for the gap between the originally given OCP (5.10) and the Rubio relaxation (5.11).

Finally let us note that a (positive) linear functional Λ in (5.11) satisfies $\Lambda(F(\cdot, \cdot, \cdot)) \geq 0$ for a continuous function $F(\cdot, \cdot, \cdot) \geq 0$. Since the equality conditions in the relaxed OCP (5.11) can be replaced by an equivalent pair of the inequalities, the optimal Rubio generalization presented above is mathematically equivalent to the celebrated abstract optimization problem of the Ky Fan theory [275,314]. This reduction makes it possible to establish the deepest relations between RT and classic analysis.

5.7 Convex Compactifications in Lebesgue Spaces

Consider a bounded domain Ω in $\mathbb{R}^n$ and a separable Banach space $\mathcal{S}$. Let

$$\mathbb{L}_p(\Omega, \mathcal{S}),\ 1 \leq p < \infty,$$

be a Lebesgue space of Bochner measurable functions (control functions) $u : \Omega \to \mathcal{S}$. In the control engineering applications we usually have

$$\mathcal{S} = \mathbb{R}^m.$$

Definition 5.2. *A triple $(K; Z; g(\cdot))$ is called a convex compactification of*

$$\mathbb{L}_p(\Omega, \mathcal{S}),\ 1 \leq p < \infty,$$

if

- *Z is a locally convex space;*
- *$g : \mathbb{L}_p(\Omega, \mathcal{S}) \to Z$ is a continuous mapping;*
- *the closure $\mathrm{cl}Z^g(B_\rho)$ is convex and compact in Z, and*

$$K = \bigcup_\rho \mathrm{cl}Z^g(B_\rho),$$

where B_ρ denotes the ball in $\mathbb{L}_p(\Omega, \mathcal{S})$ with radius ρ.

Consider a linear subspace $H \subset Car_p(\Omega, \mathcal{S})$, where $Car_p(\Omega, \mathcal{S})$ denotes the linear space of Carathéodory functions $h : \Omega \times \mathcal{S} \to \mathbb{R}$, such that

$$h(x; s)j \leq a(x) + b||s||^p_{\mathcal{S}},\ a(\cdot) \in \mathbb{L}_1(\Omega),\ b \in \mathbb{R}.$$

The dual space H^* to H is assumed to be endowed with its weak* topology. Let

$$Y^p_H(\Omega, \mathcal{S}) := \{\nu(\cdot) \in H^*, \exists \rho \in \mathbb{R},\ \exists \{u_l(\cdot)\} \subset B_\rho :\ \overset{*}{\lim_l} i_H(u_l(\cdot)) = \nu\}.$$

By i_H we next denote an embedding of $\mathbb{L}_p(\Omega, \mathcal{S})$ into the dual space H^* of H given by the following expression:

$$\langle i_H(u(\cdot))h(\cdot, \cdot)\rangle := \int_\Omega h(x, u(x))dx.$$

The next theorem gives a concrete (important) example of a convex compactification (see [127,129,132]).

Theorem 5.10. *Let $H \subset Car_p(\Omega, \mathcal{S})$. Then the triple*

$$(Y_H^p(\Omega, \mathcal{S}), H^*; i_H),$$

where i_H is an embedding of $\mathbb{L}_p(\Omega, \mathcal{S})$ into the dual space H^* (endowed with its weak topology) of H given by*

$$\langle i_H(u(\cdot))h(\cdot,\cdot)\rangle := \int_\Omega h(x, u(x))dx,$$

forms a convex compactification of $\mathbb{L}_p(\Omega, \mathcal{S})$.

Let us finally note that the presented analytic tool for the convex compactifications in Lebesgue spaces can be applied to the specific relaxations of the OCPs involving some classes of operator equations. For example, relaxation schemes for the OCPs with semilinear operator (evolution) equations and additional constraints are discussed in [246,280]. The mentioned abstractions also allow to introduce so-called $\mathbb{L}_p$-Young measures on a space of weakly measurable essentially bounded functions on Ω with values at the space of Borel measures on $\mathbb{R}^m$.

5.8 The Buttazzo Relaxation Scheme

Let us now consider a very interesting abstract relaxation approach to the specific OCPs of the following type:

$$\text{minimize } \{J(u(\cdot), y(\cdot)) + \chi_A(u(\cdot), y(\cdot)) : (u(\cdot), y(\cdot)) \in \mathcal{U} \times Y\}. \tag{5.13}$$

Here $\mathcal{U}, Y$ are the space of controls and the state space, respectively, and $\chi_A(\cdot,\cdot)$ is a characteristic function. Moreover,

$$A \subset \mathcal{U} \times Y$$

is the set of admissible pairs. Let us note that the admissible pairs mentioned above are usually determined by the ordinary differential equations and some possible additional constraints. The cost function

$$J : \mathcal{U} \times Y \to [0, \infty]$$

is assumed to be lower semicontinuous (see Chapter 3). We also assume that the given cost functional is $\tau \times \sigma$-coercive, i.e., for every $\lambda \geq 0$ there exists a $\tau \times \sigma$-compact subset K_λ of $\mathcal{U} \times Y$ such that

$$\{(u(\cdot), y(\cdot)) \in A : J(u(\cdot), y(\cdot)) \leq \lambda\} \subset K_\lambda.$$

Note that in the conventional OCPs of the type (5.1) the space $\mathcal{U}$ is a subset of a Lebesgue space endowed with its weak topology. The state space Y is usually an appropriate Sobolev space with its weak topology (or with the L_∞ topology in the one-dimensional case). The lower semicontinuity condition is a natural assumption for the practically oriented cost functionals in the classic OCPs. The coercivity condition is usually guaranteed by some boundedness assumptions related to $J(\cdot,\cdot)$ and to the set A.

Let us also note that the "specific" form of the objective functional in (5.13) is caused by the generic form of the abstract Lagrangian associated with the constrained (also in the state space) OCPs. A "similar" approach, namely, an initially given functional plus a characteristic function, is also used in the penalty function like numerical approaches to OCPs. We refer to [121,132] for some related topics.

Assume

$$\{(u_l(\cdot), y_l(\cdot))\} \subset A,\ l \in \mathbb{N},$$

is a minimizing sequence to (5.13). We now define the relaxed OCP associated with the initially given problem (5.13). The concept proposed in [120] is in fact a relaxation concept with respect to a concrete minimizing sequence $\{(u_l(\cdot), y_l(\cdot))\}$.

Definition 5.3. *The relaxed problem associated with the specific OCP (5.13) is defined as follows:*

$$\min\{F(u(\cdot), y(\cdot)) : (u(\cdot), y(\cdot)) \in \mathcal{U} \times Y\},$$

where

$$F : \mathcal{U} \times Y \to [0, \infty]$$

is determined by

$$F(u(\cdot), y(\cdot)) := \inf_{u_l(\cdot)\to u(\cdot)} \inf_{y_l(\cdot)\to y(\cdot)} \liminf_{l\to\infty} [J(u_l(\cdot), y_l(\cdot)) + \chi_A(u_l(\cdot), y_l(\cdot))].$$

The main advantage of the relaxation method proposed in [120] is a possibility to write the relaxed functional in an explicit form. In a particular case the abstract OCP (5.13) corresponds to the following classic problem:

$$\begin{aligned}
&\text{minimize } J(u(\cdot), x(\cdot)) := \int_0^{t_f} f_0(t, u(t), x(t))dt \\
&\text{subject to} \\
&\dot{x}(t) \in a(t, x(t)) + B(t, x(t))b(t, u(t)) \text{ a.e. on } [0, t_f] \\
&x(0) = x_0 \\
&u(t) \in U.
\end{aligned} \tag{5.14}$$

Here $a : [0, t_f] \times \mathbb{R}^n \to \mathbb{R}^n$ and $B : [0, t_f] \times \mathbb{R}^n \to \mathbb{R}^{n \times l}$ and

$$b : [O, t_f] \times \mathbb{R}^m \leftarrow \mathbb{R}^l$$

are Borel functions. We next assume that the basic hypotheses from [120] are fulfilled. Let $f^{**}(t, x, u, v)$ be the greatest function lower semicontinuous and convex in (u, v) which is less than or equal to $f(t, x, u, v)$, where

$$f(t, x, u, v) := f_0(t, x, u) \text{ if } v \in b(t, u),$$
$$f(t, x, u, v) := \infty \text{ otherwise.}$$

Theorem 5.11. *Assume that the basic hypotheses from [120] are satisfied. Then the relaxed functional $F(\cdot, \cdot)$ to the cost functional*

$$J(\cdot, \cdot) + \chi_A(\cdot, \cdot)$$

associated with the initially given OCP (5.14) can be written as follows:

$$F(u(\cdot), x(\cdot)) = \int_0^{t_f} \tilde{f_0}(t, u(t), x(t), \dot{x}(t))dt + \chi_{\tilde{A}}(u(t), x(t)),$$

where

$$\dot{x}(t) \in a(t, x(t)) + B(t, x(t))\tilde{b}(t, u(t)) \text{ a.e. on } [0, t_f],$$
$$b(t, u) := \{v \in \mathbb{R}^l : (u, v) \in \text{co}\{(\lambda, \mu) \in \mathbb{R}^m \times \mathbb{R}^l : \mu \in b(t, \lambda)\}\},$$
$$\tilde{f_0}(t, u, x, \dot{x}) := \inf\{f^{**}(t, u, x, v) : a(t, x) + B(t, x)v = \dot{x}\},$$
$$x(0) = x_0.$$

The proof of Theorem 5.11 can be found in [120]. Finally note that this relaxation approach can also be applied to some classes of OCPs with the elliptic-type state equations (partial differential equations).

5.9 Approximation of Generalized Solutions

As follows from the concepts of HSs and switched systems (SSs) (see Definition 1.1 and Definition 1.4 from Section 1.1) these dynamic models have state- or/and time-dependent switching mechanisms. Considering an SS and taking into consideration the "compact" dynamic system representation (1.3) we conclude that the vector $\beta(\cdot)$ constitutes a specific, namely, a piecewise constant "control" for an SS from the basic Definition 1.4. In Chapter 6 we will generalize Definition 1.4 and preserve this concept of the additional control input $\beta(\cdot)$. This fact implies a necessity to consider adequate relaxation schemes for OCPs with significantly restricted admissible control functions, namely, with the class of piecewise constant control inputs that correspond to the structure of the vector $\beta(\cdot)$ in an SS.

Motivated by the above observations we now study a nonlinear system with a switched control structure, i.e.,

$$\begin{aligned} \dot{x}(t) &= f(t, x(t), u(t)), \ t \in [t_0, t_f], \\ x(t_0) &= x_0, \end{aligned} \tag{5.15}$$

where $f(\cdot, \cdot, u)$ is a Carathéodory function, i.e., measurable in t and continuous in x. We also assume that $f(t, x, \cdot)$ is a continuously differentiable function. Let us specify formally the set of admissible controls of switched nature mentioned above. In this chapter, we study the piecewise constant control functions associated with (5.15). For each component $u_k(\cdot)$ of a feasible control input

$$u(\cdot) = [u_1(\cdot), \dots, u_m(\cdot)]^T$$

we introduce a finite set bounded values that represent admissible fixed-level controls:

$$\mathcal{Q}^{(k)} := \{q_j^{(k)} \in \mathbb{R}, j = 1, \dots, M_k\}, \ \ M_k \in \mathbb{N}, \quad k = 1, \dots, m.$$

In general, all the sets $\mathcal{Q}^{(k)}$ are different (contain different value levels) and have various numbers of elements. It is assumed that each $\mathcal{Q}^{(k)}$ possesses the following strict order property

$$q_1^{(k)} < q_2^{(k)} < \dots < q_{M_k}^{(k)}.$$

We now introduce the set of switching times related to a control function from the set of admissible inputs, i.e.,

$$\begin{aligned} &\mathcal{T}^{(k)} := \{t_i^{(k)} \in \mathbb{R}_+, i = 0, \dots, N_k\}, \\ &N_k \in \mathbb{N}, \quad k = 1, \dots, m. \end{aligned}$$

All the sets $\mathcal{T}^{(k)}$ are defined for the corresponding control components $u_k(\cdot)$, $k = 1, \dots, m$, where $\mathbb{R}_+$ denotes a nonnegative real semiaxis. Let

$$t_0^{(k)} < t_1^{(k)} < \dots < t_{N_k}^{(k)}.$$

For each $\mathcal{T}^{(k)}$ we next put

$$t_{N_1}^{(1)} = \dots = t_{N_m}^{(m)} = t_f.$$

Using the given notation of the control level sets $\mathcal{Q}^{(k)}$ and the above fixed switching times $\mathcal{T}^{(k)}$, the set of admissible controls $\mathcal{S}$ can now be specified exactly, namely, as the Cartesian product

$$\mathcal{S} = \mathcal{S}_1 \times \dots \times \mathcal{S}_m, \tag{5.16}$$

where each set $\mathcal{S}_k$, $k = 1, \ldots, m$, is defined as follows:

$$\mathcal{S}_k := \{v : [t_0, t_f] \to \mathbb{R} \mid v(t) = \sum_{i=1}^{N_k} I_{[t_{i-1}^{(k)}, t_i^{(k)})}(t) q_{j_i}^{(k)}\},$$
$$q_{j_i}^{(k)} \in \mathcal{Q}^{(k)},\ j_i \in \mathbb{Z}[1, M_k],\ t_i^{(k)} \in \mathcal{T}^{(k)}.$$

By $\mathbb{Z}[1, M_k]$ we denote here the set of all integers into the interval $[1, M_k]$ and $I_{[t_{i-1}^{(k)}, t_i^{(k)})}(t)$ is the characteristic function of the interval $[t_{i-1}^{(k)}, t_i^{(k)})$. Evidently, the set of admissible control inputs $\mathcal{S}$ can be qualitatively interpreted as the set of all functions

$$u : [t_0, t_f] \to \mathbb{R}^m,$$

such that each component $u_k(\cdot)$ of $u(\cdot)$ attains a constant level value. We have

$$q_{j_i}^{(k)} \in \mathcal{Q}^{(k)},\ t \in [t_{i-1}^{(k)}, t_i^{(k)}).$$

Moreover, the component level changes occur only at the prescribed times $t_i^{(k)} \in \mathcal{T}^{(k)}$, $i = 1, \ldots, N_k$.

Note that the general existence/uniqueness theory for nonlinear ordinary differential equations implies that for every

$$u(\cdot) \in \mathcal{S} \subset \mathbb{L}^2[t_0, t_f; \mathbb{R}^m]$$

the initial value problem (5.15) has a unique absolutely continuous solution $x(\cdot)$. Here $\mathbb{L}^2\{[t_0, t_f]; \mathbb{R}^m\}$ denotes the Lebesgue space of all square integrable control functions $u : [t_0, t_f] \to \mathbb{R}^m$. We also refer to [159,192,265] for necessary existence and uniqueness results.

In this section we study an OCP with a quadratic cost functional

$$J(u(\cdot)) = \frac{1}{2} \int_{t_0}^{t_f} (\langle Q(t)x(t), x(t)\rangle + \langle R(t)u(t), u(t)\rangle)\, dt + \frac{1}{2}\langle Gx(t_f), x(t_f)\rangle,$$

where $G \in \mathbb{R}^{n \times n}$ is a symmetric positive defined matrix and $Q(\cdot)$, $R(\cdot)$ are integrable matrix functions that satisfy standard symmetry and the positivity hypothesis; we have

$$Q(t) \geq 0,\ R(t) \geq \delta I,\ \delta > 0,$$

for all $t \in [t_0, t_f]$. Recall that in the case of system (5.15) in the presence of unrestricted control inputs we obtain the standard situation for a linear quadratic OCP. In this case the optimization problem is formally studied in the full space

$$\mathbb{L}^2[t_0, t_f; \mathbb{R}^m]$$

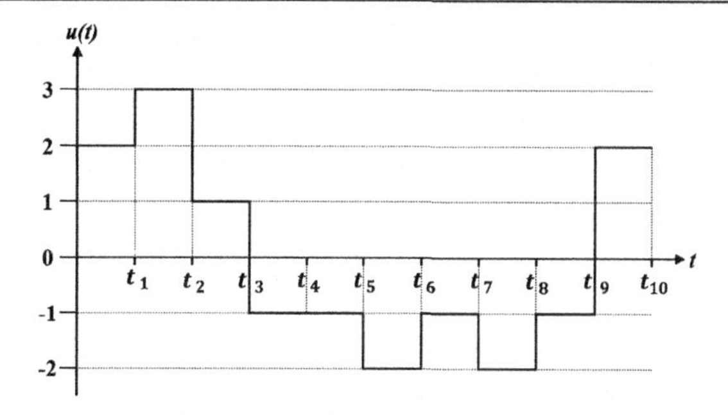

Figure 5.1: The admissible switched-type control inputs $u(\cdot)$.

of square integrable control functions. In contrast to the classic case, we consider system (5.15) in combination with the given (hard) piecewise constant admissible inputs $u(\cdot)$ (see Fig. 5.1).

The control signal $u(\cdot)$ shown in Fig. 5.1 can only take some fixed-level values within the finite feasibility set

$$\mathcal{Q} = \{-2, -1, 0, 1, 2, 3\}$$

during the time interval $[t_{i-1}, t_i)$, $i = 1, \ldots, 10$. In addition, the control signal is only allowed to change its value at the given time instants $t_0, t_1, \ldots, t_f$, so it is fixed between these times. The evident combinatorial character of the considered control constraints for the continuous time system (5.15) can be illustrated by a simple example.

Example 5.3. *Suppose $u \in \mathbb{R}^2$ and*

$$\mathcal{Q}^{(1)} = \{0, 1, 2\}, \; \mathcal{Q}^{(2)} = \{0, -1\}.$$

Furthermore, the set of switching times for each control component is assumed to be given by

$$\mathcal{T}^{(1)} = \{0, 0.5, 1\}, \; \mathcal{T}^{(2)} = \{0, 0.33, 0.66, 1\}.$$

Resulting from the above formal definitions, the set $\mathcal{S}$ in (5.16) can now be written as

$$\mathcal{S} = \mathcal{S}_1 \times \mathcal{S}_2,$$

where

$$\mathcal{S}_1 = \{v : [0, 1] \to \mathbb{R} \mid v(t) = I_{[0,0.5)}(t) q^{(1)}_{j_1} + I_{[0.5,1)}(t) q^{(1)}_{j_2}, \; q^{(1)}_{j_i} \in \mathcal{Q}^1\},$$
$$\mathcal{S}_2 = \{w : [0, 1] \to \mathbb{R} \mid w(t) = I_{[0,0.33)}(t) q^{(2)}_{j_1} + I_{[0.33,0.66)}(t) q^{(2)}_{j_2} + I_{[0.66,1)}(t) q^{(2)}_{j_3}, \; q^{(2)}_{j_i} \in \mathcal{Q}^2\}.$$

In that concrete case we have

$$M_1 = 3,\ M_2 = 2,\ N_1 = 2,\ N_2 = 3.$$

The cardinality of the control set S *in this example is given by* $|S| = 3^2 \cdot 2^3 = 72$. *In other words, we have* 72 *admissible control inputs, among which we must find the one that minimizes the performance criterion.*

Note that in general, the cardinality of the set of admissible controls $\mathcal{S}$ can be expressed as follows:

$$|\mathcal{S}| = \prod_{k=1}^{m} M_k^{N_k}.$$

Evidently, $|\mathcal{S}|$ grows exponentially if N_k increases. Motivated by various engineering applications and taking into consideration the specific input structure of an SS, we now can formulate the following constrained OCP, associated with the dynamic system (5.15):

$$\begin{aligned} &\text{minimize } J(u(\cdot)) \\ &\text{subject to } (2.1),\ u(\cdot) \in \mathcal{S}, \end{aligned} \tag{5.17}$$

where $J(\cdot)$ is the quadratic costs functional defined above. Note that $\mathcal{S}$ constitutes a nonempty subset of the real Hilbert space $\mathbb{L}^2[t_0, t_f; \mathbb{R}^m]$. Due to the highly restrictive nonlinear constraint $u(\cdot) \in \mathcal{S}$, the obtained OCP (5.17) cannot be solved by a direct application of the classic Pontryagin maximum principle. Recall that the conventional versions of the maximum principle make it possible to specify an optimal solution to an OCP in a full (nonrestricted) control space

$$\mathbb{L}^2[t_0, t_f; \mathbb{R}^m].$$

Otherwise, an effective numerical treatment of the specific variants of the maximum principle for the restricted OCPs constitutes a numerically difficult task (see, e.g., [96,154,192]). We are interested in a relatively simple and implementable computational procedure for the generic OCP (5.17). We use a specific relaxation technique for this purpose and obtain an optimal solution of the convex-type relaxed OCP. The relaxation scheme proposed in this section is next applied to the further numerical treatment of the initially given OCP (5.15).

The method we consider incorporates simple relaxed OCPs associated with the initial problem. In the context of an OCP of the type (5.17), when dealing with the minimization of $J(u(\cdot))$, the most general way of looking at relaxation is to consider the lower semicontinuous hull of $J(u(\cdot))$ determined on a convexification of the set of admissible controls in (5.17).

We refer to the results of Section 5.2 and also to [28,130,222,243,279,287,288] for relaxation procedures in general optimized control.

First let us note that under basic assumptions from Section 5.2 the following control state mapping

$$x^u(t) : \mathbb{L}^2\{[t_0, t_f]; \mathbb{R}^m\} \to \mathbb{R}^n$$

is Fréchet differentiable for every $t \in [t_0, t_f]$. Therefore, the quadratic cost functional $J(\cdot)$ in (2.3) is also Fréchet differentiable. We refer to [88,154,192,261] for the classic differentiability concepts. The original set $\mathcal{S}$ of admissible controls in (2.3) is a nonconvex set. This is an immediate consequence of the combinatorial structure of $\mathcal{S}$ given by (5.16). Motivated from this fact we next consider the following polytope:

$$\begin{aligned} \text{conv}(\mathcal{S}) := \{v(\cdot) \mid v(t) = \sum_{s=1}^{|\mathcal{S}|} \lambda_s u_s(t), \\ \sum_{s=1}^{|\mathcal{S}|} \lambda_s = 1,\ \ \lambda_s \geq 0,\ u_s(\cdot) \in \mathcal{S},\ s = 1, \ldots, |\mathcal{S}|\}. \end{aligned}$$

From the definition of $\mathcal{S}$, we conclude that the convex set $\text{conv}(\mathcal{S})$ is also closed and bounded. We can also give an easy alternative characterization of $\text{conv}(\mathcal{S})$, i.e.,

$$\text{conv}(\mathcal{S}) = \text{conv}(\mathcal{S}_1) \times \ldots \times \text{conv}(\mathcal{S}_m),$$

where

$$\text{conv}(\mathcal{S}_k)$$

is a convex hull of the partial set $\mathcal{S}_k,\ k = 1, \ldots, m$. Since

$$\text{conv}(\mathcal{Q}_k) \equiv [q_1^{(k)}, q_{M_k}^{(k)}],$$

we obtain

$$\begin{aligned} \text{conv}(\mathcal{S}_k) := \{v(\cdot) \mid v(t) = \sum_{i=1}^{N_k} I_{[t_{i-1}^{(k)}, t_i^{(k)})}(t) q_{j_i}^{(k)}, \\ q_{j_i}^{(k)} \in [q_1^{(k)}, q_{M_k}^{(k)}],\ j_i \in \mathbb{Z}[1, M_k],\ t_i^{(k)} \in \mathcal{T}^{(k)}\}. \end{aligned}$$

Roughly speaking $\text{conv}(\mathcal{S})$ contains all the piecewise constant functions $u(\cdot)$ such that the corresponding constant value $u_k(t)$ belongs to the interval $[q_1^{(k)}, q_{M_k}^{(k)}]$ for all

$$t \in [t_{i-1}^{(k)}, t_i^{(k)}).$$

Let us note that in contrast to the initially considered control set $\mathcal{S}$, the convex hull $\text{conv}(\mathcal{S})$ introduced above is an infinite-dimensional space.

We next consider the construction of the closed convex hull $\bar{co}\{J(u(\cdot))\}$ of the objective $J(u(\cdot))$. We refer to [190] for the exact concept of $\bar{co}\{J(u(\cdot))\}$. Note that the biconjugate

$$J^{**}(u(\cdot))$$

of $J(u(\cdot))$ (determined on $\mathbb{L}^2\{[t_0, t_f]; \mathbb{R}^m\}$) is equal to $\bar{co}\{J(u(\cdot))\}$ (see [190] for the formal proof).

For a given objective functional $J(\cdot)$ getting its closed convex hull is a complicated, but at the same time fascinating, operation. Note that the objective functional $J(u(\cdot))$ in (2.3) is a composite functional. This fact can be simply illustrated by the mapping $x^u(t)$ considered at the beginning of this section. This situation is a typical characterization of the general optimal control processes governed by ordinary differential equations. The numerical calculation of

$$\bar{co}\{J(u(\cdot))\}$$

is not broached here. A novel computational method for a practical evaluation of $\bar{co}\{J(u(\cdot))\}$ is proposed in [243,287,288]. This method constitutes a generalization of the celebrated McCormic relaxation scheme in the specific case of a composite functional in an OCP. The closed convexification of such a functional is realized by solving an "auxiliary control system" (see [287] for details).

Using the above convex constructions of the objective functional and of the convexified set of admissible control functions, we can now formulate the following relaxed OCP:

$$\begin{aligned} &\text{minimize } \bar{co}\{J(u(\cdot))\} \\ &\text{subject to } u(\cdot) \in \text{conv}(\mathcal{S}). \end{aligned} \tag{5.18}$$

The auxiliary problem (5.18) is in fact a simple relaxation of the initial OCP (5.17). The optimization procedure in (5.18) is determined over a convex set of admissible control inputs $\text{conv}(\mathcal{S})$. We will study this problem and use it for a constructive numerical treatment of (5.17). Note that the OCP (5.18) constitutes a convex minimization problem in a real Hilbert space. We refer to Chapter 3 and to [29,30,300] for the general theory of convex OCPs with ordinary differential equations. The approximability property of the fully relaxed OCP (5.18) can be expressed as follows [190]:

$$\inf_{u(\cdot)\in\mathbb{L}^2\{[t_0,t_f];\mathbb{R}^m\}} J(u(\cdot)) = \inf_{u(\cdot)\in\mathbb{L}^2\{[t_0,t_f];\mathbb{R}^m\}} \bar{co}\{J(u(\cdot))\}.$$

Note that the above relation is a simple consequence of the following simple facts:

$$\inf_{u(\cdot)\in\mathbb{L}^2\{[t_0,t_f];\mathbb{R}^m\}} J(u(\cdot)) = -J^*(0),$$
$$J^*(u(\cdot)) = \bar{co}\{J(u(\cdot))\},$$

where $J^*(u(\cdot))$ is a conjugate of $J(u(\cdot))$. Let us now introduce the auxiliary variable x_{n+1} and the extended state vector

$$\tilde{x} := (x^T, x_{n+1})^T$$

such that

$$\dot{x}_{n+1}(t) = \frac{1}{2}(\langle Q(t)x(t), x(t)\rangle + \langle R(t)u(t), u(t)\rangle), \quad t \in [t_0, t_f],$$
$$x_{n+1}(t_0) := 0.$$

The given OCPs (5.18) can now be equivalently rewritten using the modified terminal costs functional

$$J(u(\cdot)) = \phi(\tilde{x}(t_f)) := x_{n+1}(t_f) + \frac{1}{2}\langle Gx(t_f), x(t_f)\rangle. \tag{5.19}$$

The formal Hamiltonian associated with (5.18) for the given dynamic system (5.15) extended by the above additional differential equation has the following form:

$$H(t, x, u, p, p_{n+1}) = \langle p, f(t, x, u)\rangle + \frac{1}{2}p_{n+1}\left(\langle Q(t)x, x\rangle + \langle R(t)u, u\rangle\right), \tag{5.20}$$

where

$$p \in \mathbb{R}^n, \, p_{n+1} \in \mathbb{R}$$

are adjoint variables. Assume that $p_{n+1} \neq 0$. We use the corresponding notation $\tilde{p} := (p^T, p_{n+1})^T$. Note that the Hamiltonian introduced above does not depend on the auxiliary state variable x_{n+1}.

Assume that

$$u^*(\cdot) \in \text{conv}(\mathcal{S}) \subset \mathbb{L}^2\{[t_0, t_f]; \mathbb{R}^m\}$$

is an optimal solution of (3.1). By $x^*(\cdot)$ we denote the corresponding optimal trajectory (solution) of (2.1) generated by $u^*(\cdot)$. Since $J(u(\cdot))$ is a continuously (Fréchet) differentiable functional, the closed convex hull $\bar{co}\{J(u(\cdot))\}$ is a Fréchet differentiable functional (see [83, 190] for the formal proof). The convex structure of (3.1) makes it possible to apply powerful numerical approaches from convex programming [190,261]. In this paper, we use a variant of the projected gradient method in combination with the celebrated multiple shooting method (see, e.g., [258,295]) for a concrete numerical treatment of (3.1). The projected gradient method for (3.1) can now be expressed as follows:

$$u_{(l+1)}(\cdot) = \gamma_l \mathcal{P}_{\text{conv}(\mathcal{S})}\left[u_{(l)}(\cdot) - \alpha_l \nabla \bar{co}\{J(u_{(l)}(\cdot))\}\right] + (1-\gamma_l)u_{(l)}(\cdot),\ l \in \mathbb{N}, \tag{5.21}$$

where $\mathcal{P}_{\text{conv}(\mathcal{S})}$ is the projection operator on the convex set

$$\text{conv}(\mathcal{S}),\ \{\alpha_l\}$$

and $\{\gamma_l\}$ are sequences of some suitable step sizes. By ∇ we denote here the Fréchet derivative of the convexified functional

$$\bar{co}\{J(u_{(l)}(\cdot))\}.$$

The conventional projection operator $\mathcal{P}_{\text{conv}(\mathcal{S})}$ is next defined as follows (see Chapter 3):

$$\mathcal{P}_{\text{conv}(\mathcal{S})}[u(\cdot)] := \text{Arg} \min_{v(\cdot)\in\text{conv}(\mathcal{S})} \left(\| v(\cdot) - u(\cdot) \|_{\mathbb{L}^2\{[t_0,t_f];\mathbb{R}^m\}}\right).$$

Note that the projection here is defined in the real Hilbert space $\mathbb{L}^2\{[t_0, t_f]; \mathbb{R}^m\}$.

Recall that several choices are possible for the step sizes α_l and γ_l. Let us describe briefly the main strategies of the step size selection.

- The constant step size:

$$\gamma_l = 1,\ \alpha_l = \alpha > 0$$

  for all $l \in \mathbb{N}$.
- Armijo line search along the boundary of $\text{conv}(\mathcal{S})$:

$$\gamma_l = 1,\ l \in \mathbb{N},\ \alpha_l$$

  is determined by

$$\alpha_l := \bar{\alpha}\theta^{\chi(l)}$$

  for some $\bar{\alpha} > 0,\ \theta,\ \delta \in (0,1)$, where

$$\chi(l) := \min\{\chi \in \mathbb{N} \mid \bar{co}\{J(\mathcal{P}_{\text{conv}(\mathcal{S})}\left[u_{(l,s)}(\cdot)\right])\} \leq \bar{co}\{J(u_{(l)}(\cdot))\} - \delta\langle \nabla \bar{co}\{J(u_{(l)}(\cdot))\}, u_{(l)}(\cdot) - \mathcal{P}_{\text{conv}(\mathcal{S})}[u_{(l,s)}(\cdot)]\rangle\}$$

  and

$$u_{(l,s)}(\cdot) := u_{(l)}(\cdot) - \bar{\alpha}\theta^{\chi}\nabla \bar{co}\{J(u_{(l)}(\cdot))\}.$$

- Armijo line search along the feasible direction:

$$\{\alpha_l\} \subset [\bar{\alpha}, \hat{\alpha}]$$

for some

$$\bar{\alpha} < \hat{\alpha} < \infty$$

and γ_l is determined by the Armijo rule

$$\gamma_l := \theta^{\chi(l)}$$

for some

$$\theta,\ \delta \in (0,1),$$

where

$$\chi(l) := \min\{\chi \in \mathbb{N} \mid \bar{co}\{J(u_{(l,s)}(\cdot))\} \leq \bar{co}\{J(u_{(l)}(\cdot))\} - \theta^{\chi}\delta\langle \nabla \bar{co}\{J(u_{(l)}(\cdot))\}, u_{(l)}(\cdot) - \mathcal{P}_{\text{conv}(\mathcal{S})}[w_l(\cdot)]\rangle\}$$

and

$$u_{(l,s)}(\cdot) := \theta^{\chi}\mathcal{P}_{\text{conv}(\mathcal{S})}[w_l(\cdot)] + (1-\theta^{\chi})u_{(l)}(\cdot).$$

- Exogenous step size before projecting: $\gamma_l = 1$ for all $l \in \mathbb{N}$ and α_l given by

$$\alpha_l := \frac{\delta_l}{||\nabla \bar{co}\{J(u_{(l)}(\cdot))\}||_{\mathbb{L}^2\{[t_0,t_f];\mathbb{R}^m\}}},\quad \sum_{l=0}^{\infty}\delta_l = \infty,\ \sum_{l=0}^{\infty}\delta_l^2 < \infty.$$

Under some nonrestrictive assumptions the projected gradient iteration (5.21) generates a minimizing sequence for the relaxed optimization problem (5.18). Many useful and mathematically exact convergence theorems for the gradient iterations (3.2) can be found in [175, 178,261]. A comprehensive discussion of the weakly and strongly convergent variants of the basic gradient method can be found in [79,87]. We also refer to [150,269,301,317] for some specific convergence results associated with the gradient-based solution schemes for various hybrid and switched OCPs.

The first strategy from the ones presented above was analyzed in [178] and its weak convergence was proved under Lipschitz continuity of

$$\nabla \bar{co}\{J(\cdot)\}.$$

The main difficulty is the necessity of taking $\alpha \in (0, 2/L)$, where L is the Lipschitz constant associated with

$$\nabla \bar{co}\{J(\cdot)\};$$

see also [87].

Note that the second gradient-based strategy requires one projection onto conv$(\mathcal{S})$ for each step of the inner loop resulting from the Armijo line search. Therefore, many projections might be performed for each iteration l, making the second strategy inefficient when the projection onto the set conv$(\mathcal{S})$ cannot be computed explicitly. On the other hand, the third strategy demands only one projection for each outer step, i.e., for each iteration l. The second and third optimization strategies from the variants presented above are the constrained versions of the line search proposed in [16] for solving unconstrained optimization problems. Under existence of minimizers and convexity assumptions for a convex minimization problem, it is possible to prove, for second and third strategies, convergence of the whole sequence to a solution of the above optimization problem in finite-dimensional spaces; see [114].

The last strategy from the ones presented above, as its counterpart in the unconstrained case, fails to be a descent method. Furthermore, it is easy to show that this approach satisfies

$$||u_{(l+1)}(\cdot) - u_{(l)}(\cdot)|| \leq \delta_l$$

for all l, with δ_l given above. This reveals that convergence of the sequence of points generated by this exogenous approach can be very slow (step sizes are small). Note that the second and third strategies allow for occasionally long step sizes because both strategies employ all information available at each l-iteration. Moreover, the last strategy does not take into account the values of the objective functional for determining the "best" step sizes. These characteristics, in general, entail poor computational performance. The basic convergence results for the obtained approximating sequence $\{u_{(l)}(\cdot)\}$ can be stated as follows.

Theorem 5.12. *Assume that all hypotheses from* Section 5.2 *are satisfied and* $p_{n+1} \neq 0$. *Consider a sequence* $\{u_{(l)}(\cdot)\}$ *generated by the method* (5.21) *with a constant step size* α. *Then for an admissible initial point*

$$u_{(0)}(\cdot) \in conv(\mathcal{S})$$

the resulting sequence $\left\{u_{(l)}(\cdot)\right\}$ *is a minimizing sequence for* (5.18), *i.e.,*

$$\lim_{l \to \infty} \bar{co}\{J(u_{(l)}(\cdot))\} = \bar{co}\{J(u^*(\cdot))\}.$$

Additionally assume that $\partial f(t,x,u)/\partial u$ *is* Lipschitz *continuous with respect to* (x,u), *i.e.,*

$$||\frac{\partial f(t,x_1,u_1)}{\partial u} - \frac{\partial f(t,x_2,u_2)}{\partial u}|| \leq L_x||x_1 - x_2|| + L_u||u_1 - u_2||,$$

and $\alpha \in (0, 2/L)$, *where*

$$L := (L_x l + L_u) + \lambda, \ \ l := \max_{t \in [t_0,t_f]} \{l_t(t)\}, \ \ \lambda := \max_{t \in [t_0,t_f]} \{\lambda^R_{\max}(t)\},$$

and $\lambda^R_{\max}(t)$ is the maximal eigenvalue of the matrix $R(t)$. Here $l_t(t)$ are Lipschitz *constants of the control state mapping $x^u(t)$, for $t \in [t_0, t_f]$. Then $\{u_{(l)}(\cdot)\}$ converges $\mathbb{L}^2\{[t_0, t_f]; \mathbb{R}^m\}$-weakly to a solution $u^*(\cdot)$ of* (5.18).

Proof. As mentioned above the convexified cost functional $J(\cdot)$ in (5.17) is Fréchet differentiable. The property of $\{u_{(l)}(\cdot)\}$ to be a minimizing sequence for (3.1) is an immediate consequence of [87,261]. Following [28–30,213,300] the reduced gradient $\nabla\bar{co}\{J(\cdot)\}$ of the modified costs functional

$$\bar{co}\{J(\cdot)\},\ u(\cdot) \in \text{conv}(\mathcal{S})$$

can be computed from the following Hamilton-type boundary value problem:

$$\begin{aligned}
&\nabla\bar{co}\{J(u(\cdot))\}(t) = -\frac{\partial H(t, x(t), u(t), p(t), p_{n+1}(t))}{\partial u} = \\
&-(\frac{\partial f(t, x(t), u(t))}{\partial u})^T p(t) - \\
&R(t)u(t)p_{n+1}(t), \\
&\frac{d\tilde{p}(t)}{dt} = -\frac{\partial H(t, x(t), u(t), p(t), p_{n+1}(t))}{\partial \tilde{x}} = \\
&-((\frac{\partial H(t, x(t), u(t), p(t), p_{n+1}(t))}{\partial x})^T, 0)^T, \\
&\tilde{p}(t_f) = -\frac{\partial(\bar{co}\{\phi(\tilde{x}(t_f))\})\}}{\partial \tilde{x}} = \\
&(-(\bar{co}\{Gx(t_f)\})^T, -1)^T, \\
&\frac{d\tilde{x}(t)}{dt} = \frac{\partial H(t, x(t), u(t), p(t), p_{n+1}(t))}{\partial \tilde{p}}, \\
&\tilde{x}(t_0) = (x_0^T, 0)^T,
\end{aligned} \tag{5.22}$$

where $x(\cdot)$ and $\tilde{p}(\cdot)$ are a state variable and an adjoint variable associated with an admissible control function $u(\cdot) \in \text{conv}(\mathcal{S})$. The differentiability of $x^u(t)$ implies the Lipschitz continuity of this control state mapping on the bounded set $\text{conv}(\mathcal{S})$. Since

$$\partial f(t, x, u)/\partial u$$

is also assumed to be Lipschitz continuous and the composition of two Lipschitz continuous mappings also possesses the same property, we can establish the Lipschitz continuity of the derivative $\nabla\bar{co}J(u(\cdot))(t)$ uniformly in $t \in [t_0, t_f]$. From (3.4) we easily deduce

$$p_{n+1}(t) \equiv -1\ \forall t \in [t_0, t_f].$$

Using the explicit expression, namely, the first relation in (5.22), the Lipschitz constant for

$$\nabla \bar{co}\{J(u(\cdot))\}(t)$$

can be computed as follows:

$$L = (L_x l + L_u) + \lambda.$$

The weak convergence to $u^*(\cdot)$ of the sequence $\{u_l(\cdot)\}$ generated by (5.21) with a constant step size $\alpha \in (0, 2/L)$ follows now from [178]. The proof is completed. □

The proposed gradient-type method (5.21) provides a well-defined numerical basis for the computational treatment of (5.18). The concrete calculation of the costs function gradient is comprehensively discussed in Chapter 8. One can use the classic version of the multiple shooting method (see, e.g., [258,295]) and apply it for solving the boundary value problem in (5.22). Note that using the obtained optimal solution

$$u^*(\cdot) \in \text{conv}(\mathcal{S})$$

of the auxiliary convex OCP (5.18) one can determine a suitable approximate treatment of the original OCP (5.17).

From the computational point of view the fully (globally) convexified OCP (5.18) is related to a mathematically sophisticated procedure, namely, with the calculation of a convex envelope of a composite functional in Hilbert space. We have mentioned the corresponding techniques and will consider a related topic in Section 5.12. Motivated by this fact we now consider an alternative relaxation idea for the original OCP (5.17). This idea is related to the "local convexification" procedure associated with the infimal (prox) convolution

$$J_\lambda(u(\cdot)) + \frac{\lambda}{2}||u(\cdot)||^2_{\mathbb{L}^2\{[t_0,t_f];\mathbb{R}^m\}}$$

of the original objective functional $J(u(\cdot))$. We next adapt the general abstract concepts from [130,275] to our concrete OCP (5.17) considered in the specific Hilbert space $\mathbb{L}^2\{[t_0, t_f]; \mathbb{R}^m\}$.

Definition 5.4. *We say that $J(u(\cdot))$ is locally paraconvex around*

$$u(\cdot) \in \mathbb{L}^2\{[t_0, t_f]; \mathbb{R}^m\}$$

if the infimal convolution $J_\lambda(u(\cdot))$ is convex and continuous on a

$$\mathcal{B}_\delta(u(\cdot))$$

around $u(\cdot)$ for some $\delta > 0$ and $\lambda > 0$.

Note that the infimal convolution $J_\lambda(u(\cdot))$ from Definition 5.4 is a locally convex functional at $u(\cdot) \in \mathbb{L}^2\{[t_0, t_f]; \mathbb{R}^m\}$.

Definition 5.5. *We say that $J(u(\cdot))$ is prox-regular at*

$$\hat{u}(\cdot) \in \mathbb{L}^2\{[t_0, t_f]; \mathbb{R}^m\}$$

if there exist $\epsilon > 0$ and $r > 0$ such that

$$J(u_1(\cdot)) > J(u_2(\cdot)) + \langle \nabla J(\hat{u}(\cdot)), u_1(\cdot) - u_2(\cdot)\rangle_{\mathbb{L}^2\{[t_0,t_f];\mathbb{R}^m\}} - \frac{r}{2}||u_1(\cdot) - u_2(\cdot)||^2_{\mathbb{L}^2\{[t_0,t_f];\mathbb{R}^m\}}$$

for all $u_1(\cdot)$ from an ϵ-ball $\mathcal{B}_\epsilon(\hat{u}(\cdot))$ around $\hat{u}(\cdot)$ whenever $u_2(\cdot) \in \mathcal{B}_\epsilon(\hat{u}(\cdot))$ and

$$|J(u_1(\cdot)) - J(\hat{u}(\cdot))| < \epsilon.$$

In the case of problem (5.17) with the quadratic $J(u(\cdot))$ we evidently have

$$J_\lambda(u(\cdot)) = \frac{1}{2}\int_{t_0}^{t_f} (\langle Q(t)x(t), x(t)\rangle + \langle (R(t) + \lambda I)u(t), u(t)\rangle)\, dt + \frac{1}{2}\langle Gx(t_f), x(t_f)\rangle,$$

where I is a unit matrix. Consider now the following infimal convolution-based OCP:

$$\begin{aligned} &\text{minimize } J_\lambda(u(\cdot)) \\ &\text{subject to } u(\cdot) \in \text{conv}(\mathcal{S}) \end{aligned} \tag{5.23}$$

and assume that it possesses an optimal solution $u_\lambda^{opt}(\cdot)$. The corresponding optimal trajectory is denoted by $x_\lambda^{opt}(\cdot)$. Similar to the fully relaxed case we next introduce the auxiliary variable x_{n+1} as follows:

$$\begin{aligned} \dot{x}_{n+1}(t) &= \frac{1}{2}(\langle Q(t)x(t), x(t)\rangle + \langle (R(t) + \lambda I)u(t), u(t)\rangle), \quad t \in [t_0, t_f], \\ x_{n+1}(t_0) &:= 0. \end{aligned}$$

The infimal convolution-based OCP (5.23) is now equipped with the objective functional of the type (5.19) and the following Hamiltonian:

$$H(t, x, u, p, p_{n+1}) = \langle p, f(t, x, u)\rangle + \frac{1}{2}p_{n+1}(\langle Q(t)x, x\rangle + \langle (R(t) + \lambda I)u, u\rangle).$$

Assume $p_{n+1} \neq 0$. Evidently, (5.23) constitutes a partial relaxation (convexification) of the initial OCP (5.17). Note that in the case $\lambda = 0$ OCP (5.23) represents a relaxed variant of the initial OCP (5.17) with an original objective $J(u(\cdot))$ and convexified control set. The optimal pair for this specific problem is denoted by

$$(u_0^{opt}(\cdot), x_0^{opt}(\cdot)).$$

Recall that under some weak assumptions the prox-regularity of a functional in a Hilbert space implies paraconvexity of the corresponding infimal convolution (see [275]). Using this fact, we can finally prove a local convergence result for the sequence $\{u_l(\cdot)\}$ generated by the basic gradient method

$$u_{(l+1)}(\cdot) = \gamma_l \mathcal{P}_{\text{conv}(\mathcal{S})}\left[u_{(l)}(\cdot) - \alpha_l \nabla J_\lambda(u_{(l)}(\cdot))\right] + (1-\gamma_l)u_{(l)}(\cdot),\ l \in \mathbb{N}, \tag{5.24}$$

applied to the infimal convolution-based OCP (5.23).

Theorem 5.13. *Assume that all hypotheses from* Section 5.2 *are satisfied,* $p_{n+1} \neq 0$, *and*

$$u_0^{opt}(\cdot) \in \text{int}\{conv(\mathcal{S})\}.$$

Consider a sequence $\{u_{(l)}(\cdot)\}$ *generated by method* (5.24) *with a constant step size* α. *Then there exists an initial point*

$$u_{(0)}(\cdot) \in conv(\mathcal{S})$$

such that

$$\begin{aligned} &\lim_{\lambda \to 0} \lim_{l \to \infty} J_\lambda(u_{(l)}(\cdot)) = \\ &\min_{conv(\mathcal{S})} J(u(\cdot)) = J(u_0^{opt}(\cdot)). \end{aligned} \tag{5.25}$$

Proof. Consider an ϵ-ball $\mathcal{B}_\epsilon(u_0^{opt}(\cdot))$ around the optimal control function $u_0^{opt}(\cdot)$ and let us estimate the difference

$$J(u_2(\cdot)) - J(u_1(\cdot))$$

for some $u_1(\cdot),\ u_2(\cdot) \in \mathcal{B}_\epsilon(u_0^{opt}(\cdot))$. First note that the gradient $\nabla J(u(\cdot))(\cdot)$ in problem (5.23) with a $\lambda \geq 0$ can be calculated similar to (5.22), i.e.,

$$\begin{aligned} &\nabla J_\lambda(u(\cdot))(t) = -\frac{\partial H(t, x(t), u(t), p(t), p_{n+1}(t))}{\partial u} = \\ &-\frac{\partial f(t, x(t), u(t))}{\partial u}^T p(t) - \\ &(R(t) + \lambda I)u(t)p_{n+1}(t), \\ &\frac{d\tilde{p}(t)}{dt} = -\frac{\partial H(t, x(t), u(t), p(t), p_{n+1}(t))}{\partial \tilde{x}} = \\ &-((\frac{\partial H(t, x(t), u(t), p(t), p_{n+1}(t))}{\partial x})^T, 0)^T, \\ &\tilde{p}(t_f) = -\frac{\partial(\phi(\tilde{x}(t_f)))}{\partial \tilde{x}} = (-(Gx(t_f))^T, -1)^T, \end{aligned} \tag{5.26}$$

$$\frac{d\tilde{x}(t)}{dt} = \frac{\partial H(t, x(t), u(t), p(t), p_{n+1}(t))}{\partial \tilde{p}},$$
$$\tilde{x}(t_0) = (x_0^T, 0)^T.$$

Applying the weak version of the Pontryagin maximum principle to OCP (5.23) for

$$\lambda = 0, \quad u_0^{opt}(\cdot) \in \text{int}\{\text{conv}(\mathcal{S})\},$$

we obtain

$$\frac{\partial H(t, x_0^{opt}(t), u_0^{opt}(t), p(t), p_{n+1}(t))}{\partial u} = 0$$

(see [192]). Therefore,

$$\langle \nabla J(u_0^{opt}(\cdot)), u_1(\cdot) - u_2(\cdot) \rangle_{\mathbb{L}^2\{[t_0,t_f];\mathbb{R}^m\}} = 0 \tag{5.27}$$

for all admissible $u_1(\cdot)$, $u_2(\cdot)$. From the Lipschitz continuity of the control state mapping $x^u(\cdot)$ and also taking into consideration the boundedness of the control set $\text{conv}(\mathcal{S})$, we easily deduce

$$J(u_2(\cdot)) - J(u_1(\cdot)) < \frac{r}{2} ||u_1(\cdot) - u_2(\cdot)||^2_{\mathbb{L}^2\{[t_0,t_f];\mathbb{R}^m\}} \tag{5.28}$$

for a suitable constant $r > 0$. The combination of (5.27) and (5.28) implies the prox-regularity property of the functional $J(u(\cdot))$ at $u_0^{opt}(\cdot)$ (see Definition 5.4). Since $J(u(\cdot))$ is continuous, the prox-regularity property implies the paraconvexity of the infimal convolution

$$J_\lambda(u(\cdot))$$

of $J(u(\cdot))$ in a neighborhood of optimal solution $u_0^{opt}(\cdot)$. Using the convergence results of the gradient method for locally convex functions on a convex set (see [87,262]), we conclude that

$$\lim_{l\to\infty} J_\lambda(u_{(l)}(\cdot)) = J_\lambda(u_\lambda^{opt}(\cdot)). \tag{5.29}$$

Using the continuity property of the infimal convolution $J_\lambda(u(\cdot))$ and (5.29) we finally obtain (5.25). The proof is completed. □

As one can see Theorem 5.13 establishes the well approximating property of the infimal-based (relaxed) OCP (5.23) for the weakly relaxed variant of the problem (5.23) with $\lambda = 0$. We next use this weak relaxation for a constructive numerical treatment of the initial OCP (5.17).

The implementability issue of the proposed gradient-based methods (5.21) and (5.24) strongly depends on the constructive expressions for the functional gradients determined as solutions of the boundary value problems (5.22) and (5.26), respectively.

Existence of the nontrivial (absolutely continuous) adjoint variables

$$\left(p^T(\cdot), p_{n+1}\right)^T \in \mathbb{R}^{n+1} \setminus \{0\}$$

in (5.22) and in (5.26) that satisfy the corresponding boundary value problems is closely related to the so-called regularity (Lagrange regularity) conditions in optimal control (see, e.g., [17,96,157,192]). When solving conventional OCPs based on some necessary conditions for optimality one is often faced with two possible technical difficulties: the irregularity of the Lagrange multipliers associated with the given constraints (see, e.g., [192,198]) and the degeneracy phenomenon (see, e.g., [17,157]). Various supplementary conditions, namely, constraint qualifications, have been proposed under which it is possible to assert that the Lagrange multiplier rule holds in a "usual" constructive form (see [130]). Examples are the well-known Slater regularity condition for classic convex programming and the Mangasarian–Fromovitz regularity conditions for general nonlinear optimization problems. Let us also note here the celebrated Kurcyusz–Robinson–Zowe regularity conditions for some classes of abstract OCPs (see [198,337]). We also refer to [17,192,198,213,269] for some additional facts and mathematical details. Note that some regularity conditions for OCPs can be formulated as controllability conditions for the linearized system [198].

The last part of this section is devoted to the specific regularity conditions for the auxiliary infimal-based OCP (5.23) that guarantee existence of the nontrivial Lagrange multipliers for these optimization problems (Lagrange regularity). We restrict our consideration to the case of a stationary control system in (5.15) and correspondingly assume

$$f(t, x, u) = f(x, u).$$

The Lagrange regularity implies the consistence of the main boundary value problems in (5.22) and finally guarantees the realizability of the proposed gradient-based methods in the concrete form (5.24)–(5.26).

In parallel with the initially given nonlinear dynamic model in (5.15) we next introduce the conventional linearized system. The corresponding linearization is considered over an optimal pair

$$(u_\lambda^{opt}(\cdot), x_\lambda^{opt}(\cdot))$$

associated with the infimal-based OCP (5.23). We have

$$\begin{aligned} \dot{y}(t) = & \frac{\partial f(x_\lambda^{opt}(t), u_\lambda^{opt}(t))}{\partial x} y(t) + \\ & \frac{\partial f(x_\lambda^{opt}(t), u_\lambda^{opt}(t))}{\partial u} v(t), \\ & y(t_0) = 0, \end{aligned} \tag{5.30}$$

where $v(t) \in \mathbb{R}^m$ is a control input for (4.1).

Theorem 5.14. *Assume that all hypotheses from* Section 5.2 *are satisfied and, moreover, the linearized system* (5.30) *is controllable. Then the infimal-based* OCP (5.23) *is* Lagrange *regular.*

Proof. Let us introduce the following system operator:

$$F : \mathbb{L}^2\{[t_0, t_f]; \mathbb{R}^m\} \times \mathbb{W}^{1,\infty}\{[t_0, t_f]; \mathbb{R}^n\} \to \mathbb{W}_n^{1,\infty}\{[t_0, t_f]; \mathbb{R}^n\},$$
$$F(u(\cdot), x(\cdot)) := x(\cdot) - x_0 - \int_{t_0}^{\cdot} f(x(t), u(t)),$$

where $\mathbb{W}^{1,\infty}\{[t_0, t_f]; \mathbb{R}^n\}$ denotes a Sobolev space of all absolutely continuous functions with essentially bounded derivatives. Under the basic assumptions of Section 5.2, the system operator is Fréchet differentiable (see [192,193]). From the main result of [337] it follows that the optimization problem (5.23) is Lagrange regular, if the Fréchet derivative

$$DF(u_\lambda^{opt}(\cdot), x_\lambda^{opt}(\cdot))$$

of the system mapping $F(u(\cdot), x(\cdot))$ at $(u_\lambda^{opt}(\cdot), x_\lambda^{opt}(\cdot))$ is surjective.

Let now $z(\cdot) \in \mathbb{W}^{1,\infty}\{[t_0, t_f]; \mathbb{R}^n\}$ be an arbitrary function. The integral equation

$$y(t) = \int_{t_0}^{t} \frac{\partial f(x_\lambda^{opt}(t), u_\lambda^{opt}(t))}{\partial x} y(t) dt + z(t), \; t \in [t_0, t_f],$$

is a linear Volterra equation of the second kind. This equation has a solution

$$y(\cdot) := \zeta(\cdot)$$

from $\mathbb{W}^{1,\infty}\{[t_0, t_f]; \mathbb{R}^n\}$ [20]. We now put this specific function $y(\cdot)$ into the linearized system (5.30). The controllability assumption implies the existence of a pair

$$(\tilde{v}(\cdot), \tilde{y}(\cdot)) \in \mathbb{L}^2\{[t_0, t_f]; \mathbb{R}^m\} \times W^{1,\infty}\{[t_0, t_f]; \mathbb{R}^n\}$$

that satisfies the initial value problem (5.30). Therefore,

$$DF(u_\lambda^{opt}(\cdot), x_\lambda^{opt}(\cdot))[(\tilde{y}(\cdot) + \zeta(\cdot), \tilde{v}(\cdot))] = \tilde{y}(\cdot) + \zeta(\cdot) - \int_{t_0}^{\cdot} [\frac{\partial f(x_\lambda^{opt}(t), u_\lambda^{opt}(t))}{\partial x}(\tilde{y}(t) + \zeta(\cdot)) + \frac{\partial f(x_\lambda^{opt}(t), u_\lambda^{opt}(t))}{\partial u}\tilde{v}(t)]dt = \tilde{y}(\cdot) + z(\cdot) - \int_{t_0}^{\cdot} [\frac{\partial f(x_\lambda^{opt}(t), u_\lambda^{opt}(t))}{\partial x}\tilde{y}(t) + \frac{\partial f(x_\lambda^{opt}(t), u_\lambda^{opt}(t))}{\partial u}\tilde{v}(t)]dt = z(\cdot). \tag{5.31}$$

The final relation (5.31) characterizes the surjectivity property of

$$DF(u_\lambda^{opt}(\cdot), x_\lambda^{opt}(\cdot))$$

that implies the Lagrange regularity of (5.23). The proof is completed. □

Evidently, the controllability property of system (5.30) considered on the originally given control space $\mathcal{S}$ implies the assumption of the basic Theorem 5.14 for $v(t) \in \mathbb{R}^m$. Let us now present a simple controllability condition for the linearized system (5.30) determined on the originally given control space, namely, on the space of the fixed-levels controls $v(\cdot) \in \mathcal{S}$. We restrict here our consideration to a specific stationary case, i.e.,

$$A := \partial f(x_\lambda^{opt}(t), u_\lambda^{opt}(t))/\partial x$$

and

$$B := \partial f(x_\lambda^{opt}(t), u_\lambda^{opt}(t))/\partial u$$

for $u(\cdot) \in \mathcal{S}$.

Theorem 5.15. *Consider the linearized system* (5.30) *for* $v(\cdot) \in \mathcal{S}$ *and*

$$N_k \equiv N, \quad \mathcal{T}^k \equiv \mathcal{T}, \quad k = 1, ..., m.$$

Assume that

$$-B^T \int_{t_{i-1}}^{\cdot} e^{-A^T\tau} d\tau W(N)^{-1} \left(y(t_0) - e^{-At_f} y(t_f)\right) \in \mathcal{S}. \tag{5.32}$$

Then system (5.30) *is controllable if and only if the matrix*

$$W(N) := \sum_{i=1}^{N} \left[\int_{t_{i-1}}^{t_i} e^{-A\tau} d\tau B B^T \int_{t_{i-1}}^{t_i} e^{-A^T \tau} d\tau \right],$$

$$t_i \in \mathcal{T},$$

is nonsingular.

Proof. Let $W(N)$ be nonsingular. Then

$$y(t_f) = e^{At_f} y(t_0) + \int_{t_0}^{t_f} e^{A(t_f - \tau)} B v(\tau) d\tau,$$

or equivalently

$$y(t_f) = e^{At_f} \left[y(t_0) + \sum_{i=1}^{N} \int_{t_{i-1}}^{t_i} e^{-A\tau} d\tau B v^i \right], \tag{5.33}$$

where $v^i \in \mathbb{R}^m$ is a constant vector associated with the interval $[t_{i-1}, t_i)$. The resulting input value v^i such that

$$v(t) = v^i$$

for $t \in [t_{i-1}, t_i)$ and $y(t_0)$, $y(t_f)$ belongs to the corresponding trajectory of (5.30) generated by $v(\cdot)$ and is given by

$$v^i = -B^T \int_{t_{i-1}}^{t_i} e^{-A^T \tau} d\tau W(N)^{-1} \left(y(t_0) - e^{-At_f} y(t_f) \right) \in \mathcal{S}.$$

Substituting the last expression in (5.33), we next obtain

$$y(t_f) = e^{At_f} [y(t_0) - W(N) W(N)^{-1} (y(t_0) - e^{-At_f} y(t_f))] = y(t_f).$$

We finally conclude that the given system is controllable under piecewise constant inputs.

Let the initial system (5.30) be controllable by piecewise constant controls $v(\cdot)$ from $\mathcal{S}$. Assume that the symmetric matrix $W(N)$ is not a (strictly) positive definite matrix. This hypothesis implies the existence of a nontrivial vector $w \in \mathbb{R}^n$ such that

$$w^T W(N) w = 0,$$

or equivalently

$$0 = w^T \sum_{i=1}^{N} \left[\int_{t_{i-1}}^{t_i} e^{-A\tau} d\tau B B^T \int_{t_{i-1}}^{t_i} e^{-A^T \tau} d\tau \right] w =$$

$$\sum_{i=1}^{N} ||w^T \int_{t_{i-1}}^{t_i} e^{-A\tau} d\tau B||^2.$$

The last fact implies the following:

$$w^T \int_{t_{i-1}}^{t_i} e^{-A\tau} d\tau B = 0 \, \forall i = 1, \ldots, N.$$

Since the controllability of the system (4.1) for $v(\cdot) \in \mathcal{S}$ is assumed, there exists a sequence of values $\{v^i\}$ such that the state $y(t_0) \equiv w$ can be transferred into $y(t_f) \equiv 0$. Therefore, we next deduce

$$0 = e^{At_f} \left[w + \sum_{i=1}^{N} \left(\int_{t_{i-1}}^{t_i} e^{-A\tau} d\tau \right) B v^i \right]. \tag{5.34}$$

Evidently, (5.34) holds if and only if

$$0 = w + \sum_{i=1}^{N} \left(\int_{t_{i-1}}^{t_i} e^{-A\tau} d\tau \right) B v^i. \tag{5.35}$$

We now multiply (5.35) by w^T and obtain the contradiction with the nontriviality hypothesis $w \neq 0$. We have

$$0 = w^T w + \sum_{i=1}^{N} w^T \left(\int_{t_{i-1}}^{t_i} e^{-A\tau} d\tau \right) B v^i = w^T w.$$

Therefore, $W(N)$ is a positive definite symmetric matrix and the existence of the inverse $W(N)^{-1}$ follows immediately. The proof is completed. □

Using Theorem 5.14 and Theorem 5.15, we can now easily obtain the following specific implementability result associated with the proposed gradient-based algorithm (5.24)–(5.26).

Theorem 5.16. *Assume that all hypotheses from* Section 5.2 *and condition* (5.32) *are satisfied and, moreover, the matrix $W(N)$ associated with the linearized system* (5.30) *of the stationary original system* (5.15) *with*

$$N_k \equiv N, \;\; \mathcal{T}^k \equiv \mathcal{T}, \;\; k = 1, \ldots, m,$$

is nonsingular. Then the auxiliary infimal-based OCP (5.23) *is* Lagrange *regular and the projected gradient method for* (5.23) *can be implemented in the constructive form* (5.24)–(5.26).

Note that controllability conditions for a linearized system (5.30) equipped with controls $v(\cdot) \in \mathcal{S}$ evidently imply the controllability property of the same system considered for

$$v(\cdot) \in \mathrm{conv}(\mathcal{S}).$$

This fact is a simple consequence of the evident inclusion

$$\mathcal{S} \subset \text{conv}(\mathcal{S}).$$

Therefore, the realizability conditions for the gradient-based algorithm (5.24)–(5.26) proposed for the auxiliary OCP (5.23) with $u(\cdot) \in \text{conv}(\mathcal{S})$ can be formulated using the controllability conditions for the original dynamic system determined on the initially given fixed-level control set $\mathcal{S}$.

5.10 The β-Relaxations

Consider now the constrained OCP (5.1) from Section 5.2 and include the integral inequality constraint mentioned in the same section. We have

$$\begin{aligned}
&\text{minimize } J(x(\cdot), u(\cdot)) = \int_0^{t_f} f_0(t, x(t), u(t))dt \\
&\text{subject to } \dot{x}(t) = f(t, x(t), u(t)) \text{ a.e. on } [0, t_f], \\
&x(0) = x_0, \\
&u(t) \in U \text{ a.e. on } [0, t_f], \\
&h_j(x(t_f)) \leq 0 \; \forall j \in I, \\
&q(t, x(t)) \leq 0 \; \forall t \in [0, t_f], \\
&\int_0^{t_f} s(t, x(t), u(t))dt \leq 0.
\end{aligned} \tag{5.36}$$

Here $\mathcal{J}(u(\cdot)) := \phi(x^u(1))$, where $x^u(\cdot)$ is an absolutely continuous trajectory of the dynamic system in (5.36) generated by an admissible control input $u(\cdot)$ and $\phi(\cdot)$ is a continuous function. We assume that all the basic hypotheses from Section 5.2 are satisfied. Additionally we suppose that $f(t, \cdot, \cdot)$ is differentiable and that $f, \; f_x, \; f_u$ are continuous. Moreover, there exists a constant $C < \infty$ such that

$$||f_x(t, x, u)|| \leq C$$

for all $(t, x, u) \in [0, t_f] \times \mathbb{R}^n \times U$.

A variety of approximation schemes have been recognized as a powerful tool for theoretical studying and practical solving of the infinite-dimensional dynamic optimization problems of the type (5.36). On the other hand, theoretical approaches to the relaxed OCP that corresponds to the constrained problem (5.36) are not sufficiently advanced for the numerically tractable schemes and computer-oriented algorithms. Let us recall that some usual approximations of the differential inclusion were examined from the theoretical standpoint in [246–248,309]. As an example we refer to the proximal-based method (with respect to the state space) proposed by Mordukhovich in [246,247] for the Bolza-type OCP. An implementation of this method is directly related to a sequence of sophisticated optimization problems and cannot be used in an engineering praxis. It is necessary to stress that it is extremely difficult to generalize the method mentioned above to relaxed OCPs with some additional state and control constraints.

For the constructive approximations of the relaxed control system and the corresponding OCPs we use here the idea of the generalized system of Gamkrelidze (5.3). Our aim is to approximate this and the corresponding generalized OCP (5.36). We will approximate (5.3) from the viewpoint of numerically tractable (consistent) approximations. For this purpose we introduce (see [27,29,30]) an auxiliary control system, namely, the β-systems.

The first step in generating the relaxed β-controls consists in approximation of the admissible control set U by a finite set U_M of points (grid)

$$v^k \in U, \; k = 1, ..., M,$$

where

$$M = (n+1)\tilde{M}, \; \tilde{M} \in \mathbb{N}.$$

We assume that for a given number $\epsilon > 0$ (accuracy) there exists a natural number $\tilde{M}_\epsilon$ such that for every $v \in U$ one can find a point

$$v^k \in U_M, \; ||v - v^k||_{\mathbb{R}^m} < \epsilon,$$

where

$$M = (n+1)\tilde{M}, \; \tilde{M} > \tilde{M}_\epsilon.$$

We now consider a sequence of the accuracies $\{\epsilon_M\}$ mentioned above such that

$$\lim_{M \to \infty} \epsilon_M = 0.$$

Let us now introduce the following auxiliary system:

$$\begin{aligned} \dot{z}(t) &= \sum_{k=1}^{M} \beta^k(t) f(t, z(t), v^k) \text{ a.e. on } [0, t_f], \\ z(0) &= x_0, \end{aligned} \tag{5.37}$$

where

$$\beta^k(\cdot) \in \mathbb{L}_1^1([0, t_f]), \; \beta^k(t) \geq 0,$$

and

$$\sum_{k=1}^{M} \beta^k(t) = 1 \ \forall t \in [0, t_f].$$

We next introduce the set of admissible β-controls.

Definition 5.6. *The admissible β-controls are defined as follows:*

$$\aleph(M) := \{(\beta^1(\cdot), ..., \beta^M(\cdot))^T \ : \ \beta^k(\cdot) \in \mathbb{L}_1^1([0, t_f]), \ \beta^k(t) \geq 0,$$
$$\sum_{k=1}^{M} \beta^k(t) = 1 \ \forall t \in [0, t_f]\}, \ \beta_M(\cdot) := (\beta^1(\cdot), ..., \beta^M(\cdot))^T.$$

We now call the introduced control system (5.37) (with the β-controls) the β-control system. Note that under the basic assumptions for a fixed U_M the given β-control system (5.37) has an absolutely continuous solution $z_M^\beta(\cdot)$ for every admissible β-control

$$\beta_M(\cdot) \in \aleph(M).$$

Let us note that the set $\aleph(M)$ from Definition 5.6 is similar to the set $\aleph(n+1)$ of admissible controls in the Gamkrelidze relaxed dynamic system (5.3) (see Section 5.2).

Let $\mathbb{W}_n^{1,1}([0, T])$ be the standard Sobolev space of absolutely continuous functions

$$\varphi : [0, t_f] \to \mathbb{R}^n, \ \dot{\varphi}(\cdot) \in \mathbb{L}_n^1([0, t_f]).$$

The function space $\mathbb{W}_n^{1,1}([0, t_f])$ equipped with the norm $||\cdot||_{\mathbb{W}_n^{1,1}([0,t_f])}$ defined by

$$||\varphi(\cdot)||_{\mathbb{W}_n^{1,1}([0,t_f])} := ||\varphi(\cdot)||_{\mathbb{L}_n^1([0,t_f])} + ||\dot{\varphi}(\cdot)||_{\mathbb{L}_n^1([0,t_f])} \ \forall \varphi(\cdot) \in \mathbb{W}_n^{1,1}([0, t_f])$$

is a Banach space (see Chapter 4). Recall that $\mathbb{C}_n([0, t_f])$ is the Banach space of all continuous functions $\varrho : [0, t_f] \to \mathbb{R}^n$ equipped with the usual max-norm

$$||\varrho(\cdot)||_{\mathbb{C}_n([0,t_f])} := \max_{t \in [0,t_f]} ||\varrho(t)||_{\mathbb{R}^n}.$$

Let us first establish some basic properties of the introduced β-control systems.

Theorem 5.17. *Let the right-hand side of the dynamic system from* (5.36) *satisfy the basic assumptions. For every*

$$(\nu(\cdot), y^\nu(\cdot)), \ \nu(\cdot) \in \aleph(n+1) \times \mathcal{U}^{n+1}$$

there exists a sequence of β-controls $\{\beta_M(\cdot)\} \subset \aleph(M)$ and the corresponding sequence $\{z^{\beta}_M(\cdot)\}$ of solutions of an appropriate β-control system such that $z^{\beta}_M(\cdot)$ approximate the solution $y^{\nu}(\cdot)$ of the Gamkrelidze system (5.3) *in the following sense:*

$$\lim_{M\to\infty} ||z^{\beta}_M(\cdot) - y^{\nu}(\cdot)||_{\mathbb{C}_n([0,t_f])} = 0$$

and

$$\lim_{M\to\infty} ||z^{\beta}_M(\cdot) - y^{\nu}(\cdot)||_{\mathbb{W}^{1,1}_n([0,t_f])} = 0.$$

Let us refer to [32] for a rigorous proof of this fundamental result.

Theorem 5.18. *Let the right-hand side of the system from OCP* (5.36) *satisfy the hypothesis of Theorem 5.17. Additionally assume that*

$$\beta_M(\cdot) \in \aleph(M)$$

and $z^{\beta}_M(\cdot)$ is the corresponding solution of (5.37). *Then*

$$\dot{z}^{\beta}_M(t) \in \text{conv}\{f(t, z^{\beta}_M(t), U)\} \text{ a.e. on } [0, t_f].$$

Proof. Let δ_{v_k} be a Dirac measure at $v_k \in U_M$. For a given $\beta_M(\cdot) \in \aleph(M)$, we next define the admissible relaxed control

$$\mu(t) = \sum_{k=1}^{M} \beta_k(t)\delta_{v_k} \in \mathcal{M}^1_+(U).$$

Here $\mathcal{M}^1_+(U)$ is the space of all probability measures on the Borel sets of U. Evidently, $z^{\beta}_M(\cdot)$ is a solution of the Young relaxed system (5.4) corresponding to the relaxed control $\mu(\cdot)$ determined above. From the equivalence theorem (Theorem 5.4) we now deduce

$$\dot{z}^{\beta}_M(t) \in \text{conv}\{f(t, z^{\beta}_M(t), U)\} \text{ a.e. on } [0, t_f].$$

The proof is completed. □

Evidently the set-valued function in the right-hand side of the differential inclusion in Theorem 5.18 is compact- and convex-valued. The solvability property of this differential inclusion follows from the results of Chapter 4.

Theorem 5.17 and Theorem 5.18 show that $z^{\beta}_M(\cdot)$ is a consistent approximation to the solution of the RDI (5.5) introduced in Section 5.2. This property is understood in the strong sense, namely, with respect to the convergence in the Sobolev space $\mathbb{W}^{1,1}_n([0, t_f])$ (including the convergence of the derivatives $\dot{z}^{\beta}_{M}(\cdot)$). It constitutes a fundamental approximating result for the proposed β-controls.

However, from the point of view of numerical approaches to the OCPs we always need to construct an approximating control input from the space $\mathcal{U}$ of a "usual" (nonrelaxed) control function $u(\cdot)$. The next result gives an approximative solution to this practically motivated question.

Theorem 5.19. *Let the right-hand side of the system from OCP* (5.36) *satisfy the assumptions of Theorem 5.17. Let*

$$\beta_M(\cdot) \in \aleph(M)$$

and $z^{\beta}_{M}(\cdot)$ *be the corresponding solution of the* β*-system* (5.37). *Then there exists a piecewise constant control function*

$$\tilde{u}_M(\cdot) \in \mathcal{U}$$

such that the solution $x^{\tilde{u}_M}(\cdot)$ *of the initial system from* (5.36) *exists on the time interval* $[0, t_f]$ *and*

$$\lim_{M \to \infty} ||z^{\beta}_{M}(t) - x^{\tilde{u}_M}(t)||_{\mathbb{R}^n} = 0 \tag{5.38}$$

uniformly in $t \in [0, t_f]$.

Proof. By construction of the solution $z^{\beta}_{M}(\cdot)$ this function is a solution of the relaxed system (5.3) with the specific admissible relaxed control $\mu(\cdot)$ designed as follows:

$$\mu(t) = \sum_{k=1} \beta_k(t)\delta_{v_k}.$$

Then, there exists an admissible piecewise constant control $u_M(\cdot) \in \mathcal{U}$ such that (5.38) holds; (see [32], Theorem 13.2.1, p. 677 and Theorem 13.2.2, p. 678). □

The presented theorems make it possible to approximate the initially given constrained OCP (5.36). Let us introduce the functions

$$\bar{\mathcal{J}}_M, \bar{h}_M, \bar{s}_M : \aleph(M) \to \mathbb{R}$$

and

$$\bar{q}_M : \aleph(M) \to \mathbb{C}([0, t_f])$$

defined by

$$\bar{\mathcal{J}}_M(\beta_M(\cdot)) := \phi(z^{\beta}_M(t_f)),\ \bar{h}_M(\beta_M(\cdot)) := h(z^{\beta}_M(t_f)),$$
$$\bar{q}_M(\beta_M(\cdot))(t) := q(t, z^{\beta}_M(t))\ \forall t \in [0, t_f],$$
$$\bar{s}_M(\beta_M(\cdot)) := \int_0^{t_f} \sum_{k=1}^{M} \beta^k(t) s(t, z^{\beta}_M(t), v^k) dt.$$

We can now present our main approximability result devoted to the "comparison" of the initially given OCP (5.36) and an adequately relaxed problem, where we use the β-relaxations.

Theorem 5.20. *For every $\delta > 0$ there exists a number $M_\delta \in \mathbb{N}$ such that for all natural numbers $M > M_\delta$ the β-relaxed OCP*

$$\begin{aligned} &\text{minimize } \bar{\mathcal{J}}_M(\beta_M(\cdot)) \text{ subject to } \beta_M(\cdot) \in \aleph(M), \\ &\bar{h}_M(\beta_M(\cdot)) \leq \delta, \\ &\bar{q}_M(\beta_M(\cdot))(t) \leq \delta\ \forall t \in [0, t_f], \\ &\bar{s}_M(\beta_M(\cdot)) \leq \delta \end{aligned} \tag{5.39}$$

has an optimal solution

$$\beta^{opt}_M(\cdot) \in \aleph(M)$$

under the assumption that it has an admissible solution. Moreover,

$$|\bar{\mathcal{J}}_M(\beta^{opt}_M(\cdot)) - \bar{\mathcal{J}}(v^{opt}(\cdot))| \leq \delta,$$

where

$$v^{opt}(\cdot) \in \aleph(n+1) \times \mathcal{U}^{n+1}$$

is an optimal solution of the relaxed OCP (5.6).

Proof. By Theorem 5.17, there exists a sequence of β-controls $\{\beta_M(\cdot)\}$ such that

$$\lim_{M \to \infty} ||z^{\beta}_M(t_f) - y^{opt}(t_f)||_{\mathbb{R}^n} = 0,$$

where $y^{opt}(\cdot)$ is the optimal trajectory of system (5.3) for $v^{opt}(\cdot)$. The continuity property of the constraints functions

$$h(\cdot),\ q(\cdot),\ s(\cdot)$$

implies that, for every $\delta > 0$, there exists a number $M_{1,\delta} \in \mathbb{N}$ such that

$$\bar{h}_M(\beta_M(\cdot)) \leq \delta,\ \bar{q}_M(\beta_M(\cdot))(t) \leq \delta,\ \forall t \in [0, t_f], \bar{s}_M(\beta_M(\cdot)) \leq \delta,$$

for all

$$M > M_{1,\delta}.$$

The structure of the right-hand side of the β-system (5.37) implies convexity of the extended velocity vector (extended orientor field) in (5.39) (see, e.g., [32]). Moreover, the control set

$$\mathcal{F}(M) := \{(\beta^1, ..., \beta^M)^T : \beta^k \geq 0, \sum_{k=1}^{M} \beta^k = 1\}$$

is compact. The solutions of the β-system are uniformly bounded (see [32]). The state, target, and integral constraints in (5.39) also define a bounded and closed set. Since the set of admissible solutions to (5.39) is nonempty, the existence of an optimal solution

$$\beta_M^{opt}(\cdot) \in \aleph(M)$$

for this problem follows from the Filippov–Himmelberg theorem (see Chapter 4).

Recall that function ϕ is continuous. Then, for every $\delta > 0$, there exists a number $M_{2,\delta} \in \mathbb{N}$ such that

$$|\bar{\mathcal{J}}_M(\beta_M(\cdot)) - \bar{\mathcal{J}}(\nu^{opt}(\cdot))| \leq \delta/3,$$

for all $M > M_{2,\delta}$. Let $z_M^{opt}(\cdot)$ be the solution of the β-system (5.37) associated with $\beta_M^{opt}(\cdot)$. Evidently, $z_M^{opt}(\cdot)$ is also a solution of the Young system (5.4) with the following relaxed control:

$$\hat{\mu}(t) = \sum_{k=1}^{M} \beta^{k,opt}(t)\delta_{v_k},$$

where $\beta^{k,opt}(t)$ is the kth component of $\beta_M^{opt}(t)$.

From the equivalence theorem (see Section 5.2) we deduce that $z^{\beta_M(\cdot)}$ is also a solution of the Gamkrelidze system (5.3) for an admissible generalized control $\hat{\nu}(\cdot)$. The continuity of the functions $h(\cdot)$, $q(\cdot)$, $s(\cdot)$ implies that, for every $\delta > 0$, there exist a number $M_{3,\delta} \in \mathbb{N}$ and an admissible generalized control $\nu(\cdot)$ such that

$$|\bar{\mathcal{J}}_M(\beta_M^{opt}(\cdot)) - \bar{\mathcal{J}}(\nu(\cdot))| \leq \delta/3,$$

for all natural numbers $M > M_{3,\delta}$. Hence,

$$\bar{\mathcal{J}}_M(\beta_M^{opt}(\cdot)) = \bar{\mathcal{J}}(\nu(\cdot)) + \gamma(\delta), \ -\delta/3 \leq \gamma(\delta) \leq \delta/3.$$

Now let us observe that for all natural numbers

$$M > M_\delta := \max(M_{1,\delta},\ M_{2,\delta},\ M_{3,\delta})$$

we have

$$\begin{aligned}&\delta/3 \geq |\bar{\mathcal{J}}_M(\beta_M(\cdot)) - \bar{\mathcal{J}}(\nu^{opt}(\cdot))| \geq \\ &|\bar{\mathcal{J}}_M(\beta_M^{opt}(\cdot)) - \bar{\mathcal{J}}(\nu^{opt}(\cdot))| = \\ &\bar{\mathcal{J}}(\nu(\cdot)) - \bar{\mathcal{J}}(\nu^{opt}(\cdot)) + \gamma(\delta).\end{aligned}$$

Since

$$\bar{\mathcal{J}}(\nu(\cdot)) \geq \bar{\mathcal{J}}(\nu^{opt}(\cdot)),$$

we have

$$\delta/3 - \gamma(\delta) \geq |\bar{\mathcal{J}}(\nu(\cdot)) - \bar{\mathcal{J}}(\nu^{opt}(\cdot))|.$$

Therefore,

$$|\bar{\mathcal{J}}_M(\beta_M^{opt}(\cdot)) - \bar{\mathcal{J}}(\nu^{opt}(\cdot)) - \gamma(\delta)| \geq \delta/3 - \gamma(\delta)$$

and

$$|\bar{\mathcal{J}}_M(\beta_M^{opt}(\cdot)) - \bar{\mathcal{J}}(\nu^{opt}(\cdot)) \leq \delta.$$

The proof is now completed. □

From Theorem 5.20 it follows that the β-system (5.37) and the corresponding β-relaxed optimization problem (5.39) constitute a theoretical basis for the consistent numerical treatment of the Gamkrelidze relaxed OCP (5.6).

5.11 Generalized Solutions in Calculus of Variation

Variational problems with constraints constitute a theoretical basis for various problems of modern mechanical engineering [5,48,70,110,111,167,177,198,199,223,226,229,240,253, 314,332]. The main analytic tool for the advanced robotics is in fact the Lagrange mechanics [265]. Consider a generic variational problem with constraints,

$$\begin{aligned}&\text{minimize} \int_{t_1}^{t_2} f_0(\tau, x(\tau), \dot{x}(\tau))d\tau \\ &\text{subject to } g(t_1, t_2, t, x(t)) = 0,\end{aligned} \tag{5.40}$$

for some continuous functions $f_0(\cdot,\cdot,\cdot)$ and $g(\cdot,\cdot,\cdot,\cdot)$. Using the well-known replacement $\dot{x}(\tau) = u(t)$, we can reduce the variational problem (5.40) to a specific case of the basic OCP (5.36). Every suitable relaxation method from the family of methods discussed in this chapter can finally be applied to the resulting OCP. In this section we consider the application of the proposed β-relaxations idea proposed in Section 5.10 and study an approach that makes it possible to construct a minimizing sequence.

Consider the basic OCP (5.36) and a sequence of optimal β-controls

$$\{\beta_M^{opt}(\cdot)\},\ M \in \mathbb{R},$$

for a given sequence of approximating problems (5.39) (see Theorem 5.20). We are now interested to use this sequence,

$$\{\beta_M^{opt}(\cdot)\},$$

and construct a minimizing sequence for the originally given OCP (5.36). Taking into consideration the solution concept from Chapter 3, the expected minimizing sequence provides a numerically adequate solution to the originally given general OCP (5.36) and to the variational problem.

Using $\{\beta_M^{opt}(\cdot)\}$, we next define the associated minimizing sequence of ordinary controls $u(\cdot)$ from $\mathcal{U}$ by the approach proposed by Tikhomirov [302]. Let

$$\{\Gamma^{\omega}\},\ \Gamma^{\omega} \subset [0, t_f],\ \omega = 1, \ldots, \Omega \in \mathbb{N}$$

be an equidistant partition of the time interval $[0, t_f]$, i.e.,

$$\bigcup_{\omega=1}^{\Omega} \Gamma^{\omega} = [0, t_f].$$

We assume that every Γ^{ω} is a half-open interval closed from the left. Furthermore, for every interval Γ^{ω}, we consider an additional set of subintervals $\{T_k^{\omega}\}$ such that

$$\bigcup_{k=1}^{M} T_k^{\omega} = \Gamma^{\omega},\ |T_k^{\omega}| = \int_{\Gamma^{\omega}} \beta^{k,opt}(t)dt.$$

Here $|T_k^{\omega}|$ denotes the length of T_k^{ω}. Recall that $\beta^{k,opt}(t)$ is the kth component of the optimal β-control $\beta_M^{opt}(t)$. We next assume that every T_k^{ω} is a half-open interval closed from the left. Using the introduced sets of intervals and subintervals, we next define the following Tikhomirov-like sequence of ordinary piecewise constant controls:

$$u_{\Omega,M}(t) := \sum_{\omega=1}^{\Omega} \sum_{k=1}^{M} \chi_k^{\omega}(t) v^k, \tag{5.41}$$

where $v^k \in U$. By $\chi_k^{\omega}(t)$ we denote here a characteristic function of the interval T_k^{ω}. We next call (5.41) the Tikhomirov-like sequence associated with the given sequence of the optimal β-controls $\{\beta_M^{opt}(\cdot)\}$.

Our aim is to prove two important approximability properties of the Tikhomirov-like sequence $\{u_{\Omega,M}(\cdot)\}$ introduced above. These properties will be considered in connection with the β-system (5.37).

Theorem 5.21. *Let $\{u_{\Omega,M}(\cdot)\}$ be the* Tikhomirov-like *sequence generated by $\{\beta_M^{opt}(\cdot)\}$. Then*

$$\lim_{\Omega\to\infty} ||z_M^{\beta,opt}(\cdot) - x^{u_{\Omega,M}}(\cdot)||_{\mathbb{C}_n([0,t_f])} = 0,$$

where $z_M^{\beta,opt}(\cdot)$ is the solution of the β-control system (5.37) *associated with $\beta_M^{opt}(\cdot)$ and trajectory $x^{u_{\Omega,M}}(\cdot)$ is the solution of the initially given control system from* (5.36) *generated by the conventional control input $u_{\Omega,M}(\cdot)$.*

The proof of this result can be found in [32].

Theorem 5.22. *Let $\{u_{\Omega,M}(\cdot)\}$ be the* Tikhomirov-like *sequence generated by $\{\beta_M^{opt}(\cdot)\}$. Then for a fixed $M \in \mathbb{N}$ and for every $\Delta > 0$ there exists a number $\delta > 0$ and $\Omega_\Delta \in \mathbb{N}$ such that*

$$|\mathcal{J}(u_{\Omega,M}(\cdot)) - \bar{\mathcal{J}}_M(\beta_M^{opt}(\cdot))| \leq \Delta$$

and

$$\begin{aligned}
&\tilde{h}(u_{\Omega,M}(\cdot)) \leq \Delta + \delta, \\
&\tilde{q}(u_{\Omega,M}(\cdot))(t) \leq \Delta + \delta \;\forall t \in [0, t_f], \\
&\tilde{s}(u_{\Omega,M}(\cdot)) \leq \Delta + \delta
\end{aligned}$$

for all natural numbers $\Omega > \Omega_\Delta$.

Proof. The continuity of the function $\phi(\cdot)$ and the above Theorem 5.21 imply that, for every $\Delta > 0$, there exists a number $\Omega_\Delta \in \mathbb{N}$ such that

$$|\mathcal{J}(u_{\Omega,M}(\cdot)) - \bar{\mathcal{J}}_M(\beta_M^{opt}(\cdot))| \leq \Delta,$$

for all natural numbers

$$\Omega > \Omega_\Delta.$$

Using the continuity properties of the constraint functions

$$h(\cdot),\; q(\cdot),\; s(\cdot),$$

we obtain

$$\tilde{h}(u_{\Omega,M}(\cdot)) - \bar{h}_M(\beta_M^{opt}(\cdot)) \leq \Delta,$$
$$\tilde{s}(u_{\Omega,M}(\cdot)) - \bar{s}_M(\beta_M^{opt}(\cdot)) \leq \Delta,$$
$$\tilde{q}(u_{\Omega,M}(\cdot))(t) - \bar{q}_M(\beta_M^{opt}(\cdot))(t) \leq \Delta, \ \forall t \in [0, t_f],$$

for all natural numbers $\Omega > \Omega_\Delta$. Since $\beta_M^{opt}(\cdot)$ is assumed to be an optimal solution of the auxiliary OCP (5.39), we have

$$\bar{h}_M(\beta_M^{opt}(\cdot)) \leq \delta,$$
$$\bar{s}_M(\beta_M^{opt}(\cdot)) \leq \delta,$$
$$\bar{q}_M(\beta_M^{opt}(\cdot))(t) \leq \delta, \ \forall t \in [0, t_f].$$

Combining now the two above groups of inequalities we obtain the statement of the theorem. The proof is completed. □

Let us now introduce a basic concept of the relaxation gap (see, e.g., [246,247]).

Definition 5.7. *The initial* OCP (5.36) *is called stable with respect to the* β*-relaxation* (5.39), *if*

$$\inf_{Problem(5.36)} \mathcal{J}(u(\cdot)) = \min_{Problem(5.39)} \bar{\mathcal{J}}(\nu(\cdot)).$$

An analogous phenomenon (the relaxation gap) was considered by many authors and for various OCPs. Let us refer to [121] for some interesting examples. Note that the "relaxation stability" defined above holds for the one-dimensional OCP (5.36) in the absence of the state and integral constraints [247,248]. In general, the relaxation stability property is related to the "calmness" condition considered in [129,130]. Recall that according to Clarke's result [130], the calmness hypothesis implies that corresponding necessary optimality conditions can be taken in a "normal" form. The general result that normality implies relaxation stability for a class of OCPs has been obtained by Warga [321]. For some special classes of the basic problem (5.36) (without state and integral constraints) the relaxation stability holds with no additional calmness or normality assumptions.

Under the additional assumption of the relaxation stability, we can prove our next result.

Theorem 5.23. *Assume that the basic OCP (5.36) is stable with respect to relaxation (by Definition 5.40). Then the* Tikhomirov-like *sequence* $\{u_{\Omega,M}(\cdot)\}$ *generated by* $\{\beta_M^{opt}(\cdot)\}$ *is a minimizing sequence for the initially given OCP* (5.36). *That means that for every* $\epsilon > 0$ *there exist* $M_\epsilon \in \mathbb{N}$ *such that for all*

$$M > M_\epsilon$$

one can find a number $\Omega_\epsilon(M) \in \mathbb{N}$ *with*

$$
\begin{aligned}
&|\mathcal{J}(u_{\Omega,M}(\cdot)) - \inf_{Problem(5.36)} \mathcal{J}(u(\cdot))| \le \epsilon,\\
&\tilde{h}(u_{\Omega,M}(\cdot)) \le \epsilon,\\
&\tilde{q}(u_{\Omega,M}(\cdot))(t) \le \epsilon \ \forall t \in [0, t_f],\\
&\tilde{s}(u_{\Omega,M}(\cdot)) \le \epsilon
\end{aligned}
$$

for all $\Omega > \Omega_\epsilon(M),\ M > M_\epsilon$.

Proof. From Theorem 5.22 it follows that for a fixed $M \in \mathbb{N}$ and for every $\epsilon > 0$, there exists a natural number $\Omega_\epsilon(M)$ such that

$$|\mathcal{J}(u_{\Omega,M}(\cdot)) - \bar{\mathcal{J}}_M(\beta_M^{opt}(\cdot))| \le \epsilon/2$$

for all

$$\Omega \ge \Omega_\epsilon,\ \Omega \in \mathbb{N}.$$

Additionally Theorem 5.20 and the fact that the initial problem (5.36) is stable with respect to the relaxation imply that, for every $\epsilon > 0$, there exists a number $M_\epsilon \in \mathbb{N}$ such that

$$|\bar{\mathcal{J}}_M(\beta_M^{opt}(\cdot)) - \inf_{Problem(5.36)} \mathcal{J}(u(\cdot))| \le \epsilon/2,$$

for all

$$M > M_\epsilon,\ M \in \mathbb{N}.$$

Hence, for every $\epsilon > 0$, there exist $M_\epsilon \in \mathbb{N}$ such that for all natural numbers $M > M_\epsilon$ there exists a number

$$\Omega_\epsilon(M) \in \mathbb{N}$$

with

$$|\mathcal{J}(u_{\Omega,M}(\cdot)) - \inf_{Problem(5.36)} \mathcal{J}(u(\cdot))| \le \epsilon$$

for all natural numbers

$$\Omega > \Omega_\epsilon,\ M > M_\epsilon.$$

By Theorems 5.20 and 5.22, we now deduce

$$
\begin{aligned}
&\tilde{h}(u_{\Omega,M}(\cdot)) \le \epsilon,\\
&\tilde{q}(u_{\Omega,M}(\cdot))(t) \le \epsilon \ \forall t \in [0, t_f],\\
&\tilde{s}(u_{\Omega,M}(\cdot)) \le \epsilon
\end{aligned}
$$

for all natural numbers $\Omega > \Omega_\epsilon,\ M > M_\epsilon$. The proof is completed. □

Theorem 5.23 shows that in the case of the relaxation stability with an optimal solution of the β-relaxed OCP (5.39) can be used for the consistent numerical treatment of the initially given constrained OCP. The "numerical solution" here is in fact determined by a possibility to construct an adequate minimizing sequence for the given OCP (5.36).

5.12 The McCormic Envelopes

We start this section by introducing a general analytical concept of a factorable function.

Definition 5.8. *A function is factorable if it is defined by a finite recursive composition of binary sums, binary products, and a given library of univariate intrinsic functions.*

Recall that a convex envelope of a nonconvex function $f(\cdot)$ over some region $X \subset \mathbb{R}^n$ is the best (largest) possible convex underestimator of $f(\cdot)$ over X, i.e.,

$$\text{con}_{f,X} := \sup\{c(x) : c(x) \leq f(x) \; \forall \, x \in X\},$$

where $c(\cdot)$ is convex. A possible definition can be the same where the requirement "$c(\cdot)$ is convex" is substituted by the milder one "$c(\cdot)$ is affine." A constructive expression of the convex envelopes (convex relaxations) of the factorable functions introduced above implies some concrete rules for the relaxation of sums and products. Let us also mention here the known rules for the convex envelopes for quadratic functions over polytopes. We use here the compact definition of the celebrated McCormic envelopes (sometimes called McCormic relaxations) presented in [86,243].

Definition 5.9. *Convex relaxations of the factorable functions generated by the recursive application of the relaxation rules for the univariate composition, binary multiplication, and binary addition from convex relaxations of the univariate intrinsic functions, without the introduction of auxiliary variables, are called McCormic relaxations.*

Let $X \subset \mathbb{R}^n$ be a nonempty convex set. Consider functions $f(\cdot), \; f_1(\cdot), \; f_2(\cdot)$ on X with

$$f(x) = f_1(x) + f_2(x).$$

Assume that $\tilde{f}(\cdot)$ and $\tilde{f}_1(\cdot), \; \tilde{f}_2(\cdot)$ are convex envelopes of $f(\cdot)$ and of $f_1(\cdot), \; f_2(\cdot)$ on X, respectively. Then the McCormic envelope $\tilde{f}(\cdot)$ of $f(\cdot)$ can be easily calculated (see [243]). We have

$$\tilde{f}(x) = \tilde{f}_1(x) + \tilde{f}_2(x).$$

The property mentioned above is a simple consequence of the convexity of the sum $\tilde{f}_1(x) + \tilde{f}_2(x)$. Consider now the product

$$f(x) = f_1(x) f_2(x).$$

From the generic results of [314] we deduce that the McCormic envelope $\tilde{f}(\cdot)$ in that case is equal to

$$\max\{\alpha_1(x) + \alpha_2(x) - f_1^L f_2^L, \gamma_1(x) + \gamma_2(x) - f_1^U f_2^U\}.$$

Here

$$\begin{aligned}
\alpha_1(x) &:= \min\{f_2^L \tilde{f}_1(x), f_2^L \bar{f}_1(x)\}, \\
\alpha_2(z) &:= \min\{f_1^L \tilde{f}_2(x), f_1^L \bar{f}_2(x)\}, \\
\gamma_1(x) &:= \min\{f_2^U \tilde{f}_1(x), f_2^U \bar{f}_1(x)\}, \\
\gamma_2(x) &:= \min\{f_1^U \tilde{f}_2(x), f_1^U \bar{f}_2(x)\},
\end{aligned}$$

where $\tilde{f}_j(\cdot)$ and $\bar{f}_j(\cdot)$ are convex and concave relaxations of the functions $f_j(\cdot),\ j = 1, 2$, respectively. Moreover, the auxiliary functions $f_j^L(\cdot), f_j^U(\cdot)$, where $j = 1, 2$, are defined as follows:

$$f_1^L(x) \leq f_1(x) \leq f_1^U(x),\ f_2^L(x) \leq f_2(x) \leq f_2^U(x)$$

for all $x \in X$.

The theory of McCormic envelopes also includes an important result related to the relaxation of the composite functions.

Theorem 5.24. *Let $X_1 \subset \mathbb{R}^n$ and $X_2 \subset \mathbb{R}$ be nonempty convex sets. Consider the composite function*

$$g(\cdot) := F(\cdot) \circ f(\cdot),$$

where $f : X_1 \to \mathbb{R}$ is continuous and $F : X_2 \to \mathbb{R}$. Assume that

$$f(X_1) \subset X.$$

Suppose that a convex relaxation $\tilde{f}(\cdot)$ and a concave relaxation $\bar{f}(\cdot)$ of $f(\cdot)$ are given. Let $\tilde{F}(\cdot)$ be a convex relaxation of $F(\cdot)$ on X. Moreover, assume that $x_{\min} \in X_2$ is a solution to the following minimization problem:

$$\text{minimize}_{x \in X_2} \tilde{F}(x).$$

Then the convex envelope $\tilde{g}(\cdot)$ of the composite function $g(\cdot)$ can be calculated as follows:

$$\tilde{g}(x) = \tilde{F}(\text{mid}\{\tilde{f}(x), \bar{f}(x), x_{\min}\}).$$

The proof of the above result can be found in [86]. The simple geometrical facts presented in this section will be used in Chapter 7 for a specific relaxation approach to HSs proposed in [243].

5.13 Notes

We refer to the following works that contain the classic RT [22,32,36,61,64,86,120,121, 130–132,173,192,279–281,286,302,314,321,326].

CHAPTER 6

Optimal Control of Hybrid and Switched Systems

6.1 Main Definitions and Concepts

6.1.1 The Abstract Optimal Control Problem

We examine here a useful abstract framework for hybrid optimal control problems (HOCPs) and switched optimal control problems (SOCPs). The abstract optimal control problem (OCP) under consideration involves a generalized input variable v along with the corresponding state variable ξ. We have

$$\begin{aligned} & \text{minimize } T(\xi, v) \\ & \text{subject to } P(\xi, v) = 0, \\ & (\xi, v) \in \Omega, \end{aligned} \tag{6.1}$$

where $T : X \times Y \to R$ is an objective functional and X, Y are real Banach spaces. Moreover, by

$$P : X \times Y \to X$$

we denote here a given mapping associated with a "state equation" in the abstract form. Let us note that the minimization operation in (6.1) is considered with respect to an admissible pair (ξ, v) and treats the states and controls as unrelated. That property of the main optimization problem constitutes a conceptual difference to OCPs considered in Chapter 5. By Ω in (6.1) we denote a nonempty subset of $X \times Y$. Note that the last condition in (6.1) constitutes the specific equation-type constraints in this optimization problem.

Definition 6.1. *We say that an admissible pair*

$$(\hat{\xi}, \hat{v}) \in \Gamma := \{(\xi, v) \in \Omega \mid P(\xi, v) = 0\}$$

is a local solution of (6.1) *if*

$$T(\hat{\xi}, \hat{v}) \le T(\xi, v) \;\; \forall (\xi, v) \in W_{(\hat{\xi}, \hat{v})} \subset \Gamma,$$

where $W_{(\hat{\xi}, \hat{v})} \subset X \times Y$ *is a neighborhood of* $(\hat{\xi}, \hat{v})$.

A Relaxation-Based Approach to Optimal Control of Hybrid and Switched Systems
https://doi.org/10.1016/B978-0-12-814788-7.00012-6

Note that all derivatives considered in this chapter are Fréchet derivatives. In that sense we next assume that the given mappings T and P are Fréchet continuously differentiable and that the state equation

$$P(\xi, v) = 0$$

can be solved with respect to ξ, i.e., $\xi = \omega(v)$. Here $\omega : Y \to X$ is a differentiable function. In this case the functional $T(\xi, v)$ can be represented as a functional depending on v only, i.e.,

$$T(\xi, v) = T(\omega(v), v) = \tilde{T}(v).$$

The concrete practical realizations of the abstract framework (6.1) are of primary importance in many engineering applications. Not only the HOCPs but a dynamic optimization or an OCP with integral equations involving an ordinary differential equation and a partial differential equation can also be formulated (in various ways) as an abstract optimization problem (6.1). Moreover, the usual finite-dimensional numerically motivated approximations of an infinite-dimensional OCP have the form of the minimization problem (6.1). In the above cases the equation constraint

$$P(\xi, v) = 0$$

represents the particular "state equation" of a concrete OCP.

We next assume that the abstract problem (6.1) is Lagrange regular (see [29,36]). In that case we can define the Lagrangian of problem (6.1) as follows:

$$\mathcal{L}(\xi, v, p) := T(\xi, v) + \langle p, P(\xi, v)\rangle_X,$$

where

$$p \in X^*, \ \langle p, \cdot\rangle_X : X \to \mathbb{R}.$$

Here X^* is the (topological) dual space to the given Banach space X. For the generalized Lagrange multiplier rule and its applications in optimal control see the references at the end of this chapter. We use here the standard notation

$$T_\xi, \ T_v, \ P_\xi, P_v, \ \mathcal{L}_\xi, \ \mathcal{L}_p, \ \mathcal{L}_v$$

for the partial derivatives of the functions

$$T, \ P, \ \mathcal{L}.$$

Moreover, we introduce the adjoint operators

$$T_\xi^*, \ T_v^*, \ P_\xi^*, P_v^*, \ \mathcal{L}_\xi^*, \ \mathcal{L}_p^*, \ \mathcal{L}_v^*$$

to the corresponding derivatives (linear operators) and also consider the adjoint operator

$$\nabla \tilde{T}^*(v)$$

to $\nabla \tilde{T}(v)$. In the context of problem (6.1) we now formulate an immediate consequence of the above solvability assumption for the abstract state equation $P(\xi, v) = 0$. Note that a usual solvability criterion for this equation follows from an appropriate variant of the implicit function theorem.

Theorem 6.1. *Let T and P be continuously* Fréchet *differentiable and let the state equation in* (6.1) *be solvable. Assume that there exists the inverse operator*

$$(P_\xi^*)^{-1} \in L((X^* \times Y^*), X^*)$$

to P_ξ. Then the gradient

$$\nabla \tilde{T}^*(v)$$

can be calculated by solving the following system of equations:

$$\begin{aligned} P(\xi, v) &= \mathcal{L}_p^*(\xi, v, p) = 0, \\ T_\xi^*(\xi, v) + P_\xi^*(\xi, v)p &= \mathcal{L}_\xi^*(\xi, v, p) = 0, \\ \nabla \tilde{T}^*(v) = T_v^*(\xi, v) + P_v^*(\xi, v)p &= \mathcal{L}_v^*(\xi, v, p). \end{aligned} \tag{6.2}$$

Proof. Differentiating the functional $\tilde{T}$ and state equation in (6.1) we obtain

$$\begin{aligned} P_\xi(\xi, v)\nabla\omega(v) + P_v(\xi, v) &= 0, \\ \nabla \tilde{T}(v) &= T_v(\xi, v) + T_\xi(\xi, v)\nabla\omega(v). \end{aligned}$$

The existence of

$$(P_\xi^*)^{-1}$$

implies the following exact formula:

$$\nabla\omega(v) = -(P_\xi)^{-1}(\xi, v)P_v(\xi, v).$$

Hence

$$\nabla \tilde{T}(v) = T_v(\xi, v) - T_\xi(\xi, v)(P_\xi)^{-1}(\xi, v)P_v(\xi, v)$$

and

$$\nabla \tilde{T}^*(v) = T_v^*(\xi, v) - P_v^*(\xi, v)(P_\xi^*)^{-1}(\xi, v)T_\xi^*(\xi, v). \tag{6.3}$$

On the other hand, we calculate the adjoint variable p from the second (adjoint) equation in (6.2) and substitute it to the third (gradient) equation in (6.2). In this manner we also obtain the expected relation (6.3). The proof is completed. □

The obtained abstract result, namely, Theorem 6.1, provides a theoretical basis for a constructive reduced-gradient formalism associated with the HOCPs under consideration (see Section 6.4). Let us also note that a similar result was obtained in [3,41] for some classic OCPs.

The idea of reduced gradients discussed in Theorem 6.1 can also be used in some standard linearization procedures for the initially nonlinear OCPs. Note that various linearization techniques have been recognized for a long time as a powerful tool for solving optimization problems. The approach proposed above can be extended to some other classes of HOCPs and SOCPs. It seems to be possible to derive the constructive necessary ϵ-optimality conditions (for example, the ϵ-hybrid maximum principle) by means of the presented analytical approach.

6.1.2 Optimal Solution Concepts, Lagrangians

Recall that the solution concept presented in Definition 6.1 is usually called the "Pontryagin minimum." Different from this solution abstraction one can consider the following two definitions.

Definition 6.2. *An admissible pair*

$$(\hat{\xi}, \hat{v}) \in \Gamma := \{(\xi, v) \in \Omega \mid P(\xi, v) = 0\}$$

is a strong local solution of the abstract minimization problem (6.1) *if*

$$T(\hat{\xi}, \hat{v}) \leq T(\xi, v) \;\; \forall (\xi, v) \in W_{\hat{\xi}} \subset \Gamma_X,$$

where $W_{\hat{\xi}} \subset X$ *is a neighborhood of* $\hat{\xi}$ *in* X.

Note that the neighborhood in Definition 6.2 is considered in the sense of the norm of the Banach space X. In contrast to the above concept the Pontryagin minimum (Definition 6.1) is characterized by the neighborhood in the sense of the norm

$$||\xi||_X + ||v||_Y$$

of the product Banach space $X \otimes Y$.

In some important real-world applications of the abstract optimization problem (6.1) one has to consider a subspace $\tilde{X} \subset X$ of the Banach state space X. Such a situation is a typical scenario for the classic and hybrid OCPs. In that particular but important case we have

$$\tilde{X} \equiv \mathbb{C}^1(0, t_f)$$

and $X \equiv \mathbb{C}(0, t_f)$. The weak minimum concept corresponds to the specific case mentioned above and is expressed in the next definition.

Definition 6.3. *We say that an admissible pair*

$$(\hat{\xi}, \hat{v}) \in \Gamma := \{(\xi, v) \in \Omega \mid P(\xi, v) = 0\}$$

is a weak local solution of (6.1) *if*

$$T(\hat{\xi}, \hat{v}) \leq T(\xi, v) \ \ \forall (\xi) \in \tilde{W}_{\hat{\xi}} \subset \Gamma,$$

where $W_{\hat{\xi}} \subset \tilde{X}$ *is a neighborhood of* $\hat{\xi}$ *in the space* $\tilde{X}$.

It is easy to see that the strong local minimum for (6.1) is also a Pontryagin minimum for the same problem. For the OCPs (taking

$$\tilde{X} \equiv \mathbb{C}^1(0, t_f)$$

and $X \equiv \mathbb{C}(0, t_f)$) the Pontryagin minimum is also a minimum in the weak sense (Definition 6.2).

Note that in the general OCPs the Pontryagin minimum concept plays an intermediate role. In the next sections we study the HOCPs and SOCPs for the corresponding Pontryagin minimum definitions.

In the case of the general OCPs one can also consider the convergence in the sense of the Pontryagin metric. This consideration is of crucial importance for the numerical methods in optimal control. A sequence $\{u_k(\cdot)\}$ converges in the Pontryagin sense to an optimal control input $u^{opt}(\cdot)$, if

$$||u_k(\cdot) - u^{opt}(\cdot)||_{\mathbb{L}_1} \to 0, \ ||u_k(\cdot) - u^{opt}(\cdot)||_{\mathbb{L}_\infty} \leq \text{const}.$$

Note that the above convergence does not correspond to any topology in $\mathbb{L}_\infty$ of the Fréchet–Uryson type (when the closure of a set can be obtained by sequences). Otherwise, the Pontryagin minimum has that advantage against the weak minimum. It is invariant with respect to a broad class of the necessary transformations (reformulations) of the initial OCP, whereas the weak minimum is not [142]. This corresponds to the invariance of various versions of the Pontryagin maximum principle. Note that the Euler–Lagrange equation (considered as a necessary condition for the weak minimum) does not possess such an invariance.

On the other hand, unlike the strong minimum, which does not require restrictions on the control and hence does not allow to estimate the increment of the cost (or the complete Lagrange function) by using its expansion at the given process, the Pontryagin minimum requires some, though mild, control restrictions, which make it possible to use such expansions. The research performed up till now shows that the concept of Pontryagin minimum is a convenient and effective analytic tool. It admits a rich theory of first-order, but also of higher-order optimality conditions.

Using the formulae (given above) for the Lagrange function $\mathcal{L}(\xi, v, p)$

$$\mathcal{L}(\xi, v, p) := T(\xi, v) + \langle p, P(\xi, v) \rangle_X,$$

associated with the abstract problem (6.1), we can introduce a Weierstrass function for the same OCP (6.1). We have

$$\mathcal{E}(\xi, v, p, q) := \mathcal{L}(\xi, v, q) - \mathcal{L}(\xi, v, p) - (q - p)\mathcal{L}_p(\xi, v, p).$$

The above analogy to the classic Weierstrass function from the conventional variational analysis can be very useful for the development of the hybrid Pontryagin maximum principle.

6.2 Some Classes of Hybrid and Switched Control Systems

Consider the following initial value problem:

$$\begin{aligned} \dot{x}(t) &= \sum_{i=1}^{r} \chi_{[t_{i-1}, t_i)}(t) \; [\; A_{q_i}(t, x(t)) + B_{q_i}(t)u(t) \;] \text{ a.e. on } [0, t_f], \\ x(0) &= x_0, \end{aligned} \tag{6.4}$$

where $x_0 \in \mathbb{R}^n$ is a fixed initial state and all functions

$$A_{q_i} : (0, t_f) \times \mathcal{R} \to \mathbb{R}^n,$$

where q_i are from an index set $\mathcal{Q}$, are measurable with respect to variable $t \in (0, t_f)$ and locally Lipschitz continuous in $x \in \mathcal{R} \subseteq \mathbb{R}^n$. We also assume that for every $t \in [0, t_f]$ the values

$$B_{q_i}(t) = (b_{i,j}(t))_m^n$$

are $n \times m$ matrices with continuous components $b_{ij}(\cdot)$. We consider systems (6.4) with $r \in \mathbb{N}$ switching times $\{t_i\}$, $i = 1, ..., r$, where

$$0 = t_0 < t_1 < ... < t_{r-1} < t_r = t_f,$$

and denote by $\chi_{[t_{i-1}, t_i)}(\cdot)$ the characteristic function of the disjunct time intervals of the type

$$[t_{i-1}, t_i), \; i = 1, ..., r.$$

Motivated by numerous applications, let us examine (6.4) over a set $\mathcal{U}$ of bounded measurable control inputs. In this section we assume that the set of admissible controls $\mathcal{U}$ has the following structure:

$$\mathcal{U} := \{v(\cdot) \in \mathbb{L}^2_m(0, t_f) \mid v(t) \in U \text{ a.e. on } [0, t_f]\},$$

where $U \subseteq \mathbb{R}^m$ is a compact and convex set. By $\mathbb{L}^2_m(0, t_f)$ we denote here the Hilbert space of square integrable functions. We assume that for every admissible control function $u(\cdot) \in \mathcal{U}$ the initial value problem (1.1) has an absolutely continuous solution $x^u(\cdot)$ (on the above time interval). However, due to the presence of a nonlinearity $A_{q_i}(t, x)$ in the above equation, the global existence of a solution to (6.4) is not guaranteed, unless there are some additional conditions imposed. We refer to [154,185,265] for the corresponding details. The above technical assumptions associated with the initial value problem (6.4) are called the basic assumptions. Evidently, the control system (6.4) constitutes a specific case of the general concept of affine switched system (ASS) (see Definition 1.4, Chapter 1).

Our next definition characterizes a subclass of the ASSs of the type (6.4).

Definition 6.4. *We call the ASS* (6.4) *a convex affine system if every functional*

$$V_l(u(\cdot)) := x^u_l(t),$$
$$u(\cdot) \in \mathcal{U}, \; t \in [0, t_f], \; l = 1, ..., n,$$

is convex.

Note that Definition 6.4 is similar to the concept of the conventional convex control systems introduced in [28]. We next recall a monotonicity concept for functionals in Euclidean spaces.

Definition 6.5. *We say that a functional*

$$T : \Gamma \subset \mathbb{R}^n \to \mathbb{R}$$

is monotonically nondecreasing if $T(\xi) \geq T(\omega)$ *for all* $\xi, \omega \in \Gamma$ *such that* $\xi_l \geq \omega_l, \; l = 1, ..., n.$

Note that the presented monotonicity concept can be expressed by introducing the standard positive cone $\mathbb{R}^n_{\geq 0}$ (the positive orthant). For a similar monotonicity concept see, for example, [32,180]. A general example of a monotonically nondecreasing functional can be deduced from the mean value theorem [132], namely, a differentiable functional

$$T : \mathbb{R}^n \to \mathbb{R}, \;\; \partial T(\xi)/\partial \xi_l \geq 0 \;\; \forall \xi \in \mathbb{R}^n, \; l = 1, ..., n,$$

where $\partial T(\xi)\partial \xi_l$ is the kth component of the gradient of T, is monotonically nondecreasing. Our next result gives a constructive characterization of a convex affine system from Definition 6.4.

Theorem 6.2. *Assume that the above basic assumptions for (6.4) are satisfied. Let every component*

$$A^l_{q_i}(t, \cdot),\ l = 1, ..., n,$$

of A_{q_i}, $q \in \mathcal{Q}$, *be a convex and monotonically nondecreasing functional for every* $t \in [t_{i-1}, t_i),\ i = 1, ..., r$. *Then the ASS (6.4) is a convex affine system.*

Proof. Let

$$u_3(\cdot) := \lambda u_1(\cdot) + (1 - \lambda)u_2(\cdot),$$

where $u_1(\cdot), u_2(\cdot)$ are elements of $\mathcal{U}$ and $0 < \lambda < 1$. Consider the corresponding restrictions

$$u_{3,i}(\cdot), u_{1,i}(\cdot), u_{2,i}(\cdot)$$

of the above control functions on the time intervals $[t_{i-1}, t_i),\ i = 1, ..., r$. Let N be a (large) positive integer number and

$$G^N_i := \{t^0_i = t_{i-1}, t^1_i, ..., t^N_i = t_i\}$$

be an equidistant grid associated with every interval $[t_{i-1}, t_i],\ i = 1, ..., r$. Here we have $t^0_1 = t_0 = 0$ and $t^N_r = t_r = t_f$. Consider the Euler discretization method applied to the affine differential equation from (6.4) on every time interval $[t_{i-1}, t_i),\ i = 1, ..., r$ (for every location $q_i \in \mathcal{Q}$) with

$$u(\cdot) = u_3(\cdot),$$
$$\tilde{x}^{u_{3,i}}_l(t^{s+1}_i) = \tilde{x}^{u_{3,i}}_l(t^s_i) + \Delta_i t[(A_{q_i}(t^s_i, \tilde{x}^{u_{3,i}}(t^s_i)))_l + (B_{q_i}u_{3,i}(t^s_i))_l],$$
$$\tilde{x}^{u_{3,i}}_l(t^0_i) = \tilde{x}^{u_{3,i-1}}_l(t^N_{i-1}),$$
$$\tilde{x}^{u_{3,1}}_l(t^0_1) = (x_0)_l,$$

where

$$l = 1, ..., n,\ s = 0, ..., N - 1,$$

and

$$\Delta_i t := (t_{i-1} - t_i)/(N + 1).$$

By

$$\tilde{x}^{u_3}(t^s_i),\ t^s_i \in G^N_i,$$

we denote the corresponding Euler discretization of the solution x^{u_3} on G^N_i. From the main convexity result of [28] (Theorem 1, p. 994) it follows that the functional

$$u_q(\cdot) \to \tilde{x}_l^{u_q}(t_q^s), \ l = 1, ..., n,$$

where $q = 1$ (the first location), is convex. Using this fact and convexity of every matrix function

$$A_{q_i}(t, \cdot), \ q_i \in \mathcal{Q},$$

for $q = 2$ we obtain

$$\tilde{x}_l^{u_{3,q}}(t_q^1) = \tilde{x}_l^{u_{3,q}}(t_q^0) + \Delta_q t[(A_q(t_q^0, \tilde{x}_l^{u_{3,q}}(t_q^0))_l + (B_q(t_q^0)u_{3,q}(t_q^0)))_l] \leq$$
$$\lambda\tilde{x}_l^{u_{1,q}}(t_q^0) + (1-\lambda)\tilde{x}_l^{u_{2,q}}(t_q^0) + \lambda\Delta_q t[(A_q(t_q^0, \tilde{x}_l^{u_{1,q}}(t_q^0)))_l + (B_q(t_q^0)u_1(t_q^0)))_l]+$$
$$(1-\lambda)\Delta_q t[(A_q(t_q^0, \tilde{x}_l^{u_{2,q}}(t_q^0)))_l + (B_q(t_q^0)u_2(t_q^0)))_l] = \lambda\tilde{x}_l^{u_{1,q}}(t_q^1) + (1-\lambda)\tilde{x}_l^{u_{2,q}}(t_q^1).$$

Thus, the functional

$$u_q(\cdot) \to \tilde{x}_l^{u_q}(t_q^1), \ l = 1, ..., n,$$

where $q = 2$ is also convex. Using the convexity of every $A_{q_i}(t, \omega)$ and monotonicity properties of $(A_{q_i}(t, x))_l$ with respect to x and the convex structure of the functional

$$u_q(\cdot) \to \tilde{x}_l^{u_q}(t_q^1),$$

we deduce that the functional

$$u_q(\cdot) \to \tilde{x}_l^{u_q}(t_q^2), \ q = 2,$$

is also convex for all $l = 1, ..., n$ (see [36] for details). Applying induction on s, we obtain convexity of the functional

$$u_q(\cdot) \to \tilde{x}_k^{u_q}(t_q^s), \ q = 2,$$

for all $s = 0, ..., N-1$, $N \in \mathbb{N}$.

Now we use induction on $q \in \mathcal{Q}$ and prove the convexity of every functional. We have

$$u_{q_i}(\cdot) \to \tilde{x}_l^{u_{q_i}}(t_i^s) \ \forall s = 0, ..., N-1,$$
$$i = 1, ..., r, \ q_i \in \mathcal{Q}.$$

From the last fact we can deduce the convexity of the general functionals (the global Euler discretization)

$$u(\cdot) \to \tilde{x}_l^u(t),$$

where

$$t \in \bigcup_{i=1,...,r} G_i^N$$

and $u(\cdot) \in \mathcal{U}$.

The convexity of all A_{q_i}, $q \in \mathcal{Q}$, implies the global Lipschitz continuity of these functions on any closed subset of $\mathcal{R}$ that contains $x_l^{u_3}(t)$, $t \in [t_{i-1}, t_i)$. From this fact we deduce the uniform convergence of the considered Euler approximations for $N \to \infty$. That means

$$\lim_{N\to\infty} \sup_{t\in[t_{i-1},t_i)} ||\tilde{x}_l^{u_3}(t) - x_l^{u_3}(t)|| = 0$$

on every interval $[t_{i-1}, t_i)$. Therefore, for every number $\epsilon > 0$ there exists a number $N_\epsilon \in \mathbb{N}$ such that for all $N > N_\epsilon$ we have

$$||x_l^{u_3}(t) - \tilde{x}_l^{u_3}(t)|| \leq \epsilon,$$

where $t \in \bigcup_{i=1,\ldots,r} G_i^N$. Therefore, we obtain the inequality

$$x_l^{u_3}(t) \leq \tilde{x}_l^{u_3}(t) + \epsilon$$

for all $t \in [0, t_f]$. From convexity of $\tilde{x}_k^{u_3}(t)$ we next deduce that

$$x_l^{u_3}(t) \leq \lambda\tilde{x}_l^{u_1}(t) + (1-\lambda)\tilde{x}_l^{u_2}(t) + \epsilon, \ t \in [0, t_f].$$

That means

$$x_l^{u_3}(t) \leq \lim_{N\to\infty} (\lambda\tilde{x}_l^{u_1}(t) + (1-\lambda)\tilde{x}_l^{u_2}(t) + \epsilon) = \lambda x_l^{u_1}(t) + (1-\lambda)x_l^{u_2}(t).$$

This shows that the functional

$$u(\cdot) \to x_l^u, \ u(\cdot) \in \mathcal{U}$$

is convex for all $l = 1, \ldots, n$ and all $t \in [0, t_f]$ and the given control system (6.4) is a convex affine system. The proof is completed. □

Let us now introduce a general concept of a linear hybrid system (HS) with autonomous location transitions [19,40,41,122,123,289].

Definition 6.6. *A linear HS is a 7-tuple*

$$\{\mathcal{Q}, \mathcal{X}, U, A, B, \mathcal{U}, \mathcal{S}\},$$

where

- *$\mathcal{Q}$ is a finite set of discrete states (called locations);*
- *$\mathcal{X} = \{\mathcal{X}_q\}$, $q \in \mathcal{Q}$, is a family of state spaces such that $\mathcal{X}_q \subseteq \mathbb{R}^n$;*
- *$U \subseteq \mathbb{R}^m$ is a set of admissible control input values (called control set);*

- $A = \{A_q(\cdot)\},\ B = \{B_q(\cdot)\},\ q \in \mathcal{Q}$, *are families of continuously differentiable matrix functions*

$$A_q : \mathbb{R} \to \mathbb{R}^{n \times n},\ B_q : \mathbb{R} \to \mathbb{R}^{n \times m};$$

- $\mathcal{U}$ *is the set of all admissible control functions;*
- $\mathcal{S}$ *is a subset of* Ξ, *where*

$$\Xi := \{(q, x, q', x') \ :\ q, q' \in \mathcal{Q}, x \in \mathcal{X}_q,\ x' \in \mathcal{X}_{q'}\}.$$

A linear HS from Definition 6.6 is defined on the finite time interval $[0, t_f]$. We refer to [255, 289,296] for some abstract concepts of HSs. Let U be a convex and closed set. We assume that

$$\mathcal{U} := \{u(\cdot) \in \mathbb{L}^{\infty}_m(0, t_f) \mid u(t) \in U \text{ a.e. on } [0, t_f]\}.$$

We suppose here that affine functions

$$m_{q,q'} : \mathbb{R}^n \to \mathbb{R},\ q,\ q' \in \mathcal{Q},\ m_{q,q'}(x) = b_{q,q'}x + c_{q,q'}$$

are given such that the corresponding hyperplanes

$$M_{q,q'} := \{x \in \mathbb{R}^n \ :\ m_{q,q'}(x) = 0\}$$

are pairwise disjoint. Here

$$b_{q,q'} \in \mathbb{R}^n,\ c_{q,q'} \in \mathbb{R}$$

for every $q,\ q' \in \mathcal{Q}$. The given hyperplanes $M_{q,q'}$ represent the (affine) switching sets at which a switch from location q to location q' can take place. As usually we say that a location switch from q to q' occurs at a *switching time* $t^{switch} \in [0, t_f]$. We now consider a linear HS with $r \in \mathbb{N}$ switching times denoted as usually by

$$0 = t_0 < t_1 < ... < t_{r-1} < t_r = t_f.$$

Note that the above sequence of switching times $\{t_i\}$ is not defined a priori. A hybrid control system remains in location $q_i \in \mathcal{Q}$ for all

$$t \in [t_{i-1}, t_i),\ i = 1, ..., r.$$

In the following, we recall the notion of hybrid trajectory of the systems under consideration (see, e.g., [26]). Similar to the general case of HSs we introduce the next concept.

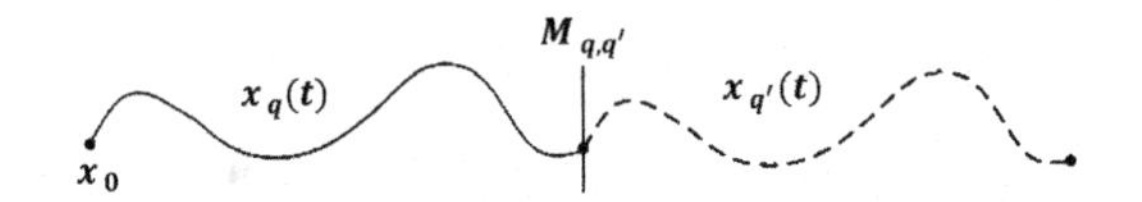

Figure 6.1: Dynamical behavior of a hybrid system.

Definition 6.7. *An admissible hybrid trajectory associated with a given linear HS from Definition 6.6 is a triple, i.e.,*

$$\mathbf{X} = (x(\cdot), \{q_i\}, \tau),$$

where $x(\cdot)$ is a continuous part of trajectory, $\{q_i\}_{i=1,...,r}$ is a finite sequence of locations, and τ is the corresponding sequence of switching times such that

$$x(0) = x_0 \notin \bigcup_{q,q' \in \mathcal{Q}} M_{q,q'}$$

and for each $i = 1, ..., r$ and every admissible control $u(\cdot) \in \mathcal{U}$ we have

- $x_i(\cdot) = x(\cdot)|_{(t_{i-1},t_i)}$ *is an absolutely continuous function on (t_{i-1}, t_i) continuously prolongable to $[t_{i-1}, t_i]$, $i = 1, ..., r$;*
- $$\dot{x}_i(t) = A_{q_i}(t)x_i(t) + B_{q_i}(t)u_i(t)$$

  for almost all times $t \in [t_{i-1}, t_i]$, where $u_i(\cdot)$ is a restriction of the chosen control function $u(\cdot)$ on the time interval $[t_{i-1}, t_i]$.

A linear HS in the sense of Definition 6.6 and Definition 6.7 that satisfies all above assumptions is denoted by LHS (see Fig. 6.1 for an illustration).

Note that the pair $(q, x(t))$ represents the hybrid state at time t, where $q \in \mathcal{Q}$ is a location and $x(t) \in \mathbb{R}^n$. Definition 6.7 describes the dynamic of an LHS. Since $x(\cdot)$ is a continuous function, the above concept describes a class of HSs without impulse components of the (continuous) trajectories. Therefore, sets $M_{q,q'}$ are defined for

$$x(t_i) = x(t_{i+1}), \ i = 1, ..., r-1.$$

Under the above assumptions, for each admissible control $u(\cdot) \in \mathcal{U}$ and for every interval $[t_{i-1}, t_i]$ (for every location $q_i \in \mathcal{Q}$) there exists a unique absolutely continuous solution of the linear differential equations from Definition 6.7. That means for each $u(\cdot) \in \mathcal{U}$ we have a unique absolute continuous trajectory of the given LHS. Moreover, the switching times $\{t_i\}$ and the discrete trajectory $\{q_i\}$ for an LHS are also uniquely defined. Note that the evolution equation for the trajectory $x(\cdot)$ of a given LHS can also be represented in the compact form, i.e.,

$$\dot{x}(t) = \sum_{i=1}^{r} \beta_{[t_{i-1},t_i)}(t) \times \\ \left(A_{q_i}(t)x_i(t) + B_{q_i}(t)u_i(t)\right) \text{ a.e. on } [0, t_f], \tag{6.5}$$

where $x(0) = x_0$ and $\beta_{[t_{i-1},t_i)}(\cdot)$ is the characteristic function of the interval $[t_{i-1}, t_i)$.

It is necessary to stress that the compact system representation (6.5) can be used for LHSs as well as for the linear-type switched systems (SSs). In the case of SSs (see Definition 1.4) the characteristic function $\beta_{[t_{i-1},t_i)}(\cdot)$ does not depend on the state $x(\cdot)$ (the switching time does not have information about the system state). If we consider an LHS the "linear" form of (6.5) has only a symbolic nature. The switching times

$$t_i, \; i = 1, ..., r,$$

are triggered by the state $x(\cdot)$. This fact implies a still nonlinear character of LHS even in the "linear-type" case (6.5). Summarizing we can conclude that an HS (even an LHS) is a strictly nonlinear dynamic system. This fact makes application of the linear control theory to general HSs impossible.

Let us now consider OCPs associated with the so-called autonomous switched mode systems [319,320]. Our approach includes an optimal sequencing scheduling as well as a simultaneous optimal timing (see [38] for the exact definitions). Note that the majority of proposed optimal control algorithms (the "mode insertion algorithm" and similar [3,18,17]) include two separate optimization steps with respect to the timing/sequencing. Otherwise, the widely used "gradient descent" methodology (see [24] and [36–38]) does not include the necessary results related to the numerical consistence. In Chapter 7, we propose a new conceptual optimization method based on the classic convex relaxation schemes for the switched mode systems.

The basic inspiration for studying switched mode control systems is the dynamical model of the form

$$\dot{x}(t) = \sum_{i=1}^{I} \beta_{[t_{i-1},t_i)}(t) \sum_{k=1}^{K} q_{k,[t_{i-1},t_i)}(t) f_k(x(t)) \\ \text{a.e. on } [0, t_f], \quad x(0) = x_0 \in \mathbb{R}^n. \tag{6.6}$$

Here $x(t) \in \mathbb{R}^n$ is a state vector, $f_k : \mathbb{R}^n \to \mathbb{R}^n$ for $k = 1, ..., K \in \mathbb{N}$ are continuously differentiable functions, and

$$\beta_{[t_{i-1},t_i)}(t) = \begin{cases} 1 & \text{if } t \in [t_{i-1}, t_i), \\ 0 & \text{otherwise} \end{cases}$$

are characteristic functions associated with disjunct time intervals $[t_{i-1}, t_i)$. Let $t_0 = 0$, $t_I = t_f$. We assume that I is a finite natural number (nonfixed a priori) and

$$q_{k,[t_{i-1},t_i)}(t) \in \{0, \ 1\}$$

for all $k = 1, ..., K, t \in [0, t_f]$ such that

$$\sum_{k=1}^{K} q_{k,[t_{i-1},t_i)}(t) = 1$$

for all $k = 1, ..., K$. Let us introduce a "vector field" $F(x)$ of the switched mode system (6.6) and a "sequencing control vector" $q_{[t_{i-1},t_i)}(t)$ (for the time interval $[t_{i-1}, t_i)$) as follows:

$$F(x) := \{f_1(x), ..., f_K(x)\},$$
$$q_{[t_{i-1},t_i)}(t) := (q_{1,[t_{i-1},t_i)}(t), ..., q_{K,[t_{i-1},t_i)}(t))^T.$$

We next use the simplified notation $q^i(t) \equiv q_{[t_{i-1},t_i)}(t)$. Note that every component

$$q_{k,[t_{i-1},t_i)}(t)$$

of $q^i(t)$ corresponds to a particular mode $f_k(x)$ of (6.6) on the time interval $[t_{i-1}, t_i)$ and hence the sequencing control vector (a vector function) $q^i(\cdot) \in \mathcal{Q}_K$,

$$\mathcal{Q}_K := \{\theta : [0, t_f] \to \{0, \ 1\}^K \mid \sum_{k=1}^{K} q_{k,[t_{i-1},t_i)}(t) = 1\}$$

represents the schedule of modes for every time instant $t \in [0, t_f]$.

System (6.6) is assumed to be associated with a sequence of switching times

$$\tau := \{t_1, ..., t_{I-1}\}.$$

Usually a technical system requires and admits a fixed maximal number of switchings on the finite time interval. In this paper we assume the absence of Zeno behavior in system (6.6) and suppose

$$I \leq I_{max} \in \mathbb{R}.$$

Every sequence τ such that the last inequality is satisfied is an admissible sequence of switching times for (6.6).

Let now

$$\beta(t) := (\beta_{[t_0,t_1)}(t), ..., \beta_{t_{I-1},t_I}(t))^T$$

be a "timing control vector." By $\mathcal{B}_I$ we next denote the set of all vectors (vector functions) $\beta(\cdot)$ such that the time intervals $[t_{i-1}, t_i)$ are disjunct and, moreover, $I \leq I_{max}$. We now are ready to describe the set of admissible switching-type control inputs for the given system (6.6): the control vector $u(t)$ in (6.6) includes the sequencing control and the timing control vectors. It has the following specific structure:

$$u(t) := (\beta^T(t), q^T(t))^T \in \mathcal{B}_I \bigotimes (\mathcal{Q}_K)^I,$$

where $q := \{q^1, ..., q^I\}$ is a family of sequencing controls for every time interval $[t_{i-1}, t_i)$. We assume that for every admissible control function $u(\cdot)$ the initial value problem (6.6) has an absolutely continuous solution $x^u(\cdot)$ (on the given time interval). However, due to the presence of the conventional and switched-type nonlinearities in the above equation, the global existence of a solution to (6.6) is not guaranteed, unless there are some additional conditions imposed. We refer to [19,23] for the corresponding details. Considering the vector field $F(x)$ as an appropriate vector, we can rewrite system (6.6) in the compact form, i.e.,

$$\dot{x}(t) = \langle\, \beta(t),\ \big(\langle q^i(t), F(x(t))\rangle_K\big)_{i=1,...,I}\, \rangle_I, \tag{6.7}$$

where $\langle\cdot,\cdot\rangle_N$ denotes a scalar product in the N-dimensional Euclidean space. Note that Eq. (6.7) constitutes a conventional dynamic system with the Carathéodory right-hand side (see, e.g., [20]). We denote this class of dynamic models as "a system with piecewise constant inputs" (see [8,9]). A dynamic system (6.7) constitutes a specific HS such that a control design in this hybrid model includes an optimal switching time selection ("timing") as well as an optimal mode sequence scheduling ("sequencing").

Given a switched mode system (6.6) we next formulate the main OCP associated with this model. We have

$$\begin{aligned} &\text{minimize } J(u(\cdot)) \\ &\text{subject to (2.2)},\ u(\cdot) \in \mathcal{B}_I \bigotimes (\mathcal{Q}_K)^I, \end{aligned} \tag{6.8}$$

with the cost functional

$$J(u(\cdot)) := \int_0^{t_f} f_0(x(t))dt$$

that involves a continuously differentiable function (integrand) $f_0 : \mathbb{R}^n \to \mathbb{R}$. We next assume that problem (6.8) possesses an optimal solution (pair) $(u^{opt}(\cdot), x^{opt}(\cdot))$.

Observe that the set

$$\mathcal{B}_I \bigotimes (\mathcal{Q}_K)^I$$

of admissible control functions in (6.8) constitutes a nonempty subset of the generic (real) Hilbert space

$$\mathbb{L}^2\{[0, t_f]; \mathbb{R}^{I+K\times I}\}.$$

Taking account of the highly restrictive nonlinear constraint for inputs $u(\cdot)$, the OCP under consideration cannot be solved by a direct application of the classic Pontryagin maximum principle. Recall that the conventional versions of the maximum principle make it possible to determine an optimal solution to an OCP in a full (nonrestricted) control space

$$\mathbb{L}^2\{[0, t_f]; \mathbb{R}^{I+K\times I}\}.$$

Moreover, an effective numerical treatment of the extended variant of the maximum principle for restricted OCPs constitutes a numerically difficult task. Motivated by that fact, we will propose a class of relatively simple and implementable relaxation-based computational procedures for OCP (6.8).

Let us recall that the general concept of the important class of HSs, namely, of the IHS, was introduced in Section 4.4. Let us consider here a specific case of an IHS given by the following definition.

Definition 6.8. *An autonomous impulsive HS (AIS) is a collection*

$$H = (Q, E, X, F, G, R),$$

where

$$Q = \{q_0, q_1, ..., q_Q\}$$

is a finite set of locations,

$$E \subset Q \times Q$$

is a set of edges, and

$$X = \{Xq\},\ q \in Q,$$

is a collection of state spaces where, for all $q \in Q$, X_q is an open subset of $\mathbb{R}^n$. Moreover,

$$F = \{f_q(\cdot)\},\ q \in Q,$$

is a collection of vector fields such that, for all $q \in Q$, function $f_q : X_q \to \mathbb{R}^n$ and

$$G = \{G_e\},\ e \in E,$$

is a collection of guards. For all possible transitions

$$e = (q_i, q_j) \in E, G_e \subset X_{q_i}.$$

Additionally

$$R = \{R_e\},\ e \in E,$$

is a collection of reset maps. For all $e = (q_i, q_j) \in E$,

$$R_e : G_e \rightarrow 2^{X_{q_j}},$$

where $2^{X_{q_j}}$ *denotes the power set of* X_{q_j}. *Finally assume that the vector fields* $f_q(\cdot)$ *are smooth enough and that the sets* G_e *are nonempty for all* $e \in E$.

An execution is as follows: starting from an initial condition (x_0, q_{i_0}) the continuous state evolves according to the autonomous differential equations

$$\dot{x}(t) = f_{q_{i_0}}(x(t)).$$

The discrete state q_{i_0} remains constant as long as the trajectory does not reach a guard $G_{(q_{i_0}, q_{i_1})}$. Once this guard is reached, $q(\cdot)$ will switch from q_{i_0} to $q_{i_{1,1}}$ at the same time the continuous state gets reset to some value according to the map $R_{(q_{i_0}, q_{i_1})}$ and the whole process is repeated. Usually one supposes that the guards can be described by smooth $(n-1)$ dimensional surfaces in $\mathbb{R}^n$,

$$G_e = \{x : S_e(x) = 0\},$$

for all $e \in E$, and that the reset maps are linear maps for all $e \in E$ of the form

$$R(q_{i_{k-1}}, q_{i_k})(x) = x + \theta_k.$$

Here θ_k belongs to Θ (a compact subset of $\mathbb{R}^n$). In our setup, the variables θ_k are degrees of freedom that can be selected by the designed controller. Let us give an illustration of a qualitative behavior of the trajectory $x(\cdot)$ associated with the AIS from Definition 6.8 (see Fig. 6.2).

Let us close this section by a simple concept of a useful class of SSs.

Definition 6.9. *An SS with controllable location transitions is a collection,*

$$\{\mathcal{Q}, \mathcal{F}, \tau, \mathcal{S}\},$$

where:

- $\mathcal{Q}$ *is a finite set of indices (called locations);*

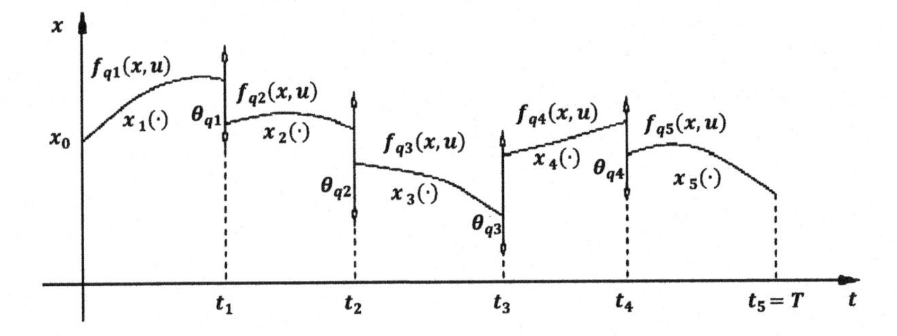

Figure 6.2: Dynamical behavior of an impulsive hybrid system.

- $\mathcal{F} = \{f_q\}_{q \in \mathcal{Q}}$ *is a family of vector fields* $f_q : \mathbb{R}^n \to \mathbb{R}^n$*;*
- $\tau = \{t_i\}$, $i = 1, ..., r$, *is an admissible sequence of switching times such that*

$$0 = t_0 < t_1 < ... < t_{r-1} < t_r = t_f;$$

- $\mathcal{S} \subset \Xi := \{(q, q) \; : \; q, q \; \in \mathcal{Q}\}.$

Usually one can assume that all functions f_q, $q \in \mathcal{Q}$, are continuously differentiable and the corresponding derivatives are bounded. Note that the elements of $\mathcal{F}$ here do not contain any conventional control parameter. An input of an SS from the above definition is represented by a sequence τ of the length $r \in \mathbb{N}$. An admissible sequence τ determines a partitioning of the time interval $[0, t_f]$ by the adjoint intervals $[t_{i-1}, t_i)$ associated with locations $q_i \in \mathcal{Q}$, $i = 1, ..., r$. A switched control system under consideration remains in a location q_i for all $t \in [t_{i-1}, t_i)$. The dynamics of the SS in every location are given by the differential equation

$$\dot{x}_i(t) = f_{q_i}(x_i(t))$$

for almost all times $t \in [t_{i-1}, t_i]$, where

$$x_i(\cdot) = x(\cdot)|_{(t_{i-1}, t_i)}$$

is an absolutely continuous function on (t_{i-1}, t_i) continuously prolongable to $[t_{i-1}, t_i]$, $i = 1, ..., r$. By $x(\cdot)$ we denote here an admissible trajectory of the given SS such that $x(0) = x_0 \in \mathbb{R}^n$.

Let us note that a general concept of an SS can be generated using the simple Definition 6.9 and considering a family $\mathcal{F}$ as a family of sufficiently smooth vector fields, i.e., $f_q : \mathbb{R}_+ \times \mathbb{R}^n \times \mathbb{R}^m \to \mathbb{R}^n$.

The classes of HSs and SSs considered in this section constitute usual dynamic restrictions for the corresponding classes of HOCPs and SOCPs we study in the next section.

6.3 Optimal Control Theory for Hybrid and Switched Systems

6.3.1 Linear Quadratic Hybrid and Switched Optimal Control Problems

Let

$$S_f : \mathbb{R} \to \mathbb{R}^{n \times n},$$
$$S_q : \mathbb{R} \to \mathbb{R}^{n \times n},$$
$$R_q : \mathbb{R} \to \mathbb{R}^{m \times m},$$

where $q \in \mathcal{Q}$. Assume that S_f is symmetric and positive semidefinite and that for every time instant $t \in [0, t_f]$ and every $q \in \mathcal{Q}$ the matrix $S_q(t)$ is also a symmetric and positive semidefinite matrix. Moreover, let $R_q(t)$ be symmetric and positive definite for every $t \in [0, t_f]$ and every $q \in \mathcal{Q}$. We also assume that the given matrix functions $S_q(\cdot)$, $R_q(\cdot)$ are continuously differentiable. Given an LHS (6.5) we consider the following hybrid linear quadratic (HLQ) problem:

$$\text{minimize } J(u(\cdot), x(\cdot)) := \frac{1}{2}(x_r^T(t_f) S_f x_r(t_f)) +$$
$$\frac{1}{2}\sum_{i=1}^{r} \int_{t_{i-1}}^{t_i} \left(x_i^T(t) S_{q_i}(t) x_i(t) + u_i^T(t) R_{q_i}(t) u_i(t)\right) dt \tag{6.9}$$

over all admissible trajectories $\mathbf{X}$ of LHS.

Evidently, (6.9) is the problem of minimizing the quadratic Bolza cost functional $J(\cdot)$ over all trajectories of the given linear HS. Note that we study the hybrid OCP (6.9) in the absence of possible target and state constraints. Throughout our consideration we assume that the HLQ problem (6.9) has an optimal solution,

$$(u^{opt}(\cdot), \mathbf{X}^{opt}(\cdot)),$$

where $u^{opt}(\cdot) \in \mathcal{U}$ and $\mathbf{X}^{opt}(\cdot)$ belongs to the set of admissible trajectories from Definition 6.7. It is necessary to stress that the existence of an optimal pair

$$(u^{opt}(\cdot), \mathbf{X}^{opt}(\cdot))$$

for an HLQ problem of the above type follows from the general existence theory for linear quadratic (LQ) OCPs with a convex closed control set U (see, e.g., [154]). We now apply the Pontryagin hybrid maximum principle (HMP) (see [34,35,38,39]) to the HLQ problem under consideration and formulate the corresponding necessary optimality conditions. For general optimality conditions in the form of a Pontryagin HMP see also [39,123,255,289,296].

Theorem 6.3. *Let*

$$(u^{opt}(\cdot), \mathbf{X}^{opt}(\cdot))$$

be an optimal solution of the regular OCP (6.9). Then there exist absolutely continuous functions $\psi_i(\cdot)$ on the time intervals $(t_{i-1}^{opt}, t_i^{opt})$, where $i = 1, ..., r$ and a nonzero vector of Lagrange multipliers

$$a = (a_1, ..., a_{r-1})^T \in \mathbb{R}^{r-1}$$

such that

$$\begin{aligned} &\dot{\psi}_i(t) = -A_{q_i}^T(t)\psi_i(t) + \\ &S_{q_i}(t)x_i^{opt}(t) \text{ a.e. on } [t_{i-1}^{opt}, t_i^{opt}], \\ &\psi_r(t_f) = -S_f x_r^{opt}(t_f), \end{aligned} \tag{6.10}$$

and

$$\begin{aligned} &\psi_i(t_i^{opt}) = \psi_{i+1}(t_i^{opt}) + \\ &a_i \frac{dm_{q_i,q_{i+1}}(x_i^{opt}(t_i^{opt}))}{dx_i} = \psi_{i+1}(t_i^{opt}) + a_i b_{q_i,q_{i+1}}, \end{aligned} \tag{6.11}$$

where $i = 1, ..., r-1$. Moreover, for every admissible control $u(\cdot) \in \mathcal{U}$ the partial Hamiltonian

$$\begin{aligned} &H_{q_i}(t, x, u, \psi) := \langle \psi_i, A_{q_i}(t)x_i + B_{q_i}(t)u_i \rangle - \\ &\frac{1}{2}\left(x_i^T S_{q_i}(t)x_i + u_i^T R_{q_i}(t)u_i\right) \end{aligned}$$

satisfies the following maximization conditions:

$$\begin{aligned} &\max_{u \in U} H_{q_i}(t, x^{opt}(t), u, \psi(t)) = \\ &H_{q_i}(t, x^{opt}(t), u^{opt}(t), \psi(t)), \ t \in [t_{i-1}^{opt}, t_i^{opt}), \end{aligned} \tag{6.12}$$

where $i = 1, ..., r$ and

$$\psi(t) := \sum_{i=1}^{r} \beta_{[t_{i-1}^{opt}, t_i^{opt})}(t)\psi_i(t)$$

for all $t \in [0, t_f]$.

Note that the adjoint variable $\psi(\cdot)$ in (6.10)–(6.12) is an absolutely continuous function on every open time interval $(t_{i-1}^{opt}, t_i^{opt})$ for $i = 1, ..., r$ but discontinuous at the switching points

$t_i^{opt} \in \tau^{opt}$. On the other hand, we are able to establish the continuity properties of the "full" optimal Hamiltonian

$$\tilde{H}^{opt}(t) := \sum_{i=1}^{r} \beta_{[t_{i-1}^{opt}, t_i^{opt})}(t) H_{q_i}(t, x^{opt}(t), u^{opt}(t), \psi(t))$$

computed for optimal pair $(u^{opt}(\cdot), \mathbf{X}^{opt}(\cdot))$ and for the corresponding adjoint variable $\psi(\cdot)$.

Theorem 6.4. *Under assumptions of Theorem 6.3, the "full" time-dependent optimal Hamiltonian $\tilde{H}^{opt}(\cdot)$ introduced above is a continuous function on* $[0, t_f]$.

Proof. Consider the time interval $[t_{i-1}^{opt}, t_i^{opt}]$ and the associated partial Hamiltonian $H_{q_i}(t, x, u, \psi)$. Evidently, the function $\tilde{H}^{opt}(\cdot)$ is continuous on the open time intervals

$$(t_{i-1}^{opt}, t_i^{opt}), \; i = 1, ..., r.$$

For an optimal pair $(u^{opt}(\cdot), \mathbf{X}^{opt}(\cdot))$ we have

$$H_{q_i}(t_i^{opt}, x^{opt}(t_i^{opt}), u^{opt}(t_i^{opt}), \psi(t_i^{opt})) = \\ -\frac{\partial J(u^{opt}(\cdot), x^{opt}(\cdot))}{\partial t_i^{opt}}.$$

Using (6.10)–(6.12) and the well-known formula for variation of the costs functional $J(\cdot)$ (see, e.g., [110,289]), we can compute

$$\begin{aligned}
&H_{q_i}(t_i^{opt}, x^{opt}(t_i^{opt}), u^{opt}(t_i^{opt}), \psi(t_i^{opt})) = \\
&H_{q_{i+1}}(t_i^{opt}, x^{opt}(t_i^{opt}), u^{opt}(t_i^{opt}), \psi(t_i^{opt})) + \\
&a_i \frac{\partial m_{q_i, q_{i+1}}(x^{opt}(t_i^{opt}))}{\partial t_i^{opt}} = \\
&H_{q_{i+1}}(t_i^{opt}, x^{opt}(t_i^{opt}), u^{opt}(t_i^{opt}), \psi(t_i^{opt})) + \\
&a_i \frac{\partial [b_{q_i, q_{i+1}} x^{opt}(t_i^{opt}) + c_{q_i, q_{i+1}}]}{\partial t_i^{opt}} = \\
&H_{q_{i+1}}(t_i^{opt}, x^{opt}(t_i^{opt}), u^{opt}(t_i^{opt}), \psi(t_i^{opt})),
\end{aligned} \tag{6.13}$$

where $i = 1, ..., r - 1$. Clearly, from the obtained relation (6.13) follows the continuity of the introduced function $\tilde{H}^{opt}(t)$ not only on the open time intervals

$$(t_{i-1}^{opt}, t_i^{opt}), \; i = 1, ..., r,$$

but also for all switching times $t_i^{opt} \in \tau^{opt}$, where τ^{opt} is the optimal sequence of switching times from $\mathbf{X}^{opt}$. The proof is completed. □

Note that a similar result is obtained in [289] for general HSs with controlled location transitions and for some classes of nonlinear HSs with autonomous location transitions. From Theorem 6.4 follows the continuity of the Hamiltonian for the HOCP (6.9) computed for optimal state and control variables and for the corresponding discontinuous adjoint variables. Note that a similar result can also be proved for general nonlinear optimal control processes governed by HSs with autonomous location transitions (see [39,289]). The corresponding proof is based on a generalization of the classic needle variations and on the associated formula for variation of the costs functional in the hybrid OCP under consideration.

Recall that the conventional LQ theory includes the celebrated Riccati formalism (see, e.g., [96,111]). We now consider a similar theoretic development for the HOCP (6.9) involved the given LHS (6.5). We will extend the well-known Bellman DP techniques for conventional LQ problems to the hybrid LQ (HLQ) optimization problems of the type (6.9).

Let us consider the linear boundary value problem (6.10)–(6.12) for the specific (but classic) case

$$U \equiv \mathbb{R}^m.$$

The maximization condition (6.12) from the above Pontryagin HMP (Theorem 6.3) implies that

$$u_i^{opt}(t) = R_{q_i}^{-1}(t) B_{q_i}^T(t) \psi_i(t)$$

for $t \in [t_{i-1}^{opt}, t_i^{opt})$. Using this representation of an optimal control and the basic facts from the theory of linear differential equations, we now compute (similarly to [110,154]) an optimal control $u^{opt}(\cdot)$ for (6.9) in the form of an optimal partially linear feedback control law

$$\begin{aligned} u^{opt}(t) &= -C(t)x^{opt}(t) = \\ &- \sum_{i=1}^{r} \beta_{[t_{i-1},t_i)}(t) C_i(t) x_i^{opt}(t), \end{aligned} \tag{6.14}$$

where

$$C_i(t) := R_{q_i}^{-1}(t) B_{q_i}^T(t) P_i(t)$$

is a partial gain matrix and $P_i(\cdot)$ is the partial Riccati matrix associated with every location $q_i^{opt} \in \mathcal{Q}$. Analogously to the classic case, for every location $q_i^{opt} \in \mathcal{Q}$ and for almost all

$$t \in (t_{i-1}^{opt}, t_i^{opt})$$

we obtain the differential equation

$$\begin{aligned} &\dot{P}_i(t) + P_i(t) A_{q_i}(t) + A_{q_i}^T(t) P_i(t) - \\ &P_i(t) B_{q_i}(t) R_{q_i}^{-1}(t) B_{q_i}^T(t) P_i(t) + S_{q_i}(t) = 0, \end{aligned} \tag{6.15}$$

known as the Riccati matrix differential equation. We call this equation the partial Riccati equation. Evidently, every matrix $C_i(\cdot)$ and every matrix $P_i(\cdot)$ and the corresponding partial Riccati equation (6.15) are associated with a current location $q_i \in \mathcal{Q}$ of the given HS. We also can deduce the usual relations

$$\psi_i(t) = -P_i(t)x_i^{opt}(t) \tag{6.16}$$

for $t \in [t_{i-1}^{opt}, t_i^{opt})$ and $i = 1, ..., r$. A symmetric (for all variables $t \in [0, t_f]$) hybrid Riccati matrix

$$P(t) := \sum_{i=1}^{r} \beta_{[t_{i-1}^{opt}, t_i^{opt})}(t) P_i(t)$$

which satisfies all equations (6.15) and the boundary (terminal) condition

$$P(t_f) = S_f$$

gives rise to the optimal feedback dynamics of (6.9) determined by the above partially linear feedback control function (6.14).

It is necessary to stress that the partial Riccati equation (6.15) can also be derived with the help of the general Bellman equation (see [123]). Analogously to the classic optimal control theory it is possible to determine a partial value function associated with every location of an HS. Replacing the control variable by $u_i^{opt}(t)$ in the abovementioned hybrid Bellman equation, one can obtain a hybrid version of the well-known differential equation for the partial value function in an LQ problem (see [110] for details). One can also prove that this partial value function for the HLQ under consideration can be chosen (similarly to the classic LQ problem) as a quadratic function with a shifting vector defined for every location $q_i \in \mathcal{Q}$.

The further investigation of the family of equations (6.15) on the full time interval $[0, t_f]$ involves the continuity question associated with the above introduced hybrid Riccati matrix $P(\cdot)$. Evidently, the continuity/smoothness of a value function is a question of general interest also in the context of other classes of OCPs governed by linear or nonlinear HSs. Related to the above presented optimization theory for an LHS we are now able to formulate our main theoretical result, namely, the discontinuity of the hybrid Riccati matrix $P(\cdot)$.

Theorem 6.5. *Under assumptions of Theorem 6.3, the hybrid Riccati matrix $P(\cdot)$ is a discontinuous function on $[0, t_f]$.*

Proof. Assume that $P(\cdot)$ is continuous on the time interval $[0, t_f]$. In particular, this means that

$$P_i(t_i^{opt}) = P_{i+1}(t_i^{opt})$$

for all numbers $i = 1, ..., r-1$. Using the above continuity assumption for $P(\cdot)$, the continuity of $x(\cdot)$, and the formula for the adjoint variable, we deduce that

$$\psi_i(t_i^{opt}) = -\lim_{t \uparrow t_i^{opt}} P_i(t) x_i^{opt}(t),$$

$$\psi_{i+1}(t_i^{opt}) = -\lim_{t \downarrow t_i^{opt}} P_{i+1}(t) x_{i+1}^{opt}(t).$$

Then from (6.16) and from the jump conditions (6.11) for the adjoint variables $\psi(\cdot)$ we obtain the following relation:

$$-P_i(t_i^{opt}) x^{opt}(t_i^{opt}) = -P_{i+1}(t_i^{opt}) x^{opt}(t_i^{opt}) + a_i b_{q_i, q_{i+1}},$$

where $i = 1, ..., r-1$. Hence

$$\left[P_{i+1}(t_i^{opt}) - P_i(t_i^{opt})\right] x_i(t_i^{opt}) = a_i b_{q_i, q_{i+1}}. \tag{6.17}$$

Since $x^{opt}(\cdot)$ is continuous and the (optimal) Lagrange multipliers

$$a = (a_1, ..., a_{r-1})^T$$

are nontrivial, the function $P(\cdot)$ is a discontinuous function on $[0, t_f]$. The obtained contradiction completes the proof. □

From (6.17) it follows that the *jump* of the hybrid Riccati matrix $P(\cdot)$ to an optimal switching time $t_i^{opt} \in \tau^{opt}$ is proportional to the associated Lagrange multiplier a_i and to the vector $b_{q_i, q_{i+1}}$ which characterizes the corresponding switching hyperplane $M_{q_i, q_{i+1}}$.

Note that Theorem 6.5 is a consequence of the continuity of the optimal Hamiltonian $\tilde{H}^{opt}(\cdot)$ (Theorem 6.4). Similarly to the above proof let us assume that the function $P(\cdot)$ is continuous. Consider now the continuity condition for two partial Hamiltonians $H_{q_i}^{opt}$ and $H_{q_{i+1}}^{opt}$, namely, the result of Theorem 6.4 for some locations $q_i, q_{i+1} \in Q$. We have

$$\begin{aligned} &H_{q_i}(t_i^{opt}, x^{opt}(t_i^{opt}), u^{opt}(t_i^{opt}), \psi(t_i^{opt})) = \\ &H_{q_{i+1}}(t_i^{opt}, x^{opt}(t_i^{opt}), u^{opt}(t_i^{opt}), \psi(t_i^{opt})). \end{aligned} \tag{6.18}$$

Using (6.18) and the above relations for $\psi_i(t_i^{opt})$ and $\psi_{i+1}(t_i^{opt})$, we deduce that

$$\begin{aligned} &-\langle P_i(t_i^{opt}) x^{opt}(t_i^{opt}), A_{q_i}(t_i^{opt}) x^{opt}(t_i^{opt}) + \\ &B_{q_i}(t_i^{opt}) u^{opt}(t_i^{opt})\rangle - \\ &\frac{1}{2}((x^{opt})^T(t_i^{opt}) S_{q_i}(t_i^{opt}) x^{opt}(t_i^{opt}) + \end{aligned}$$

$$\begin{aligned}&(u^{opt})^T(t_i^{opt})R_{q_i}(t_i^{opt})u^{opt}(t_i^{opt})) = \\ &-\langle P_{i+1}(t_i^{opt})x^{opt}(t_i^{opt}), A_{q_{i+1}}(t_i^{opt})x^{opt}(t_i^{opt}) + \\ &B_{q_{i+1}}(t_i^{opt})u^{opt}(t_i^{opt})\rangle - \\ &\frac{1}{2}((x^{opt})^T(t_i^{opt})S_{q_{i+1}}(t_i^{opt})x^{opt}(t_i^{opt}) + \\ &(u^{opt})^T(t_i^{opt})R_{q_{i+1}}(t_i^{opt})u^{opt}(t_i^{opt})).\end{aligned} \tag{6.19}$$

Since $P(\cdot)$ is assumed to be continuous in general, this function is also continuous for a special case of an LHS indicated by the following relations:

$$\begin{aligned}&S_{q_{i+1}}(t_i^{opt}) = S_{q_i}(t_i^{opt}),\ R_{q_{i+1}}(t_i^{opt}) = R_{q_i}(t_i^{opt}), \\ &B_{q_i}(t_i^{opt}) = B_{q_{i+1}}(t_i^{opt}),\ A_{q_i}(t_i^{opt}) \neq A_{q_{i+1}}(t_i^{opt})\end{aligned} \tag{6.20}$$

for all $i = 1, ..., r-1$. Note that under assumptions (6.20) we have a continuous (on the full time interval $[0, t_f]$) optimal control function $u^{opt}(\cdot)$ for (6.9). In particular, $u^{opt}(\cdot)$ has no jumps at $t = t_i^{opt}$. Then from (6.19) we deduce the following relations:

$$P_i(t_i^{opt})A_{q_i}(t_i^{opt}) = P_{i+1}(t_i^{opt})A_{q_{i+1}}(t_i^{opt})$$

and

$$A_{q_i}(t_i^{opt}) = A_{q_{i+1}}(t_i^{opt}).$$

This is a contradiction with respect to assumed conditions. This means that even in the special case of an HOCP (6.9) given by assumptions (6.20) we have discontinuity conditions

$$P_i(t_i^{opt}) \neq P_{i+1}(t_i^{opt})$$

for some $i = 1, ..., r-1$ and the contradiction with the continuity assumption for $P(\cdot)$ on the full time interval $[0, t_f]$. It is necessary to stress that the hybrid Riccati matrix $P(\cdot)$ is a discontinuous function considered on the full time interval $[0, t_f]$. On the other hand, for some (but not for all) locations

$$q_i^{opt}, q_{i+1}^{opt} \in Q$$

we can have

$$P_i(t_i^{opt}) = P_{i+1}(t_i^{opt}).$$

In this case from (6.19) it follows that the corresponding Lagrange multiplier

$$a_i \in a \neq 0 \in \mathbb{R}^{r-1}$$

is equal to zero. Let us note that the hybrid Riccati matrix $P(\cdot)$ is a completely (for all $q_i \in \mathcal{Q}$, $i = 1, \ldots, r$) continuous function only in the case of a special HLQ problem characterized by

$$\begin{aligned} S_{q_{i+1}}(t_i^{opt}) = S_{q_i}(t_i^{opt}), \ R_{q_{i+1}}(t_i^{opt}) = R_{q_i}(t_i^{opt}), \\ B_{q_i}(t_i^{opt}) = B_{q_{i+1}}(t_i^{opt}), \ A_{q_i}(t_i^{opt}) = A_{q_{i+1}}(t_i^{opt}) \end{aligned} \tag{6.21}$$

for all $i = 1, \ldots, r-1$. It is evident that under conditions (6.21) the given LHS can be rewritten as a conventional linear control system (by introduction of the new continuous system matrices and new continuous matrices in the costs functional). Therefore, the corresponding HLQ problem (6.9) with (6.21) is equivalent (in this special case) to the classic LQ-type OCP.

Let us note that similar to the conventional LQ problems the closed-loop LHS associated with the partially linear feedback (6.14) also possesses stability properties (in the sense of the classic Lyapunov concept) on the infinite time horizon. This fact can be established using the abovementioned quadratic partial value functions as candidate Lyapunov functions for the corresponding stability analysis (see [103] for details).

Our main theoretical result, namely, Theorem 6.5, provides a theoretic basis for the constructive design of the optimal feedback control strategy in the framework of the above formulated LQ-type HOCP (6.9). Evidently, in the context of the presented advanced Riccati formalism the main difficulties in computing the optimal partially linear feedback control (6.14) are caused by jumps of the hybrid Riccati matrix $P(\cdot)$ at some $t_i^{opt} \in \tau^{opt}$. Note that the discontinuity property of the hybrid Riccati matrix is a new effect in relation to the conventional LQ theory. Let us now study this discontinuity effect from a numerical point of view. By $u_i^{opt}(\cdot)$ we denote the restriction of the control function $u^{opt}(\cdot)$ on the time interval $[t_{i-1}, t_i)$. Evidently,

$$u_i^{opt}(t) = \beta_{[t_{i-1},t_i)}(t) C_i(t) x_i^{opt}(t).$$

Assume that for two given locations $q_i^{opt}, q_{i+1}^{opt} \in \mathcal{Q}$ we can compute the value of the partial Riccati matrix $P_{i+1}(t)$ for every time instant t from the closed interval $[t_i^{opt}, t_{i+1}^{opt}]$, $i = 0, \ldots, r-1$. Using the continuity property of the function $\tilde{H}^{opt}(\cdot)$ (Theorem 6.4), we obtain the following nonspecific algebraic Riccati equation with respect to the unknown matrix $P_i(t_i^{opt})$:

$$\begin{aligned} &\frac{3}{2} P_i(t_i^{opt}) B_{q_i}(t_i^{opt}) R_{q_i}^{-1}(t_i^{opt}) B_{q_i}^T(t_i^{opt}) P_i(t_i^{opt}) + \\ &P_i(t_i^{opt}) A_{q_i}(t_i^{opt}) - \frac{1}{2}(S_{q_{i+1}}(t_i^{opt}) - S_{q_i}(t_i^{opt})) - \\ &\frac{3}{2} P_{i+1}(t_i^{opt}) B_{q_{i+1}}(t_i^{opt}) R_{q_{i+1}}^{-1}(t_i^{opt}) \times \\ &B_{q_{i+1}}^T(t_i^{opt}) P_{i+1}(t_i^{opt}) - P_{i+1}(t_i^{opt}) A_{q_{i+1}}(t_i^{opt}) = 0. \end{aligned} \tag{6.22}$$

If we now transpose (6.22) and combine it with itself we get the system of the linear (Lyapunov-type) equation and the symmetric Riccati equation

$$\begin{aligned}
&P_i(t_i^{opt})A_{q_i}(t_i^{opt}) - A_{q_i}^T(t_i^{opt})P_i(t_i^{opt})- \\
&P_{i+1}(t_i^{opt})A_{q_{i+1}}(t_i^{opt}) + A_{q_{i+1}}^T(t_i^{opt})P_{i+1}(t_i^{opt}) = 0, \\
&3P_i(t_i^{opt})B_{q_i}(t_i^{opt})R_{q_i}^{-1}(t_i^{opt})B_{q_i}^T(t_i^{opt})P_i(t_i^{opt})+ \\
&P_i(t_i^{opt})A_{q_i}(t_i^{opt}) + A_{q_i}^T(t_i^{opt})P_i(t_i^{opt}) + S_{q_i}(t_i^{opt})- \\
&3P_{i+1}(t_i^{opt})B_{q_{i+1}}(t_i^{opt})R_{q_{i+1}}^{-1}(t_i^{opt})B_{q_{i+1}}^T(t_i^{opt})P_{i+1}(t_i^{opt})- \\
&P_{i+1}(t_i^{opt})A_{q_{i+1}}(t_i^{opt}) - A_{q_{i+1}}^T(t_i^{opt})P_{i+1}(t_i^{opt}) - S_{i+1}(t_i^{opt}) = 0.
\end{aligned} \tag{6.23}$$

This system (6.23) defines the value of the partial Riccati matrix, namely, $P_i(t_i^{opt})$, which can be used as the necessary start condition for solving the Riccati matrix differential equation (6.15) on the next time interval $[t_{i-1}^{opt}, t_i^{opt})$. Note that for $i = r - 1$ we have the final condition for the last partial Riccati matrix

$$P_r(t_r) = P(t_f) = S_f.$$

From this terminal condition, we can obtain the inverted time solution of the differential Riccati equation (6.15) for the interval $[t_{r-1}^{opt}, t_f]$ and determine the value $P_r(t_{r-1}^{opt})$. This value is used in system (6.23) for $i = r - 1$ and one calculates

$$P_{r-1}(t_{r-1}^{opt}).$$

It is necessary to stress that the jumps in the Riccati matrices at the time instants $t = t_i^{opt}$ are given by the solutions of system (6.23) and the resulting optimal feedback control $u^{opt}(\cdot)$ from (6.14) is a discontinuous piecewise linear control function.

Let us now summarize a general conceptual computational algorithm for the numerical treatment of the optimal partially linear feedback control in the given HLQ problem (6.9). Note that in the algorithm presented below an approximating trajectory $x^{appr}(\cdot)$ to $x^{opt}(\cdot)$ and the corresponding sequence τ^{appr} to τ^{opt} are assumed to be given. The elements of τ^{appr} approximate the optimal switching times $t_i^{opt} \in \tau^{opt}$ for every $i = 1, ..., r - 1$. A trajectory $x^{appr}(\cdot)$, a sequence τ^{appr}, and the associated sequence of the corresponding locations can be obtained in various ways, for instance with the help of the *gradient-based* algorithms proposed in [19], or using the optimality zone algorithms from [122,289].

Algorithm 6.1.

1) *Consider an approximating trajectory $x^{appr}(\cdot)$, the corresponding sequence τ^{appr}, the sequence of locations, and the terminal condition*

$$P(t_f) = S_f$$

for a given LHS. Set

$$k = 1, \ l = 1.$$

2) *With the help of the inverted time integrating procedure, compute the value* $P_r(t_{r-1}^{opt})$ *of the partial Riccati matrix* P_r. *Using (6.23) calculate the Riccati matrix*

$$P_{r-1}(t_{r-1}^{opt}).$$

3) *By the inverted time integrating solution define*

$$P_{r-k}(t_{r-k-1}^{opt}),$$

increase k by one. If $k = r - 1$, *then go to Step 4. Otherwise go to Step 2.*

4) *Complete all partial Riccati matrices* $P_i(\cdot)$ *and define the corresponding partial gain matrices*

$$C_i(t) = R_{q_i}^{-1}(t) B_{q_i}^T(t) P_i(t).$$

Compute the quasioptimal (in the sense of the above approximations) piecewise feedback control function from (6.14).

5) *Using the obtained quasioptimal feedback control low, compute the corresponding trajectory* $x^l(\cdot)$ *of the* $\mathcal{LHS}$ *under consideration. Determine the new approximating sequence* τ^l *from the conditions*

$$t_i^l := \min\{t \in [0, t_f] \ : \ x^l(t) \bigcap M_{q_i, q_{i+1}} \neq \emptyset\},$$

where $i = 1, ..., r - 1$. *Finally, increase l by one and go to Step 2.*

We are also able to prove the following convergence result for the presented conceptual Algorithm 6.1.

Theorem 6.6. *Under assumptions of Theorem 6.3, there exists an initial approximating trajectory* $x^{appr}(\cdot)$ *such that the sequence of hybrid trajectories generated by Algorithm 6.1 is a minimizing sequence for (6.9).*

The proof of Theorem 6.6 is based on the convexity arguments (see [36]) for the HLQ problem under consideration and on the convergence properties of the general first-order optimization methods in real Hilbert spaces.

6.3.2 Optimization of Impulsive Hybrid Systems

Consider now an IHS from Definition 4.7 and formulate the following HOCP:

$$\begin{aligned}&\text{minimize} \sum_{i=1}^{r} \int_{t_{i-1}}^{t_i} f^0(t, x(t), u(t))dt \\ &\text{over all trajectories } X \text{ of IHS.}\end{aligned} \tag{6.24}$$

Here $f^0 : \mathbb{R} \times \mathbb{R}^n \times \mathbb{R}^m \to \mathbb{R}$ is a continuously differentiable function. Recall that

$$\theta := (\theta_{q_1} ... \theta_{q_r})$$

is a matrix of the jump magnitudes associated with an IHS. The Mayer-type HOCP for an IHS can be written as follows:

$$\begin{aligned}&\text{minimize } \phi(x(t_f)) \\ &\text{over all trajectories } X \text{ of IHS.}\end{aligned} \tag{6.25}$$

Here $\phi(\cdot)$ is a sufficiently smooth objective function.

Using the equivalent representation (4.6) of an IHS (Chapter 4, Section 4.4), we can rewrite (6.25) in the following form:

$$\begin{aligned}&\text{minimize } \phi(y(t_f) + \sum_{i=1}^{r} \theta_{q_i}) \\ &\text{over all trajectories } Y \text{ of HS (4.6).}\end{aligned} \tag{6.26}$$

The basic relation between HOCPs (6.25) and (6.26) is given by the next result.

Theorem 6.7. *Suppose that HOCPs (6.25) and (6.26) have optimal solutions. Under the basic technical assumptions from Section 4.4, every optimal solution*

$$(u^{opt}(\cdot), \theta^{opt}, Y^{opt})$$

of problem (6.26) determines the corresponding optimal solution

$$(u^{opt}(\cdot), \theta^{opt}, \mathbf{X}^{opt}(\cdot))$$

for the initial impulsive problem (6.25). Moreover,

$$\{q_i\}^a = \{q_i\}, \ \tau^a = \tau,$$

and

$$x^{opt}(t) = y^{opt}(t) + \sum_{i=1}^{r} \theta_{qi}^{opt} \eta(t - t_i^{opt}).$$

Here t_i^{opt} are elements of the optimal sequence τ^{opt} and θ_{qi}^{opt} are optimal jumps in the original HOCP (6.25).

Recall that $\eta(\cdot)$ in Theorem 6.7 denotes the classic Heaviside step function. The proof of the above result can be found in [159] and is an immediate consequence of the main results of Section 4.4. The paper cited above also includes a constructive gradient-based computational algorithm for a numerical treatment of the initial (HOCP (6.25)) and auxiliary (HOCP (6.26)) problems.

Note that a similar result can also be obtained for the initially given Bolza-type impulsive HOCP and the following equivalent HOCP:

$$\text{minimize} \sum_{i=1}^{r} \int_{t_{i-1}}^{t_i} f^0\big(t, y(t) + \sum_{i=1}^{r} \theta_{q_i} \eta(t - t_i), u(t)\big)dt \tag{6.27}$$

over all trajectories Y of HS (4.6).

Finally let us present an interesting OCP associated with an SS from Definition 6.8: under a fixed switching sequence of locations $\{q_{i_k}\}k$, solve the following optimization problem:

$$\begin{aligned} &min_\theta J(\theta) := \sum_{k=0}^{K} \int_{t_k}^{t_{k+1}} f_0(x(s))ds \\ &\text{subject to} \\ &\dot{x}(t) = f_{q(t)}(x(t)), \\ &q(t) = q_{i_k},\ t \in [t_k, t_{k+1}), \\ &x(t_0^+) = x(t_0) + \theta_0, \\ &x(t_{k+1}^+) = x(t_{k+1}) + \theta_{k+1},\ S_{(q_{i_k}, q_{i_{k+1}})}(x(t_{k+1})) = 0, \\ &\dot{x}(t) = f_{q_{i_K}}(x(t)),\ t \in [t_K, t_{K+1}], \end{aligned} \tag{6.28}$$

with $x(t_0) = x_0$ and where $K \in \mathbb{N}$ is the total number of switches. The vector θ denotes here the $n \times (K+1)$-dimensional vector

$$(\theta_0^T, ..., \theta_K^T)^T,$$

and $f_0 : \mathbb{R}^n \to \mathbb{R}_+$ is a cost function. Note that the initial and finite time instants are assumed to be given. The necessary optimality conditions for the above OCP are deeply discussed in [40].

6.3.3 On the Convex Switched Optimal Control Problems

An effective application of computational methods can be usually realized under some additional technical assumptions. The main condition for a successful implementation of the gradient method studied in this section is the convexity of the minimization problem under consideration. To put it another way, the original SOCP needs to have a structure of the abstract convex optimization problem (see Chapter 3) in a suitable Hilbert space. Evidently, not every OCP of the type is equivalent to the mentioned convex problem. Therefore, we now restrict the class of SOCPs, consider convex ASSs of the type (6.4), and formulate the corresponding SOCPs. We have

$$\begin{aligned} &\text{minimize } \phi(x(t_f)) \\ &\text{subject to (6.4)}, \\ &q_i \in \mathcal{Q},\ u(\cdot) \in \mathcal{U}, \end{aligned} \tag{6.29}$$

where $\phi : \mathbb{R}^n \to \mathbb{R}$ is a continuously differentiable function. Note that the switched-type OCP (6.29) has a quite general nature. Consider some functions

$$\tilde{A}_i : \mathbb{R}_+ \times \mathbb{R}^n \to \mathbb{R}, \quad i = 1, ..., r,$$

and assume that $\tilde{A}_i(\cdot, \cdot)$ are uniformly bounded on the set $(0, t_f) \times \mathcal{R}$, measurable with respect to $t \in (0, t_f)$ and uniformly Lipschitz continuous in $x \in \mathcal{R} \subseteq \mathbb{R}^n$. Moreover, let us assume that

$$\tilde{B}_i : \mathbb{R}_+ \to \mathbb{R}^m, \quad i = 1, ..., r,$$

are continuous functions. An initially given alternative HOCP with the performance criterion

$$\tilde{\phi}(x(t_f)) + \sum_{i=1}^{r} \int_{t_{i-1}}^{t_i} [\tilde{A}_{q_i}(t, x(t)) + \tilde{B}_{q_i}(t)u(t)]dt,$$

where $q_i \in \mathcal{Q}$ and $\tilde{\phi} : \mathbb{R}^n \to \mathbb{R}$ is a continuously differentiable function, can be reduced to an SOCP of the type (6.29) via simple state augmentation. We have

$$\dot{x}_{n+1}(t) := \sum_{i=1}^{r} \chi_{[t_{i-1},t_i)}(t)[\tilde{A}_{q_i}(t, x(t)) + \tilde{B}_{q_i}(t)u(t)] \text{ a.e. on } [0, t_f],$$

where x_{n+1} is considered as a new component of the extended $(n+1)$-dimensional state vector $\tilde{x} := (x^T, x_{n+1})^T$. The correspondingly transformed performance criterion can now be written as a terminal criterion, i.e.,

$$\phi(\tilde{x}(t_f)) := \tilde{\phi}(x(t_f)) + x_{n+1}(t_f).$$

Note that the above hybrid state space extension is similar to the usual state augmentation technique for the conventional OCPs (see, e.g., [192]).

We now are ready to introduce a specific class of SOCPs.

Definition 6.10. *If the infinite-dimensional minimization problem* (6.29) *is equivalent to a convex optimization problem* (2.1), *then we call* (6.29) *a convex SOCP.*

Our aim now is to find some constructive characterizations of the elements of the hybrid problem (6.29) that make it possible to establish the convexity of this OCP in the sense of Definition 6.10.

Theorem 6.8. *Let the control system from* (6.29) *be a convex affine system and let the function $\phi(\cdot)$ be convex and monotonically nondecreasing. Then the corresponding* SOCP (6.29) *is convex.*

Proof. Using the convexity and monotonicity of the function ϕ and the convexity of

$$V_l(u(\cdot)),\ u(\cdot) \in \mathcal{U},\ t \in [0, t_f],\ l = 1, ..., n,$$

we can prove that the functional $J(\cdot)$, where

$$J(u(\cdot)) = \phi(x^u(t_f)),$$

is convex. Since U is a convex subset of $\mathbb{R}^m$, the set $\mathcal{U}$ is also convex. Therefore, problem (1.2) is equivalent to a classical convex program of the type (2.1). The proof is completed. □

Evidently, Theorem 6.8 provides a basis for a numerical consideration of convex SOCPs from the point of view of the classical convex programming (see Chapter 3).

Let us now discuss shortly the applicability of the obtained convexity results. Consider the following simple initial value problem involving a one-dimensional Riccati differential equation:

$$\dot{x}(t) = \sum_{i=1}^{r} \chi_{[t_{i-1},t_i)}(t)(a_{q_i} t x^2(t) + b_{q_i} u(t)),$$
$$x(0) = x_0 > 0,$$

where $x(t) \in \mathbb{R}$, $u(t) \in \mathbb{R}$, $0 \le u(t) \le 1$, and the coefficients a_{q_i}, b_{q_i} are positive constants. Note that in this case the basic assumptions for (6.29) are fulfilled. Since

$$A_{q_i}(t,x) = a_{q_i} t x^2$$

is a locally Lipschitz function, the given initial value problem is locally solvable. Under the above positivity assumptions, $A_{q_i}(t,x)$ can be defined on the open set $\mathbb{R}_+ \times \mathbb{R}_+$. This nonlinear function is convex and monotonically nondecreasing. Since the conditions of Theorem 6.8 are satisfied, the control system under consideration is a convex affine system. Note that the same consideration can also be made for a more general controllable Riccati-type system (1.1) that admits positive solutions. Next we examine an example of a simplified controllable "explosion" model (see [34]). We have

$$\dot{x}(t) = \sum_{i=1}^{r} \chi_{[t_{i-1},t_i)}(t)(a^1_{q_i} e^{a^2_{q_i} x(t)} + a^3_{q_i} x(t) + b_{q_i} u(t)),$$
$$x(0) = x_0 > 0, \quad 0 \leq u(t) \leq 1$$

with some positive scalar parameters $a^1_{q_i}$, $a^2_{q_i}$, $a^3_{q_i}$, and b_{q_i}, $q_i \in \mathcal{Q}$. This dynamic model also constitutes a convex affine system.

Finally note that our convexity result, namely, Theorem 6.8, primarily involves the nonlinearly affine systems (6.4). The convexity properties of a linear SS and the associated OCP of the type (6.4) can be established under some weaker assumptions and are in fact simple consequences of the theory developed in [54].

6.3.4 Pontryagin-Type Maximum Principle for Hybrid and Switched Optimal Control Problems

Consider now an IHS from Definition 4.7 and the corresponding HOCP (6.24). For both HSs, namely, for the originally given IHS (4.5) and (4.6), we introduce the extended control vector

$$v(\cdot) := (u(\cdot), \theta),$$

where $\theta := (\theta_{q_1}, ..., \theta_{q_r})$. An admissible extended control vector $v(\cdot)$ satisfies the natural restrictions

$$u(\cdot) \in \mathcal{U}, \; ||\theta_{q_i}|| \leq \Theta_{q_i}, \; i = 1, ..., r.$$

We next denote an optimal extended control vector (if it exists) by $v^{opt}(\cdot)$ and the corresponding components by $u^{opt}(\cdot)$ and θ^{opt}. The aim of this section is to formulate a Pontryagin maximum principle for the initially given impulsive HOCP (6.24). From the analytic results of [39] we firstly obtain a Pontryagin maximum principle for the auxiliary (Bolza-type) HOCP (6.27).

Theorem 6.9. *Let the functions f^0, f_{q_i} be continuously differentiable and the auxiliary Bolza-type HOCP (6.27) be Lagrange regular. Then there exist a function $\psi_i(\cdot)$ from $\mathbb{W}_n^{1,\infty}(0, t_f)$ and a vector*

$$a = (a_1 \dots a_{r-1})^T \in \mathbb{R}^{r-1}$$

such that

$$\begin{aligned} \dot{\psi}_i(t) &= -\frac{\partial H_{q_i}(y_i^{opt}(t), v^{opt}(t), \psi(t))}{\partial (y + \sum_{j=i}^r \theta_{q_j})} \text{ a. e. on } (t_{i-1}^{opt}, t_i^{opt}), \\ \psi_r(t_f) &= 0, \end{aligned} \tag{6.30}$$

and

$$\psi_i(t_i^{opt}) = \psi_{i+1}(t_i^{opt}) + \left(a_i \frac{\partial m_{q_i, q_{i+1}}(y^{opt}(t_i^{opt}) + \sum_{j=i}^r \theta_{q_j}^{opt})}{\partial (y + \sum_{j=i}^r \theta_{q_j})} \right), i = 1, ..., r-1. \tag{6.31}$$

Moreover, for every admissible control $v(\cdot)$ the following inequality is satisfied:

$$\begin{aligned} &\left(\frac{\partial H_{q_i}(y^{opt}(t), v^{opt}(t), \psi(t))}{\partial v}, (v(t) - v^{opt}(t)) \right) \leq 0 \\ &\text{a.e. on } [t_{i-1}^{opt}, t_i^{opt}],\ i = 1, ..., r, \end{aligned} \tag{6.32}$$

where

$$H_{q_i}(y, v, \psi) := \left(\psi_i, f_{q_i}\left(t, y + \sum_{i=1}^r \theta_{q_i} \eta(t - t_i), u\right)\right) - f_{q_i}^0\left(t, y + \sum_{i=1}^r \theta_{q_i} \eta(t - t_i), u\right)$$

is a "partial" Hamiltonian for the location $q_i \in \mathcal{Q}$ and $(\cdot, \cdot)$ denotes the corresponding scalar product.

Using the one-to-one correspondence between the solutions $x^{opt}(\cdot)$ and $y^{opt}(\cdot)$ of the initial value problems (4.5) and (4.6) established by Theorem 4.13 (see Section 4.4), we are now able to formulate the necessary optimality conditions for the original impulsive (Bolza-type) HOCP (6.24).

Theorem 6.10. *Let functions f^0, f_{q_i} be continuously differentiable and the originally given HOCP (6.24) be regular. Then there exist a function*

$$p_i(\cdot) \in \mathbb{W}_n^{1,\infty}(0, t_f)$$

and a vector

$$b = (b_1 \dots b_{r-1})^T \in \mathbb{R}^{r-1}$$

such that

$$\dot{p}_i(t) = -\frac{\partial H_{q_i}(x_i^{opt}(t), v^{opt}(t), p(t))}{\partial x} \text{ a.e. on } (t_{i-1}^{opt}, t_i^{opt}),$$
$$p_r(t_f) = 0, \tag{6.33}$$

and

$$p_i(t_i^{opt}) = p_{i+1}(t_i^{opt}) + \left(b_i \frac{\partial m_{q_i,q_{i+1}}(x^{opt}(t_i^{opt}))}{\partial x} \right), i = 1, ..., r-1. \tag{6.34}$$

Moreover, for every admissible control $v(\cdot)$ the following inequalities are satisfied:

$$\left(\frac{\partial H_{q_i}(x^{opt}(t), v^{opt}(t), p(t))}{\partial u}, (u(t) - u^{opt}(t)) \right) \le 0,$$
$$\left(\frac{\partial H_{q_i}(x^{opt}(t), v^{opt}(t), p(t))}{\partial \theta}, (\theta - \theta^{opt}) \right) \le 0 \tag{6.35}$$
$$\text{a.e. on } [t_{i-1}^{opt}, t_i^{opt}], \; i = 1, ..., r,$$

where

$$H_{q_i}(y, v, p) := \left(p_i, f_{q_i}\left(t, x, u\right) + \theta_{q_i}\delta(t - t_i)\right) - f_{q_i}^0\left(t, x, u\right)$$

is a "partial" Hamiltonian for the location $q_i \in \mathcal{Q}$ and $(\cdot, \cdot)$ denotes the corresponding scalar product.

Note that Theorem 6.10 is a formal consequence of Theorem 6.9. A complete proof of Theorem 6.10 can be found in [255].

We have derived the necessary optimality conditions (6.30)–(6.32) and (6.33)–(6.35) (Theorem 6.9 and Theorem 6.10) and obtained the generic Hamiltonian minimization conditions. These conditions, namely, relations (6.32) and (6.35), are written in the form of variational inequalities. It is well known that variational inequalities play an important role in optimization theory. We refer to [70] for details. For the numerical treatment of variational inequality see also [202]. It is also well known that the variational inequality (6.32) is equivalent to the following equation:

$$v^{opt}(t) = \Pi_W \left(v^{opt}(t) - \alpha \frac{\partial H_{q_i}(x_i^{opt}(t), v^{opt}(t), p(t))}{\partial v} v^{opt}(t) \right), \tag{6.36}$$

where $\alpha > 0$ and Π_W is a projection operator on the set $U \times U_\theta$. Here U_θ is the set of admissible jumps given by the inequality constraints

$$||\theta_{q_i}|| \le \Theta_{q_i}, \; i = 1, ..., r.$$

To solve (6.36) one can use a variety of gradient-type algorithms with a projection procedure. Let N be a sufficiently large positive integer number and

$$G_N := \{t^0 = 0, t^1, \ldots, t^N = T\}$$

be a (possibly nonequidistant) partition of $[0, T]$ with

$$\max_{0 \le k \le N-1} |t^{k+1} - t^k| \le \epsilon$$

for a given accuracy constant ϵ. For every control function $u(\cdot) \in \mathcal{U}$ we introduce the piecewise constant control signals $u^N(\cdot)$ such that

$$u^n(t) := \sum_{k=0}^{N-1} \eta_k(t) u^k, \ u^k = u(t^k), \ k = 0, \ldots, N-1, \ t \in [0, t_f],$$

$$\eta_k(t) = \begin{cases} 1, & \text{if } t \in [t^k, t^{k+1}], \\ 0, & \text{otherwise.} \end{cases}$$

Then for an approximate solution of Eq. (6.36) we can consider the following finite-dimensional gradient method:

$$\begin{aligned} & u^{N,0} \in U, \ \theta^{N,0} \in U_\theta, \\ & u^{N,(s+1)} = \Pi^1_U \left(u^{N,s} - \alpha_1 \frac{\partial H_{q_i}(x_i^{N,s}(t), (u^{N,s}(t), \theta^{N,s}), p^{N,s}(t))}{\partial u} u^{N,s} \right), \\ & \theta^{N,s+1} = \Pi^2_{U_\theta} \left(\theta^{N,s} - \alpha_2 \frac{\partial H_{q_i}(x_i^{N,s}(t), (u^{N,s}(t), \theta^{N,s}), p^{N,s}(t))}{\partial \theta} \theta^{N,s} \right), \end{aligned} \tag{6.37}$$

where $\alpha_1, \ \alpha_2$ are some positive constants and

$$s = 0, \ldots, \ x_i^{N,s}(\cdot), \ p^{N,s}(\cdot)$$

are solutions of the corresponding initial and boundary value problems in the actual location q_i. Here Π^1 and Π^2 are partial projection operators on the set U and U_θ, respectively. The iteration of the extended control vector is denoted as $(u^{N,s}(\cdot), \theta^{N,s})$. Evidently, U and U_θ are convex sets. Note that in every step of the gradient algorithm (6.37) we need to solve the corresponding boundary value problem from the main Theorem 6.10 (the Pontryagin-type maximum principle).

6.4 Numerical Approaches to Optimal Control Problems of Hybrid and Switched Systems

6.4.1 The Mayer-Type Hybrid Optimal Control Problem

Given a hybrid control system from Definition 1.1 we now consider a "discrete" trajectory $\mathcal{R}$ from Definition 1.2. We next call $\mathcal{R}$ an admissible discrete trajectory associated with the given HS if there exists an open bounded set

$$\mathcal{V} \subset \mathcal{U}$$

such that for every $u(\cdot)$ from $\mathcal{V}$ we obtain the same discrete trajectory $\mathcal{R}$. Assume now that the HS under consideration possesses an admissible discrete trajectory $\mathcal{R}$. Moreover, let $\phi : \mathbb{R}^n \to \mathbb{R}$ be a continuously differentiable function. We now formulate the following Mayer-type HOCP:

$$\begin{aligned} &\text{minimize } \phi(x(1)) \\ &\text{subject to } \dot{x}(t) = f_{q_i}(t, x(t), u(t)) \text{ a.e. on } [t_{i-1}, t_i] \\ &q_i \in \mathcal{R},\ i = 1, ..., r+1,\ x(0) = x_0 \in M_{q_1}, \\ &u(\cdot) \in \mathcal{U}. \end{aligned} \tag{6.38}$$

Note that (6.38) deals with the minimizing of the Mayer cost functional

$$J(\mathcal{X}) := \phi(x(1))$$

over all "continuous" trajectories $x(\cdot)$ of the HS under consideration. Note that we study an HOCP in the absence of possible target (endpoint) and additional state constraints. For some related general necessary optimality conditions in the form of a HMP we refer to [122,255].

The abstract results from Section 6.1 make it possible to formulate first-order necessary optimality conditions for the class of HOCPs under consideration. These results also provide a basis for creating the related computational algorithms. Consider an HS from (6.38). For an admissible control function $u(\cdot) \in \mathcal{U}$ and a fixed admissible discrete trajectory $\mathcal{R}$ we obtain (by integration) the complete hybrid trajectory $\mathcal{X}^u$. For every interval $[t_{i-1}, t_i]$, where $t_i \in \tau$, we can define the characteristic function associated with $\mathcal{R}$ (see also Chapter 1), i.e.,

$$\beta_{[t_{i-1},t_i)}(t) = \begin{cases} 1 & \text{if } t \in [t_{i-1}, t_i), \\ 0 & \text{otherwise.} \end{cases}$$

The differential equations from Definition 1.2 can now be written in the compact form

$$\dot{x}(t) = \sum_{i=1}^{r+1} \beta_{[t_{i-1},t_i)}(t) f_{q_i}(t, x(t), u(t)), \tag{6.39}$$

where $x(0) = x_0$ and $q_i \in \mathcal{R}$. We consider the OCP (6.38) and the dynamic system (6.39) on the fixed time interval $[0, 1]$. Under the above assumptions for the family of vector fields F (from the basic Definition 1.1), the right-hand side of the obtained differential equation (6.39) satisfies the conditions of the extended Carathéodory theorem (see, e.g., [9,25,32] and Chapter 5). Therefore, there exists a unique (absolutely continuous) solution of (6.39). We now apply the abstract Theorem 6.1 to the HOCP (6.38). In the case of the hybrid control system under consideration we can evidently specify the concrete Banach spaces

$$X = \mathbb{W}_n^{1,\infty}([0, 1]), \; Y = \mathbb{L}_m^{\infty}([0, 1]).$$

Recall that by $\mathbb{W}_n^{1,\infty}([0, 1])$ we denote the Sobolev space of all absolutely continuous functions with essentially bounded derivatives (see Chapter 5). Let us introduce the following system operator:

$$P : \mathbb{W}_n^{1,\infty}([0, 1]) \times \mathbb{L}_m^{\infty}([0, 1]) \to \mathbb{W}_n^{1,\infty}([0, 1]) \times \mathbb{R}^n,$$

where

$$P(x(\cdot), u(\cdot))\Big|_t := \begin{pmatrix} \dot{x}(t) - \sum_{i=1}^{r+1} \beta_{[t_{i-1}, t_i)}(t) f_{q_i}(t, x(t), u(t)) \\ x(0) - x_0 \end{pmatrix}.$$

The corresponding operator equation

$$P(x(\cdot), u(\cdot)) = 0$$

is evidently consistent with the abstract state equation from the general optimization problem (6.1) in the abstract Banach spaces $X, \; Y$.

Under some mild assumptions the trajectories of a conventional control system described by ordinary differential equations with sufficiently smooth right-hand sides are Fréchet differentiable mappings with respect to control functions. We next establish the same differentiability property for a "continuous" trajectory of the given HS (under the assumption of a prescribed "discrete" trajectory $\mathcal{R}$).

Theorem 6.11. *Assume that all above hypotheses are satisfied. Then the functional mapping*

$$u(\cdot) \to x(\cdot),$$

where

$$u(\cdot) \in \mathcal{V}$$

and $x(\cdot)$ is the corresponding solution of (6.39) for a prescribed $\mathcal{R}$, is Fréchet differentiable on $\mathcal{V}$.

Proof. Consider the first location $q_1 \in \mathcal{R}$ and the vector field f_{q_1}. Let

$$u(\cdot) \in \mathcal{V}$$

and $\{u^k(\cdot)\}$ be a sequence of admissible controls from $\mathcal{U}$ such that

$$\lim_{k \to \infty} ||u^k(\cdot) - u(\cdot)||_{\mathbb{L}^{\infty}_m([0,1])} = 0.$$

The corresponding solutions to the initial value problems from (6.38) are denoted by $x(\cdot)$ and $x^k(\cdot)$, respectively. The continuity property of the vector fields from F (see Definition 1.1) implies

$$\int_{I_1} ||f_{q_1}(t, x(t), u^k(t)) - f_{q_1}(t, x(t), u(t))|| \to 0 \tag{6.40}$$

as

$$u^k(\cdot) \to u(\cdot)$$

in the norm of $\mathbb{L}^{\infty}_m([0, 1]) = 0$. Here

$$I_1^k := \max\{[t_0, t_1), [t_0, t_1^k)\}$$

and

$$t_1^k := \min\{t \in [0, 1] \ : \ x^k(t) \bigcap M_{q_1} \neq \emptyset\}$$

is the first switching time for the trajectory $x^k(\cdot)$. Considering (6.40) and the basic assumptions from this section, we can next apply the classic theorem on continuous dependence of solutions of differential equations on a parameter (on the topological space $\mathbb{L}^{\infty}_m([0, 1])$). This fundamental fact (see Chapter 5) and the smoothness property of the switching manifolds from M (see Definition 1.1) imply the convergence of the above trajectories and the associated switching times

$$\sup_{t \in I_1^k} ||x^k(t) - x(t)|| \to 0, \quad t_1^k \to t_1 \tag{6.41}$$

as $k \to \infty$. Moreover, we have

$$x^k(t_1^k) \to x(t_1).$$

We now can repeat the above theoretic arguments for the next location q_2 from $\mathcal{R}$ and additionally use the classic theorem on continuous dependence of solutions of differential equations on the initial data $(t_1^k, x^k(t_1^k))$. This makes it possible to establish the convergence property on the next time interval (similar to (6.41))

$$\sup_{t \in I_2^k} ||x^k(t) - x(t)|| \to 0,$$

where $k \to \infty$ and I_2^k is defined similarly to I_1^k.

Applying the proof by induction over the locations

$$q_i \in \mathcal{R}_{r+1}, \ i = 1, ..., r+1,$$

we finally establish the continuity of the full mapping

$$u(\cdot) \to x(\cdot)$$

defined on $\mathcal{V}$. The Fréchet differentiability of $x(\cdot)$ with respect to $u(\cdot) \in \mathcal{V}$ can now be proved similarly to the conventional (nonhybrid) case (see, e.g., [29] for details). □

Consider now a regular OCP (6.38) and introduce the associated Hamiltonian

$$H(t, x, u, p) = \langle p, \sum_{i=1}^{r+1} \beta_{[t_{i-1}, t_i)}(t) f_{q_i}(t, x, u) \rangle,$$

where $p \in \mathbb{R}^n$. Since every admissible control $u(\cdot)$ determines a unique hybrid trajectory $\mathcal{X}^u$ with a prescribed $\mathcal{R}$, the objective functional

$$\tilde{J} : \mathcal{U} \to \mathbb{R},$$

where

$$\tilde{J}(u(\cdot)) := J(\mathcal{X}^u) = \phi(x(1)),$$

is well defined. Under conditions of Theorem 6.11, the differentiability of the given function $\phi(\cdot)$ implies the differentiability of $\tilde{J}(\cdot)$ (as a superposition of two Fréchet differentiable functions). The corresponding derivative is denoted by $\nabla \tilde{J}$. In the particular case of the given HOCP (6.38) the necessary calculation of the adjoint operator $\nabla \tilde{J}^*$ to the gradient $\nabla \tilde{J}$ is relatively easy. We now present a concrete result that constitutes a formal consequence of Theorem 6.11.

Theorem 6.12. *Consider a regular HOCP* (6.38). *The gradient* $\nabla \tilde{J}^*(u(\cdot))$ *can be found as a solution to the following system of equations:*

$$\begin{aligned} &\dot{x}(t) = H_p(t, x(t), u(t), p(t)), \\ &x(0) = x_0, \\ &\dot{p}(t) = -H_x(t, x(t), u(t), p(t)), \\ &p(1) = -\phi_x(x(1)), \\ &\nabla \tilde{J}^*(u(\cdot))(t) = -H_u(t, x(t), u(t), p(t)), \end{aligned} \tag{6.42}$$

where $p(\cdot)$ *is an absolutely continuous function (an "adjoint variable").*

Proof. The Lagrangian

$$\mathcal{L} : \mathbb{W}_n^{1,\infty}([0,1]) \times \mathbb{L}_m^{\infty}([0,1]) \times \mathbb{R}^n \times \mathbb{W}_n^{1,\infty}([0,1]) \to \mathbb{R}^n$$

associated with the regular problem (6.38) can be written as

$$\mathcal{L}(x(\cdot), u(\cdot), \hat{p}, p(\cdot)) = \phi(x(1)) + \langle \hat{p}, x(0) - x_0 \rangle + \langle p(t), \dot{x}(t) -$$
$$\sum_{i=1}^{r+1} \beta_{[t_{i-1}, t_i)}(t) f_{q_i}(t, x(t), u(t)) \rangle,$$

where the adjoint variable contains two components $\hat{p} \in \mathbb{R}^n$ and $p(\cdot)$. If we differentiate the above Lagrange function with respect to the adjoint variable, then we obtain the first equation from (6.42), i.e.,

$$\dot{x}(t) = \sum_{i=1}^{r+1} \beta_{[t_{i-1}, t_i)}(t) f_{q_i}(t, x(t), u(t)) = H_p(t, x(t), u(t), p(t)).$$

Here $x(0) = x_0$.

Consider now the term

$$\int_0^1 \langle p(t), \dot{x}(t) \rangle dt.$$

From integration by parts we have

$$\int_0^1 \langle p(t), \dot{x}(t) \rangle dt = \langle p(1), x(1) \rangle -$$
$$\langle p(0), x(0) \rangle - \int_0^1 \langle \dot{p}(t), x(t) \rangle.$$

Hence

$$\begin{aligned} \mathcal{L}(x(\cdot), u(\cdot), \hat{p}, p(\cdot)) &= \phi(x(1)) + \langle p(1), x(1) \rangle + \langle \hat{p} - p(0), x(0) \rangle - \\ &\langle \hat{p}, x_0 \rangle - \int_0^1 \langle \dot{p}(t), x(t) \rangle dt \\ &+ \int_0^1 \langle p(t), \sum_{i=1}^{r+1} \beta_{[t_{i-1}, t_i)}(t) f_{q_i}(t, x(t), u(t)) \rangle dt. \end{aligned} \tag{6.43}$$

If we differentiate $\mathcal{L}$ in (6.43) with respect to $x(\cdot)$, we can use the abstract Theorem 6.11 and compute the partial derivative $\mathcal{L}_x$. We have

$$\mathcal{L}_x(x(\cdot), u(\cdot), \hat{p}, p(\cdot))h(\cdot) = \langle \phi_x(x(1))h(1) \rangle + \langle p(1), h(1) \rangle -$$
$$\int_0^1 \langle \dot{p}(t), h(t) \rangle dt$$
$$- \int_0^1 \langle h(t), \sum_{i=1}^{r+1} \beta_{[t_{i-1}, t_i)}(t) \frac{\partial f_{q_i}(t, x(t), u(t))}{\partial x} p(t) \rangle dt,$$

where $h(\cdot)$ is an element of the (topological) dual space to the Sobolev space $\mathbb{W}_n^{1,\infty}([0, 1])$. Therefore, we finally have

$$\mathcal{L}_x^* = (\phi_x(x(1)) + p(1)) - \int_0^1 (\dot{p}(t) + H_x(t, x(t), u(t), p(t)))dt.$$

Put now

$$\mathcal{L}_x^* = 0$$

and derive the second differential equation and the corresponding boundary condition in (6.42).

Using the obtained relation (6.43), we also obtain

$$\mathcal{L}_u(x(\cdot), u(\cdot), \hat{p}, p(\cdot))v(\cdot) = - \int_0^1 H_u(t, x(t), u(t))v(t)dt$$

for every

$$v(\cdot) \in \mathbb{L}_m^\infty([0, 1]).$$

By Theorem 6.11, we obtain the last relation in (6.42), i.e.,

$$\nabla \tilde{J}^*(u(\cdot))(t) = \mathcal{L}_u^*(x(\cdot), u(\cdot), \hat{p}, p(\cdot)) = -H_u(t, x(t), u(t)).$$

The proof is finished. □

Summarizing this section we can note that the obtained theoretic result allows an explicit computation of the gradient $\nabla \tilde{J}$ (or $\nabla \tilde{J}^*$) in sophisticated HOCPs of the type (6.38).

6.4.2 Numerics of Optimal Control

In Section 6.4.1 we have developed an explicit formalism for the reduced gradient of the cost functional in HOCPs (6.38). To make a step forward in the study of this dynamic optimization we discuss the necessary optimality conditions and some related numerical aspects. Let us formulate an easy consequence of Theorem 6.12.

Theorem 6.13. *Assume that the HOCP* (6.38) *has an optimal solution* $(u^{opt}(\cdot), x^{opt}(\cdot))$. *Then this solution satisfies the following:*

$$\begin{aligned}
&\dot{x}^{opt}(t) = H_p(t, x^{opt}(t), u^{opt}(t), p(t)),\\
&x(0) = x_0,\\
&\dot{p}(t) = -H_x(t, x^{opt}(t), u^{opt}(t), p(t)),\\
&p(1) = -\phi_x(x^{opt}(1)),\\
&H_u(t, x^{opt}(t), u^{opt}(t), p(t)) = 0, \;\; u^{opt}(\cdot) \in \mathcal{V}.
\end{aligned} \tag{6.44}$$

The formal proof of Theorem 6.13 can be obtained as a combination of the result of the basic Theorem 6.12 and the standard facts from the classic optimization theory (see [29,34]).

Evidently, the conditions (6.44) from Theorem 6.13 constitute the necessary optimality conditions for the specific case of problem (6.38). These conditions are similar to the exact Pontryagin-type Maximum Principle (the HMP; see [255,296]). Note that the last equation in (6.44) represents in fact the usual first-order optimality condition

$$\nabla \tilde{J}^*(u(\cdot))(t) = 0, \;\; t \in [0, 1]$$

if the optimal control takes values in an open bounded control set. Note that Theorem 6.12 and Theorem 6.13 provide an analytic basis for a wide class of the gradient-based optimization algorithms for (6.38). Let us assume (in a specific case) that the control set $\mathcal{V}$ possesses the usual box form, namely,

$$\mathcal{V} := \{u \in \mathbb{R}^m \; : \; b_-^j \leq u_j \leq b_+^j, \;\; j = 1, ..., m\},$$

where $b_-^j, b_+^j, j = 1, ..., m$, are some given constants. The generic gradient algorithm in $\mathbb{L}_m^\infty([0, 1])$ can now be easily expressed. We have

$$\begin{aligned}
&u^{k+1}(t) = u^k(t) + \gamma_k H_u(t, x(t), u(t), p(t)), \;\; t \in [0, 1],\\
&b_-^j \leq u_j^{k+1}(t) \leq b_+^j, \;\; j = 1, ..., m, \;\; k = 0, 1, ...,\\
&u^0(\cdot) \in \mathcal{V},
\end{aligned} \tag{6.45}$$

where γ_k is a step size of the gradient algorithm and

$$\{u^k(\cdot)\} \subset \mathbb{L}_m^\infty([0, 1])$$

is the sequence of iterations. Note that a practical realization of (6.45) can be used for a numerical treatment of the HOCPs with bounded admissible control inputs. As usual a feasible control function $u^{k+1}(\cdot)$ can be obtained by a projection

$$u^{k+1}(t) = P_{\mathcal{V}}[u^k(t) + \gamma_k H_u(t, x(t), u(t), p(t))], \tag{6.46}$$

where $P_{\mathcal{V}}$ is a projection operator on the (convex) box-type set $\mathcal{V}$.

For the projected gradient algorithm and for convergence properties of (6.46) and of some related gradient-type optimization procedures see the references at the end of this chapter. We also refer to Chapter 3. Note that there are several rules for a practical selection of a suitable step size γ_k in numerical schemes (6.45) and (6.46). In this section we consider a conceptual computational scheme that possesses a numerical consistence property.

Algorithm 6.2. *Make the following steps.*

1) *Choose an admissible initial control* $u^0(\cdot) \in \mathcal{V}$ *and the corresponding continuous trajectory* $x^0(\cdot)$. *Set* $l = 0$.
2) *For the given* $\mathcal{R}$ *determine the hybrid trajectory*

$$\mathcal{X}^{u^l} = (\tau^l, x^l(\cdot), \mathcal{R}),$$

solve the system of equations (6.44), *and compute*

$$\nabla \tilde{J}^*(u^l(\cdot))(t) \ \forall t \in [0, 1].$$

3) *Using* $\nabla \tilde{J}^*(u^l(\cdot))$, *compute the iteration* $u^{l+1}(t)$ *as in (6.46). Increase* l *by one and go to* Step 2.

Algorithm 6.2 constitutes a conceptual computational scheme. This approach need to be concretized for the concrete HOCPs. For example, in the presented optimization model the switching times, number of switches, and switching sets in the given HOCP (6.38) are assumed to be unknown. Using the iterative structure of the proposed Algorithm 6.2, one can compute the corresponding approximations of the optimal trajectory, optimal switching times, and optimal switching sets.

Note that the most critical point of the procedure described in Algorithm 6.2 is a constructive characterization of the feasible set $\mathcal{V}$. This characterization determines the projection operator $P_{\mathcal{V}}$ and the initial iteration $u^0(\cdot)$. Note that a suitable selection of the initial data is a general problem of many effective iterative algorithms. The usual complexity of the HSs and the associated HOCPs makes it impossible to elaborate a general selection rule for the initial approximation $u^0(\cdot)$. The calculation of this initial iteration is a heuristical part of the above algorithm and it needs to be determined for a specific OCP.

Taking into consideration the above theoretical results, namely, Theorem 6.12 and Theorem 6.13, the practical implementation of the gradient-based procedure can be rewritten as follows:

$$u^{k+1}(t) = P_U[u^k(t) + \gamma_k H_u(t, x(t), u(t), p(t))], \tag{6.47}$$

where P_U is a projection operator on the convex control set $\mathcal{U}$ and γ_k is sufficiently small. Since $\mathcal{V}$ is assumed to be open and $u^k(\cdot) \in \mathcal{V}$, there exists a step size γ_k such that $u^{k+1}(\cdot) \in \mathcal{V}$. Let us now present the following convergence result for the proposed Algorithm 6.2.

Theorem 6.14. *Assume that the regular HOCP* (6.38) *satisfies all the basic hypotheses and that* $(u^{opt}(\cdot), x^{opt}(\cdot))$ *is a unique optimal solution of this problem. Let us additionally assume that every function*

$$\partial f_q(t, x, \cdot)/\partial u, \; q \in \mathcal{Q},$$

is a Lipschitz continuous function. Consider a sequence of hybrid trajectories $\{\mathcal{X}^{u^k}\}$ *generated by* Algorithm 6.2 *associated with (6.47). Then there exist admissible initial data* $(u^0(\cdot), x^0(\cdot))$ *and a sequence of the step sizes* $\{\gamma_k\}$ *for scheme (6.46) such that* $\{\mathcal{X}^{u^k}\}$ *is a minimizing sequence for the HOCP* (6.38), *i.e.,*

$$\lim_{k\to\infty} J(\mathcal{X}^{u^k}) = \phi(x^{opt}(1)).$$

We now sketch a proof of Theorem 6.14. The Lipschitz continuity of the functions from the family F (see Definition 1.1) and the last equation in (6.44) guarantee the Lipschitz property of the gradient $\nabla \tilde{J}^*(u(\cdot))$. Moreover, from the boundedness of the control set $\mathcal{V}$ follows the boundedness of the functional $\tilde{J}^*(u(\cdot))$. Using the dominated convergence theorem and the convergence properties of the standard gradient algorithm in $\mathbb{L}^p$ spaces (see, e.g., [5,20,34,41], Chapter 3), we obtain the result of Theorem 6.14. Evidently, the above result establishes the consistency of the presented first-order computational scheme.

From the general necessary optimality conditions discussed in Section 6.3 naturally follows the corresponding "indirect" numerical scheme for various types of HOCPs.

Recall that computational methods based on the Bellman optimality principle were among the first proposed for OCPs [78,110,111]. These methods are especially attractive when an OCP is discretized a priori and the discrete version of the Bellman equation is used to solve it. Dynamic programming algorithms are widely used specifically in engineering and operations research (see [87,230]). The relation between the Pontryagin-type HMP and Bellman methodology for HSs is discussed in [40,123]. Different approximation schemes have been proposed to cope with the "curse of dimensionality" (Bellman's own phrase) with rather limited success. However, the Bellman-type numerical methods are in fact inefficient for OCPs with constraints.

Application of necessary conditions of optimal control theory, specifically various applications of the celebrated Pontryagin maximum principle, yields a boundary value problem for ordinary differential equations. The necessary optimality conditions and the corresponding boundary value problems play an important role in numerics of optimal control. There is a body of work focusing on the numerical treatment of the boundary value problem for the Hamilton–Pontryagin system of differential equations (see [155,244,284] and the references therein). Some of the first numerical methods for OCPs were based on the classical shooting method and on the advanced modification of the generic shooting method [112,113,253]. This family of numerical techniques, which was extended to a class of the problems with state constraints (the multiple shooting method) (see [112,113,253]), usually guarantees very accurate numerical solutions provided that good initial guesses for the adjoint variable $p(0)$ are available. In practice, one needs to integrate the Hamilton–Pontryagin system (for the classic as well as the hybrid OCPs) and then adjust the chosen initial value of the adjoint variable such that the computed $p(\cdot)$ satisfies all terminal conditions imposed on $p(t_f)$. In the view of convergence properties and numerical consistence, the multiple shooting scheme has all advantages and disadvantages of the family of Newton methods. The general disadvantage of the indirect numerical methods (methods based on the Hamilton–Pontryagin system) is the possible local nature of the obtained optimal solution. On the other hand there is also a body of work focusing on the so-called Sakawa-type algorithms (see, e.g., [284,101]). This group of methods uses, in fact, the idea of augmented Hamiltonian for regularizing the generic Pontryagin maximizing problem. For the closely related family of Chernousko methods see also [128,286].

An OCP (classic or hybrid) with state constraints can also be solved by using some modern numerical algorithms of nonlinear programming. For example, the implementation of the interior point method is presented in [322]. In [82] the proximal point methods are used for solving stochastic and in [27,29,30,32,34] for solving some classes of deterministic OCPs. For a general survey of applications of nonlinear programming to optimal control see [90]. Although there is extensive literature on computational optimal control, relatively few results have been published on the application of a sequential quadratic programming (SQP)-type optimization algorithm to OCPs (see [252] and references therein). Note that calculation of second-order derivatives of the objective functional and of constraints in (1.2) can be avoided by applying an SQP-based scheme in which these derivatives are approximated by quasi-Newton formulae. It is evident that in this case one deals with all usual disadvantages of the Newton-based methods.

The gradient-type algorithms (see [259]) are the methods based on the evaluation of the gradient for the objective functional in an OCP. The first applications of the gradient methods in optimal control are discussed in [72]. See also the work of Bryson and Denham [111] and the book of Teo, Goh, and Wong [300]. The gradient of the objective functional can be computed by solving the state equations such that the obtained trajectory is then used to integrate the adjoint equations backward in time. The state and the adjoint variables are used to calculate the gradient of the functional with respect to control variable only. This is possible due to the fact that the trajectories of a control system are uniquely determined by controls. The gradient methods are widely used and are regarded as the most reliable, though not very accurate, methods for OCPs [149,204]. There are two major sources of the inaccuracies in the gradient-type methods: errors caused by integration procedures applied to the state equations and errors introduced by the optimization algorithm. In the works [27,29,32] of the author we also applied the evaluation of the gradient for the objective functional and the gradient method to OCPs with ordinary differential equations. In [36] we use gradients in a solution procedure for OCPs with partial differential equations. Note that the computational gradient-type methods are also studied in connection with a suitable numerical approach to relaxed OCPs [301]. The computational aspect of the theory of β-relaxation proposed in [36] also contains the technique of reduced gradients.

When solving an OCP with ordinary differential equations we deal with functions and systems which, except in very special cases, are replaced by numerically tractable approximations. The implementation of our numerical schemes for original and relaxed control problems is based on the discrete approximations of the control set and on the finite-difference approximations. Finite-difference methods turn out to be a powerful tool for theoretical and practical analysis of control problems (see, e.g., [247,259,148]). Discrete approximations can be applied directly to the problem at hand or used in the solution procedure [246,145,269].

The numerical methods for OCPs with constraints (with the exception of the works [155,269]) are either the methods based on the full discretizations (parametrization of state and control variables), or function space algorithms. The first group of methods assumes a priori discretization of system equations. The second group of methods is, in fact, theoretical work on the convergence of algorithms. The major drawback of some numerical schemes from the first group is the lack of the corresponding convergence analysis. This is especially true in regard to the multiple shooting method [112,113] and to the *collocation* method of Stryk [294]. For the alternative model of collocations see, e.g., [89]. Note that collocation methods are similar to the gradient algorithms applied to a priori discretized problems with the exception that gradients are not calculated with the help of adjoint equations. Instead both variables $x(\cdot)$ and $u(\cdot)$ are regarded as optimization parameters. We refer to [146,183] for high-order schemes applied to unconstrained problems and for error estimates for discrete approximations in optimal control. The discrete approximations to dual OCPs are studied in [184]. There are a number of results scattered in the literature on discrete approximations that are very often closely related, although apparently independent. For a survey of some recent works on computational optimal control, including discrete approximations, see [145] and [32,36].

In some works one deals with a priori discrete approximations of OCPs governed by ordinary differential equations. In [36] the problem is convex-linear; this work introduces a new class of control systems, namely, the convex control systems. We attach much importance to the convergence properties of the discretizations of the initial OCP (1.2). In this case we examine not only the usual regularization technique for the abovementioned ill-posed OCPs, but also use the proximal point approach for "regularizing" the sequence of discrete *convex approximations* to (1.2). Note that the convergence analysis of discrete approximations is of primary importance in computational optimal control (see, e.g., [149,204,239]). For example, in the works mentioned above one considers the "consistent" interplay between approximations such as discretizations and linearizations of an OCP. Finally note that the easy structure of the introduced convex control systems makes it possible to apply the standard gradient methods for concrete numerical computations. For this family of OCPs it is not necessary to consider the sophisticated numerical procedures like multiple shooting or the collocation method of Stryk.

As mentioned above the necessary optimality conditions are not only a main theoretical tool of the general optimization theory but also provide a basis for several numerical approaches in mathematical programming. The same is true in connection with various numerical solution schemes associated with an HOCP or an SOCP. We refer to [44,118,124,125,238,262,284, 301] for theoretical and some computational aspects. The necessary optimality conditions for convex minimization problems (2.1) and (6.29) are also sufficient. In this case a local solution of a convex minimization problem coincides with a global solution. Consider the convex SOCP (6.29). In this case one can also formulate the necessary optimality conditions in the form of the specific Pontryagin maximum principle (see, e.g., [154,192] and Section 6.3). Note that in this case we need to use a special form of the maximum principle, namely, a hybrid version of this optimality criterion (see [289]). The initial SOCP (6.29) can be easily rewritten in the following equivalent form (see [192]):

$$\begin{aligned} &\text{minimize} \sum_{i=1}^{r} \int_{t_{i-1}}^{t_i} f^0_{q_i}(t, x(t))dt \\ &\text{subject to (6.4)},\ q_i \in \mathcal{Q},\ u(\cdot) \in \mathcal{U}, \end{aligned} \tag{6.48}$$

where $f^0_q : \mathbb{R} \times \mathbb{R}^n \to \mathbb{R}$, $q \in \mathcal{Q}$, are continuously differentiable functions. We also assume that the control set has a simple box structure, i.e.,

$$U := \{u \in \mathbb{R}^m \ : \ v^j_- \leq u_j \leq v^j_+,\ j = 1, ..., m\},$$

where v^j_-, v^j_+, $j = 1, ..., m$, are some constants. The application of the hybrid Pontryagin maximum principle (from Section 6.3) to problem (6.48) involves the existence of the adjoint variable $\psi(\cdot)$ that is a solution of the corresponding boundary value problem. We have

$$\begin{aligned} &\dot{\psi}(t) = -\sum_{i=1}^{r} \chi_{[t_{i-1},t_i)}(t) \frac{\partial H_{q_i}(t, x^{opt}(t), u^{opt}(t), \psi(t))}{\partial x} \quad \text{a.e. on } [0, t_f], \\ &\psi_r(t_f) = 0, \end{aligned} \tag{6.49}$$

where

$$H_{q_i}(t, x, u, \psi) := \langle \psi, A_{q_i}(t, x) + B_{q_i}(t)u \rangle - f^0_{q_i}(t, x)$$

is a "partial" Hamiltonian for the location $q_i \in \mathcal{Q}$, $x^{opt}(\cdot)$ is a solution to the initial value problem (6.4) associated with an admissible input $u^{opt}(\cdot)$, and $\langle \cdot, \cdot \rangle$ denotes the scalar product in $\mathbb{R}^n$. Note that in difference to conventional OCPs the adjoint function $\psi(\cdot)$ determined in (6.49) is not an absolutely continuous function (it has "jumps" at switching times t_i, $i = 1, ..., r$). We assume that problem (6.48) is regular (nonsingular) in the following sense:

$$\psi(t) \neq 0 \; \forall t \in [0, t_f] \setminus \Upsilon,$$

where Υ is a subset of $[0, t_f]$ of measure zero. When solving sophisticated OCPs based on some necessary optimality conditions one can obtain a singular solution. Recall that there are two possible scenarios for a singularity: the irregularity of the Lagrange multiplier associated with the cost functional [132,192] and the irregularity of the Hamiltonian. In the latter case the Hamiltonian is not an explicit function of the control function during a time interval. Various supplementary conditions (constraint qualifications) have been proposed under which it is possible to assert that the Lagrange multiplier rule (and the corresponding Pontryagin HMP) holds in "normal" form, i.e., that the first Lagrange multiplier is nonequal to zero. In this case the corresponding minimization problem is called regular.

Evidently, the differential equation from (6.49) can be rewritten as follows:

$$\begin{aligned} \dot{\psi}(t) = -\sum_{i=1}^{r} \chi_{[t_{i-1},t_i)}(t)\Big(\Big[\frac{\partial A_{q_i}(t, x^{opt}(t))}{\partial x}\Big]^T + \\ (u^{opt}(t))^T \Big[\frac{\partial B_{q_i}(t)}{\partial x}\Big]^T \psi(t) + \frac{\partial f^0_{q_i}(t, x^{opt}(t))}{x}\Big). \end{aligned} \tag{6.50}$$

The maximality condition from the abovementioned HMP (see [29]) implies the "bang-bang" structure of the optimal control $u^{opt}(\cdot)$ given by components

$$\begin{aligned} & u_j^{opt}(t) = \mathbf{1}([\psi^T(t) B_{q_i}(t)]_j) v_+^j + (1 - \mathbf{1}([\psi^T(t) B_{q_i}(t)]_j)) v_-^j, \\ & \forall t \in [t_{i-1}, t_i), \; j = 1, ..., m, \end{aligned} \tag{6.51}$$

where $\mathbf{1}(z) \equiv 1$ if $z \geq 0$ and $\mathbf{1} \equiv 0$ if $z < 0$ for a scalar variable z. As we can see the optimal control is a function of $\psi(\cdot)$ and functions $B_{q_i}(\cdot)$, $q_i \in \mathcal{Q}$. A practical solution of the boundary value problem (6.49) is usually based on an iterative numerical scheme. We refer to [265] for some iterative computational approaches. Using (6.50), the $(l+1)$-iteration $\psi^{(l)}$ can now be written as

$$\begin{aligned} \psi^{S,(l+1)}(t) = -L_S\Big(\sum_{i=1}^{r} \chi_{[t_{i-1},t_i)}(t)\Big(\Big[\frac{\partial A_{q_i}(t, x^{opt}(t))}{\partial x}\Big]^T + \\ (u^{opt}(t))^T \Big[\frac{\partial B_{q_i}(t)}{\partial x}\Big]^T \psi^{S,(l)}(t) + \frac{\partial f^0_{q_i}(t, x^{opt}(t))}{x}\Big)\Big), \end{aligned} \tag{6.52}$$

where components of $u^{opt}(\cdot)$ are given by (6.51) replacing $\psi(t)$ by

$$\psi^{S,(l)}(t),\ l \in \mathbb{N}.$$

Here $L_S(w(\cdot))$ is a sequence of Riemann sums

$$L_S(w(\cdot)) := \frac{t}{S}\sum_{s=1}^{S} w(\frac{t}{S}s),\ s \in \mathbb{N},$$

for the integral

$$L(w(\cdot)) := \int_0^t w(t)dt$$

of a piecewise continuous function $w(\cdot)$. It is well known (see, e.g., [20]) that a sequence $\{L_S\}(\cdot)$ converges pointwise (weakly) to L. To put it another way, we have a pointwise convergence of the approximations given by (6.52) to an exact solution $\psi(\cdot)$ of (6.49). Since a weak convergence in the space of piecewise continuous functions coincides with the weak convergence, we conclude that a sequence of controls $\{u^{S,(l)}(\cdot)\}$, where every j-component of $u^{S,(l)}(\cdot)$ is defined as

$$\begin{aligned} &u_j^{S,(l)}(t) = \mathbf{1}([(\psi^{S,(l)}(t))^T B_{q_i}(t)]_j)v_+^j + (1 - \mathbf{1}([(\psi^{S,(l)}(t))^T B_{q_i}(t)]_j))v_-^j, \\ &\forall t \in [t_{i-1}, t_i),\ j = 1, ..., m, \end{aligned}$$

converges weakly to $u^{opt}(\cdot)$. Using some additional facts (which will be presented in Chapter 7) we can deduce the strong convergence of the sequence $x^{S,(l)}$ (the trajectories associated with $\{u^{S,(l)}(\cdot)\}$) to an optimal trajectory x^{opt}. That means a numerical consistence (in the sense of a strong convergence of trajectories) of the usual computational schemes applied to the boundary value problem (6.49) associated with the optimality conditions for the SOCP (6.29) (and (6.48)).

Additionally, the convex structure of an infinite-dimensional minimization problem (6.48) allows us to write necessary and sufficient optimality conditions in the form of the Karush–Kuhn–Tucker theorem or in the form of an easy variational inequality. We refer to Chapter 3 and also to [192,259] for some concrete forms of the optimality conditions in the convex case. Let now (6.48) be a convex minimization problem with a (Fréchet) differentiable cost functional $J(\cdot)$. Similar to the general Mayer HOCP (6.38) we now describe a gradient-based approach for this purpose (see, e.g., [28,262,300] and Section 6.1). Using this representation, $\nabla J(u^{opt}(\cdot))$ can be expressed as

$$
\begin{aligned}
&\nabla J(u^{opt}(\cdot))(t) = -H_u(t, x^{opt}(t), u(t), p^{opt}(t)),\\
&\dot{p}^{opt}(t) = -H_x(t, x^{opt}(t), u^{opt}(t), p^{opt}(t)),\\
&p^{opt}(t_f) = -\phi_x(x^{opt}(t_f)),\\
&\dot{x}^{opt}(t) = H_p(t, x^{opt}(t), u^{opt}(t), p^{opt}(t)),\\
&x(0) = x_0,
\end{aligned}
\tag{6.53}
$$

where

$$
H(t, x, u, p) = \langle p, \sum_{i=1}^{r} \chi_{[t_{i-1}, t_i)}(t) \; [\; A_{q_i}(t, x(t)) + B_{q_i}(t)u(t) \;] \rangle_{\mathbb{R}^n}
$$

is the joint Hamiltonian of the SOCP problem (6.48). By

$$
H_u, \; H_p, \; H_x
$$

we denote here the partial derivatives of H with respect to u, p, and x. Moreover, $x^{opt}(\cdot)$ and $p^{opt}(\cdot)$ are the optimal state and the optimal adjoint variable corresponding to the optimal control function $u^{opt}(\cdot) \in \mathcal{U}$. Note that the gradient $\nabla J(\cdot)$ in (6.53) is computed using the above joint Hamiltonian H to the SOCP (6.48). We do not use here a sequence of "partial" Hamiltonians associated with every location for an HS/SS (see, e.g., [27,29,289]).

We may use the notation $\langle \cdot, \cdot \rangle_{\mathbb{L}^2_m([0,t_f])}$ to denote the inner product of the space $\mathbb{L}^2_m([0, t_f])$ and formulate the following optimality criterion.

Theorem 6.15. *Consider a regular convex* SOCP (6.29) *and assume that the corresponding basic assumptions are satisfied. Then*

$$
u^{opt}(\cdot) \in \mathcal{U}
$$

is an optimal solution of (6.29) *if and only if*

$$
\langle \nabla J(u^{opt}(\cdot)), u(\cdot) - u^{opt}(\cdot) \rangle_{\mathbb{L}^2_m([0,t_f])} \geq 0 \;\; \forall u(\cdot) \in \mathcal{U}. \tag{6.54}
$$

Proof. Under the basic assumptions associated with the convex problem (6.29) the mapping

$$
x^u(t_f) : \mathbb{L}^2_m([0, t_f]) \to \mathbb{R}^n
$$

is Fréchet differentiable [192]. Since $\phi(\cdot)$ is a differentiable function, the composition

$$
J(u(\cdot)) = \phi(x^u(t_f))
$$

is also (Fréchet) differentiable. The set of admissible controls $\mathcal{U}$ is convex, closed, and bounded. Hence, we obtain the necessary and sufficient optimality condition for the given SOCP (6.29) in the form (6.54) (see [192]). The proof is completed. □

Generally, the reduced gradient $\nabla J(u(\cdot))$ at $u(\cdot) \in \mathbb{L}^2_m([0,t_f])$ can be computed as follows:

$$\begin{aligned}
&\nabla J(u(\cdot))(t) = -H_u(t, x^u(t), u(t), p^u(t)),\\
&\dot{p}^u(t) = -H_x(t, x^u(t), u(t), p^u(t)),\\
&p^u(t_f) = -\phi_x(x^u(t_f)),\\
&\dot{x}^u(t) = H_p(t, x^u(t), u(t), p^u(t)),\\
&x(0) = x_0,
\end{aligned}$$

where $x^u(\cdot)$ and $p^u(\cdot)$ are the state and the adjoint variable corresponding to $u(\cdot)$. The presented formalism provides a basis for a variety of useful gradient-type algorithms.

6.4.3 Some Examples

In this section we consider some examples of linear and impulsive HOCPs introduced in the previous parts of this chapter.

Example 6.1. *First, let us examine the classic example of a linear HOCP (see [171]). Consider the following simple HS:*

$$\begin{aligned}
\dot{x} &= u, \text{ for } q_1\\
\dot{x} &= -x + u \text{ for } q_2,
\end{aligned}$$

where $x_0 = 0.9$ and the switching manifold has the affine-linear structure

$$b_{1,2}x + c_{1,2} = 0$$

with $b_{1,2} = 1$ and $c_{1,2} = -1$. The quadratic cost functional has the following easy form:

$$J(u(\cdot), x(\cdot)) = \frac{1}{2}\int_0^1 (x^2(t) + u^2(t))dt.$$

The above HOCP is evidently a simple example of an HLQ problem (6.9). Using the numerical approaches proposed in Section 6.3, we calculate the optimal trajectory and the corresponding optimal control for the HLQ problem under consideration. The optimal behavior of the given LHS is presented in comparison with the classic optimal dynamics of the separate LQ strategy for the first subsystem (indicated by $(x_0(t), u_0(t))$) (see Fig. 6.3).

As one can see, an optimal HLQ dynamics is given here by a discontinuous (in time) partially linear feedback strategy. Unlike the classic Riccati matrix function $P_0(\cdot)$ (for the first subsystem), the corresponding hybrid Riccati matrix $P_1(\cdot)$ is a discontinuous function (see Fig. 6.4).

For the computed switching time instant $t_1^{opt} = 0.1066$ we apply the developed hybrid Riccati formalism and obtain the following values of the hybrid Riccati "matrix":

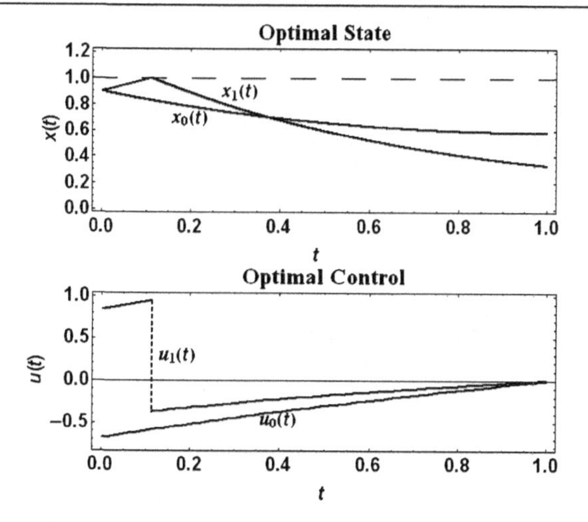

Figure 6.3: Optimal behavior of the LHS.

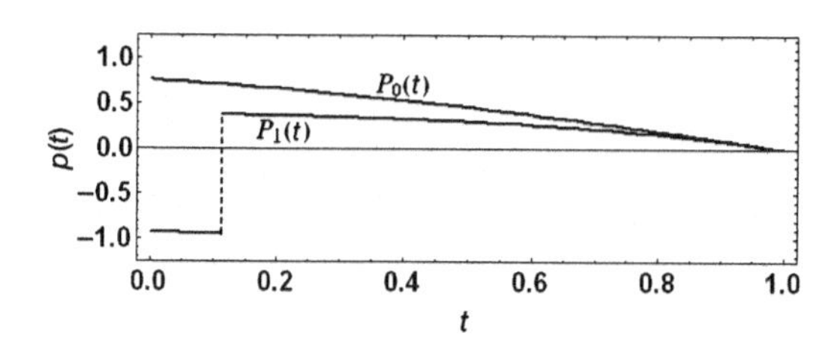

Figure 6.4: Optimal Riccati matrix.

$$P_1^-(t_1^{opt}) = -0.9897, \;\; P_1^+(t_1^{opt}) = 0.2315.$$

The computed optimal cost in the given LQ-type HOCP problem is $J^{opt} = 0.4215$. *Finally note that the optimal cost in the above classic LQ problem formulated for the first location is equal to* 0.6169.

We now present another example of an HLQ-type HOCP. We consider here an LHS with three locations.

Example 6.2. *The dynamics of the HS is given by the following linear equations associated with the corresponding locations:*

$$\begin{aligned}
\dot{x}_1 &= -x_1(t) + 4u_1(t) && \forall t \in [0, t_1], \\
\dot{x}_2 &= -2x_2(t) + 2u_2(t) && \forall t \in [t_1, t_2], \\
\dot{x}_3 &= -4x_3(t) - 5u_3(t) && \forall t \in \left[t_2, t_f\right],
\end{aligned}$$

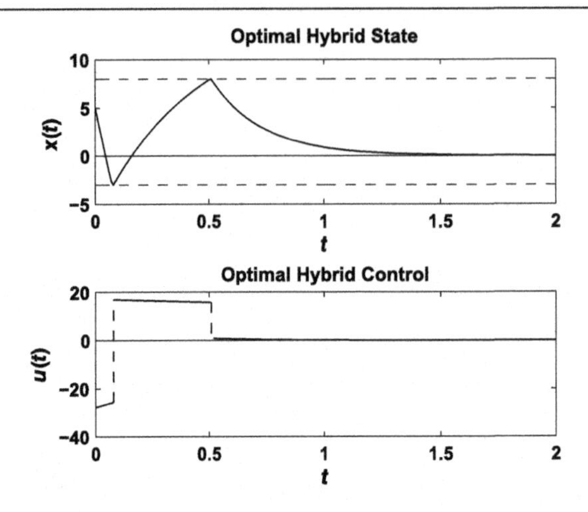

Figure 6.5: Optimal state and control vectors of the LHS.

where $x_0 = 5$ and t_1, t_2 are (unknown) switching times. The switching manifolds are affine-linear manifolds; we have

$$M_{1,2}(x) = x + 3, \; M_{2,3}(x) = x - 8.$$

Our aim is to minimize the quadratic cost function $J(\cdot)$, where

$$S_f = 0, \; S_{q_1} = S_{q_2} = S_{q_3} = 1, \; R_{q_1} = R_{q_2} = R_{q_3} = 1.$$

Applying the numerical results from Section 6.3 we obtain a trajectory shown in Fig. 6.5 and the computed optimal cost is $J^{opt} = 168.6059$.

Unlike the conventional LQ theory the first closed-loop subsystem from Example 6.1 is an unstable system (the Riccati "matrix" $P_1(t)$ is negative for all $t \in [0, t_1^{opt})$). Otherwise the "full" closed-loop systems from both examples given above are asymptotically stable in the sense of Lyapunov.

We next combine the basic IHS definition from Section 4.4 and Definitions 6.6–6.7 of an LHS and consider an LQ-type HOCP with impulses.

Example 6.3. *Let us consider a linear IHS with one switching. The dynamics of the system is given by two linear equations associated with the corresponding location, i.e.,*

$$\begin{aligned} \dot{x}_1 &= x_1(t) + u_1(t) + \theta\delta(t - t_1) & \forall t \in [0, t_1], \\ \dot{x}_2 &= -x_2(t) + 2u_2(t) & \forall t \in [t_1, t_f], \end{aligned}$$

where t_1 is a switching time and

$$x(0) = -0.04.$$

Consider here various switching manifolds which are affine-linear manifolds to show the different values that can take a jump θ. We write

$$M_{1,2}^{(1)}(x) = x + 0.3,\ M_{1,2}^{(2)}(x) = x + 0.1,\ M_{1,2}^{(3)}(x) = x + 0.05.$$

The value of the state jump is determined by the condition

$$||\theta|| \leq 0.1.$$

Our aim is to minimize the quadratic cost function

$$J(u(\cdot), x(\cdot)) = \frac{1}{2}\sum_{i=1}^{2}\int_{t_i-1}^{t_i}(x_i^2(t) + u_i^2(t))dt.$$

Since the presented example deals with the one-dimensional model, the computational algorithm can be sufficiently simplified. We write

$$J^{opt^{(1)}} = 0.0791,\ J^{opt^{(2)}} = 0.0054,\ J^{opt^{(3)}} = 4.8455 \times 10^{-4}$$

with optimal jumps

$$\theta^{opt^{(1)}} = 0.1,\ \theta^{opt^{(2)}} = 0.1,$$

and $\theta^{opt^{(3)}} = 0.05$, respectively. Comparison between each of the trajectories is shown in Fig. 6.6 and the optimal control for the first switching rule is presented in Fig. 6.7.

Our examples illustrate the numerical effectiveness of the general gradient-based computational scheme, namely, of the proposed Algorithm 6.2.

6.5 Approximations Based on the Optimal Control Methodology

6.5.1 Approximations of the Zeno Behavior in ASSs

In the face of a recent progress in hybrid and SS theory, there are a number of fundamental properties of these systems that have not been investigated in sufficient detail. Among others, these include a self-closed formal description of the so-called Zeno executions (see [10,105, 150,199,218,219,306,333]). Recall that a behavior of a system is called Zeno dynamics, if it takes infinitely many discrete transitions on a finite time interval. The real-world engineering

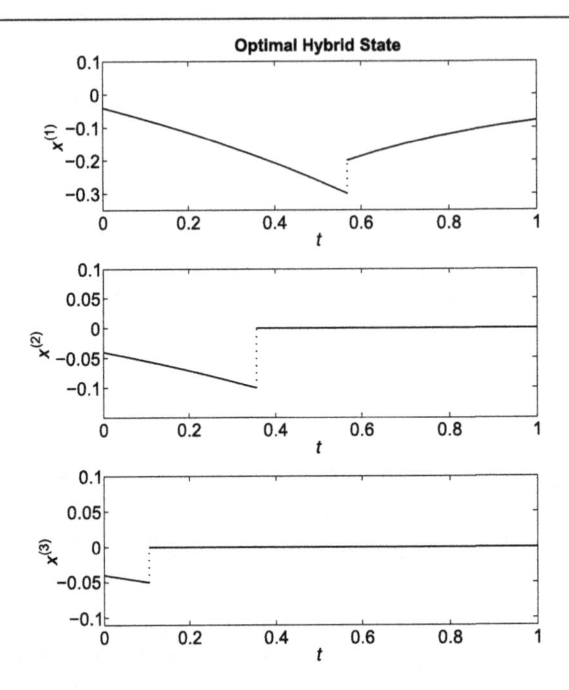

Figure 6.6: Optimal trajectory for three different switching rules.

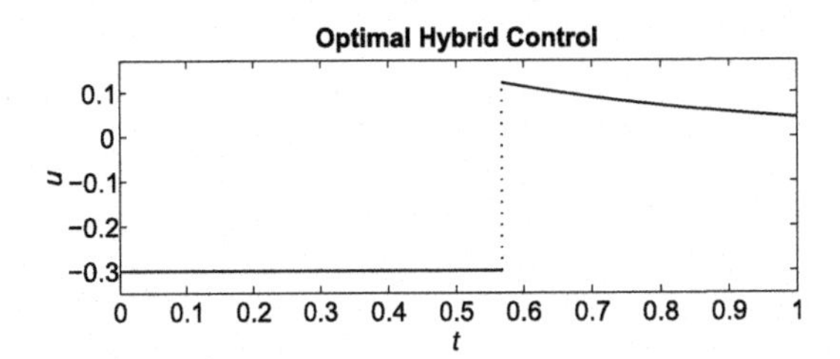

Figure 6.7: Optimal control for $M^{(1)}_{1,2}$.

and physical systems are, of course, not Zeno, but a switched-type mathematical model of a physical process may be Zeno, due to the usual modeling overabstractions. Since an adequate abstraction provides an inevitable theoretic basis for a concrete control design procedure, understanding when it leads to a Zeno effect in an SS is essential for an adequate numerical implementation of the designed control algorithms. Let us note that a Zeno-like switched dynamic behavior can cause imprecise computer simulations and imply the corresponding calculating and modeling errors. It is explainable that the main computer tools and numerical packages developed for SSs (see, e.g., [199,333]) get stuck when a large number of discrete transitions take place within a short time interval. A possible improvement of the mentioned simulation technique can be realized on a basis of adequate extensions of the available theory of the Zeno effects including some approximative approaches. Recall that the Zeno hybrid automata have been usually examined from the point of view of modern computer science [7, 103] such that the necessary investigations of the continuous part of a system under consideration and the discrete-continuous interplay are quite underrated. The general Zeno hybrid executions under assumption of a family of admissible vector fields have been deeply investigated in [10,218,219,333]. On the other hand, an analytic approach based on the abstract systems theory and approximation techniques have not been sufficiently advanced to the Zeno SS setting, nor to the corresponding control design procedures.

Note that usually, one avoids the examination of a Zeno effect in a work on switched control design by some specific additional assumptions. The aim of this section is to give a constructive optimal control–based characterization of the Zeno executions in a specific case of ASSs. We propose an analytic technique that eliminates the Zeno effect from the consideration. These approximations are consistent in the sense of the resulting systems trajectories.

Consider an ASS (see Definition 1.4). Let $x(\cdot)$ be a trajectory of an ASS associated with an admissible τ_r for a given (finite) number $r \in \mathbb{N}$. This trajectory is an absolutely continuous solution of the following initial value problem:

$$\begin{aligned} \dot{x}(t) &= \sum_{i=1}^{r} \beta_{[t_{i-1},t_i)}(t)\Big(a_{q_i}\big(x(t)\big) + b_{q_i}\big(x(t)\big)u(t)\Big), \\ x(0) &= x_0, \end{aligned} \tag{6.55}$$

where (as usual) $\beta_{[t_{i-1},t_i)}(\cdot)$ is a characteristic function of the interval $[t_{i-1}, t_i)$ for $i = 1, ..., r$ and $u(\cdot) \in \mathcal{U}$. Recall that in contrast to the general HSs, the switching times from a sequence τ^r do not depend on the state vector. The complete (switched-type) control input $\rho^r(\cdot)$ associated with the ASS under consideration can now be expressed as follows:

$$\rho^r(\cdot) := \{\beta(\cdot), u(\cdot)\}, \;\; \beta(\cdot) := \{\beta_{[t_0,t_1)}(\cdot), ..., \beta_{[t_{r-1},t_r)}(\cdot)\},$$

where $r \in \mathbb{N}$, $u(\cdot) \in \mathcal{U}$, and the set Γ of all admissible sequences $\beta(\cdot)$ is characterized by the generic conditions

$$\beta(t) \subset \{0,1\}^r \;\; \forall t \in [0, t_f], \;\; \sum_{i=1}^{r} \beta_{[t_{i-1},t_i)}(t) = 1.$$

Let us now introduce some main concepts.

Definition 6.11. *For an admissible control input $\rho^r(\cdot)$, where*

$$\beta(\cdot) \subset \Gamma, \; u(\cdot) \in \mathcal{U},$$

an execution of the ASS *is defined as a collection*

$$(\{q_i\}, \tau^r, x^r(\cdot)),$$

where $q_i \in \mathcal{Q}$ *and* $x^r(\cdot)$ *is a solution to* (6.55). *Here* τ^r *is the corresponding sequence of switching times.*

Let us assume that every admissible control $\rho^r(\cdot)$ generates a unique execution $(\{q_i\}, \tau^r, x^r(\cdot))$ of the given ASS. Following [10,218,219,333,334] we now introduce a formal concept of a Zeno execution.

Definition 6.12. *An execution*

$$(\{q_i\}^Z, \tau^Z, x^Z(\cdot))$$

of an ASS *is called* Zeno *execution if the corresponding switching sequence* $\tau^Z := \{t_i^Z\}_{i=0}^{\infty}$ *is an infinite sequence such that*

$$\lim_{r\to\infty} t_r^Z = \lim_{r\to\infty} \sum_{i=0}^{r} (t_i^Z - t_{i-1}^Z) = t_f < \infty.$$

An admissible input $\rho^Z(\cdot) := \{\beta^Z(\cdot), u^Z(\cdot)\}$, *where*

$$\begin{aligned} &\beta^Z(\cdot) := \{\beta^Z_{[t_0,t_1)}(\cdot), ..., \beta^Z_{[t_{r-1},t_r)}(\cdot), ...\} \subset \mathbb{L}^1(0, t_f), \\ &u^Z(\cdot) \in \mathcal{U}^Z \subseteq \mathcal{U}, \end{aligned}$$

that generates the Zeno *execution* $(\{q_i\}^Z, \tau^Z, x^Z(\cdot))$ *is called a* Zeno *input.*

The above characteristic functions $\beta^Z_{[t_{i-1},t_i)}(\cdot)$ correspond to the Zeno time sequence

$$\tau^Z := \{t_i^Z\}_{i=0}^{\infty}.$$

Due to the specific character of $\beta^Z_{[t_{i-1},t_i)}(\cdot)$ (elements of the Zeno sequence $\beta^Z(\cdot)$) and taking into account Definition 6.12, we can express $x^Z(\cdot)$ as a solution to the initial value problem (6.55) with $r = \infty$. By

$$\mathcal{U}^Z \subseteq \mathcal{U}$$

we denote here a set of the "conventional" control inputs $u^Z(\cdot)$ that involve a Zeno execution. The existence of the Zeno dynamics for various classes of SSs and HSs is deeply discussed in [333]. Note that the Zeno behavior is a typical effect that occurs in a switched dynamic system. It appears in mathematical models of many real engineering systems, such as water tank models, mechanics of a bouncing ball, aircraft conflict resolution models, and the dynamic models for safety control. We refer to [10,105,199,218,219,306,333] for some further switched-type mathematical models that admit Zeno executions. Moreover, in the following we will discuss the sliding mode behavior and the celebrated Fuller OCP that constitute typical examples of the Zeno behavior (see [332]).

Let us introduce some additional mathematical concepts and facts. We use here the generic notation $\mathbb{C}_n(0, t_f)$ for the Banach space of all n-valued continuous functions on $[0, t_f]$ equipped with the usual maximum-norm $||\cdot||_{\mathbb{C}_n(0,t_f)}$.

Definition 6.13. *Consider an* ASS *given by system* (6.55) *and assume that all the above technical conditions for the given ASS are satisfied. We say that this* ASS *possesses a strong approximability property (with respect to the admissible control inputs) if there exists a* $\mathbb{L}^1(0, t_f) \times \mathbb{L}^\infty_m(0, t_f)$*-weakly convergent sequence*

$$\rho^r(\cdot) := \chi^r(\cdot) \times \{v^r(\cdot)\},\ r = 1, ..., \infty$$

of functions

$$\chi^r(\cdot) \subset \mathbb{L}^1(0, t_f),\ \{v^r(\cdot)\} \subset \mathbb{L}^\infty_m(0, t_f)$$

such that the initial value problem (6.55) *with the control input* $\rho^r(\cdot)$ *has a unique solution* $x^r(\cdot)$ *for every* $r \in \mathbb{N}$ *and* $\{x^r(\cdot)\}$ *is a* $\mathbb{C}_n(0, t_f)$*-convergent (uniformly convergent) sequence.*

We now formulate an interesting analytic result that guarantees the strong approximability property for the ASS given affine systems (6.55) (see [42,50] for the additional mathematical details).

Theorem 6.16. *Let all the assumptions for the given ASS be satisfied and let*

$$(\{q_i\}^Z, \tau^Z, x^Z(\cdot))$$

be a Zeno *execution generated by a* Zeno *input* $\rho^Z(\cdot)$. *Consider the initial value problem* (6.55) *associated with an* $\mathbb{L}^1(0, t_f) \times \mathbb{L}^\infty_m(0, t_f)$*-weakly convergent sequence* $\{\rho^r(\cdot)\}$ *of control inputs*

$$\rho^r(\cdot) = \chi^r(\cdot) \times \{v^r(\cdot)\},$$

where

$$\chi^r(\cdot) \subset \mathbb{L}^1(0, t_f),\ \{v^r(\cdot)\} \subset \mathbb{L}^\infty_m(0, t_f)$$

and $\rho^Z(\cdot)$ *is a*

$$\mathbb{L}^1(0, t_f) \times \mathbb{L}^\infty_m(0, t_f)$$

weak limit of $\{\rho^r(\cdot)\}$. *Then, for all* $r \in \mathbb{N}$ *the corresponding initial value problem* (6.55) *has a unique absolutely continuous solution* $x^r(\cdot)$ *determined on* $[0, t_f]$ *and*

$$\lim_{r\to\infty} \|x^r(\cdot) - x^Z(\cdot)\|_{\mathbb{C}_n(0,t_f)} = 0.$$

Proof. The $\mathbb{L}_m^\infty(0, t_f)$-weak convergence implies the $\mathbb{L}_m^\infty(0, t_f)$-weak* convergence (see, e.g., [20] and Chapter 2). Therefore, $\{\rho^r(\cdot)\}$ also converges $(\mathbb{L}^1(0, t_f) \times \mathbb{L}_m^\infty(0, t_f))$-weakly* to the Zeno input $\rho^Z(\cdot)$.

From the equivalent definition of the $\mathbb{L}_m^\infty(0, t_f)$-weak* convergence (see [20]) we next deduce

$$\lim_{r\to\infty} \int_0^{t_f} v^r(t)w(t)dt = \int_0^{t_f} u^Z(t)w(t)dt, \quad \forall w(\cdot) \in \mathbb{L}_m^1(0, t_f). \tag{6.56}$$

Here $u^Z(\cdot) \in \mathcal{U}^Z$ is the corresponding "conventional" part of the given Zeno input $\rho^Z(\cdot)$. Using the basic properties of the Lebesgue spaces ($\mathbb{L}^p$-spaces) on the finite-measure sets, we obtain

$$(\mathbb{L}_m^1(0, t_f))' = \mathbb{L}_m^\infty(0, t_f) \subset \mathbb{L}_m^1(0, t_f).$$

By $(\mathbb{L}_m^1(0, t_f))'$ we denote here the topologically dual space to $\mathbb{L}_m^1(0, t_f)$. Consequently, (6.56) is also true for all $w(\cdot)$ from $(\mathbb{L}_m^1(0, t_f))'$. This implies the $\mathbb{L}^1(0, t_f) \times \mathbb{L}_m^1(0, t_f)$-weak convergence of the given sequence of control inputs

$$\rho^r(\cdot) \subset \mathbb{L}^1(0, t_f) \times \mathbb{L}_m^\infty(0, t_f) \subset \mathbb{L}^1(0, t_f) \times \mathbb{L}_m^1(0, t_f)$$

to the same function

$$\rho^Z(\cdot) \subset \mathbb{L}^1(0, t_f) \times \mathbb{L}_m^\infty(0, t_f) \subset \mathbb{L}^1(0, t_f) \times \mathbb{L}_m^1(0, t_f).$$

The expected result

$$\lim_{r\to\infty} \|x^r(\cdot) - x^Z(\cdot)\|_{\mathbb{C}_n(0,t_f)} = 0$$

now follows from the main result of [50] (Theorem 3.3, p. 5). The proof is completed. □

The proved result, namely, Theorem 6.16, provides an analytic basis for the constructive approximations of the Zeno executions. This approximation tools can not only be applied for a possible simplified analysis of the Zeno involving ASSs but also be used on the systems modeling phase. Note that the real-world engineering processes do not present an exact Zeno behavior. On the other side the corresponding mathematical models of some engineering interconnected systems may be Zeno due to the associated formal abstraction. This situation evidently implies a specific modeling conflict. The proposed approximations of the possible Zeno dynamics in ASSs can help to restore the adequateness of the abstract dynamic model under consideration.

Now let us assume that all the conditions of Theorem 6.16 are fulfilled. Since $\rho^Z(\cdot)$ is bounded and $\rho^r(\cdot)$ is weakly convergent, $\rho^r(\cdot)$ is a uniformly integrable sequence. This is a simple consequence of the celebrated Dunford–Pettis theorem (see, e.g., [21]). Therefore one can choose an (existing) subsequence

$$v^r(\cdot) \in \mathbb{L}_m^\infty(0, t_f)$$

of the approximating sequence $\rho^r(\cdot)$ such that

$$\sup_{v \in \{v^r(\cdot)\}} \int_\Omega v(t)dt < \epsilon, \ \epsilon > 0,$$

for every measurable set $\Omega \subset [0, t_f]$ with measure

$$\text{mes}\Omega < \delta,$$

where

$$\delta = \delta(\epsilon) > 0.$$

For example, $\{v^r(\cdot)\}$ can be chosen as a sequence of step functions. A concrete approximation

$$\chi^r(t), \ t \in [t_{i-1}, t_i), \ t_i \in \tau^r,$$

of elements $\beta^Z_{[t_{i-1},t_i)}(\cdot)$ of a Zeno input $\rho^Z(\cdot)$ can be found from a suitable class of numerically tractable approximations. In the context of Theorem 6.16 one can consider, for example, the Fourier series. Then we have

$$\chi^r(t)\big|_{t\in[t_{i-1},t_i)} = \sum_{l=1}^{r} \frac{2}{l\pi}(\cos\frac{l\pi t_{i-1}}{t_i} - \cos l\pi)\sin\frac{l\pi t}{t_i}, \quad t \in [0, t_f]. \tag{6.57}$$

The Fourier series converges (as $r \to \infty$) in the sense of the $\mathbb{L}^2(0, t_f)$-norm to a characteristic function $\beta_{[t_{i-1},t_i)}(\cdot)$ of the chosen time interval $[t_{i-1}, t_i)$. This fact is a simple consequence of the well-known Parseval equality (see, e.g., [20,282]). The given $\mathbb{L}^2(0, t_f)$-convergence implies the $\mathbb{L}^1(0, t_f)$-convergence of (6.57). Let now

$$\tau^r \to \tau^Z$$

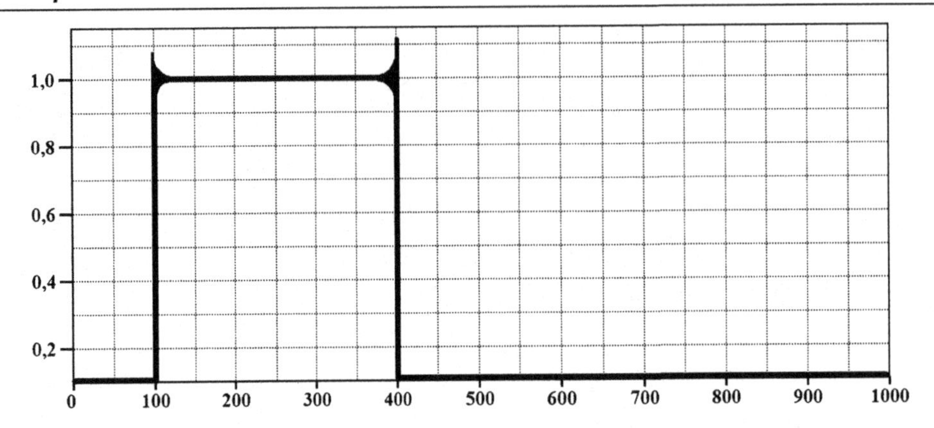

Figure 6.8: Approximation of an element $\beta^Z_{[t_{i-1},t_i)}(\cdot)$ of a Zeno input $\rho^Z(\cdot)$ by Fourier series.

as $r \to \infty$ (the simple pointwise convergence). This construction implies the $\mathbb{L}^1(0,t_f)$-convergence of the sequence $\chi^r(\cdot)$ to $\beta^Z(\cdot)$ (see Fig. 6.8).

Evidently, the component $\beta^Z(\cdot)$ of a Zeno input $\rho^Z(\cdot)$ takes its values in the set of vertices $\text{ver}(T_r)$ of the following r-simplex:

$$T_r := \left\{ \beta \in \mathbb{R}^r \mid \beta_i \geq 0, \ \sum_{i=1}^{r} \beta_i = 1 \right\}.$$

The $\mathbb{L}^1(0,t_f)$-weakly convergent component $\chi^r(\cdot)$ of an approximating sequence

$$\rho^r(\cdot) = \chi^r(\cdot) \times \{v^r(\cdot)\}$$

can also be defined as a function that takes its values from the above simplex $\chi^k(t) \in T_r$. In that case $\chi^r(\cdot)$ possesses the additional useful properties summarized in the following result.

Theorem 6.17. *Assume that all the conditions of* Theorem 6.16 *are satisfied. Let $\chi^r(\cdot)$ be an $\mathbb{L}^1(0,t_f)$-weakly convergent component of the approximating sequence $\rho^r(\cdot)$ and let*

$$\chi^r(t) \subset T_r$$

for almost all $t \in [0,t_f]$. Then $\chi^r(\cdot)$ also converges strongly to $\beta^Z(\cdot)$.

Proof. Since T_r is a convex polyhedron, the tangent cone to this simplex at every vertex is a pointed cone (does not contain a line). Therefore, for any e from

$$\text{ver}(T_r)$$

there exist a number $c > 0$ and a unit vector $\vartheta \in \mathbb{R}^r$ such that for all $d \in \mathbb{R}^r$ we have

$$\|d - e\|_{\mathbb{R}^r} \leq c\langle \vartheta, d - e\rangle_{\mathbb{R}^r}.$$

By $\|\cdot\|_{\mathbb{R}^r}$ and $\langle \cdot, \cdot \rangle_{\mathbb{R}^r}$ we denote here the norm and the scalar product in $\mathbb{R}^r$. The above estimation and the $\mathbb{L}^1(0, t_f)$-weak convergence of $\chi^r(\cdot)$ imply the following:

$$\|\chi^r(\cdot) - \beta^Z(\cdot)\|_{\mathbb{L}^1(0,t_f)} \leq$$
$$c\int_0^{t_f} \langle \vartheta(t), \chi^r(t) - \beta^Z(t)\rangle_{\mathbb{R}^r} dt \leq \epsilon(r),$$

where

$$\lim_{r\to\infty} \epsilon(r) = 0.$$

This completes the proof. □

Evidently, Theorem 6.17 brings out the best choice of the element $\chi^r(\cdot)$ of an approximating sequence $\rho^r(\cdot)$ associated with a Zeno execution.

We now consider a smooth manifold

$$\mathcal{S} := \{y \in \mathbb{R}^n \mid h(y) = 0\},$$

with

$$s := \dim(\mathcal{S}) < n.$$

Here $h : \mathbb{R}^n \to \mathbb{R}^s$, $s \in \mathbb{N}$, is a continuously differentiable function. Following [132,205] we call $\mathcal{S}$ an invariant manifold for a dynamic system

$$\begin{aligned} \dot{y}(t) &= \phi(y(t)), \ t \in \mathbb{R}_+, \\ y(0) &= y_0 \end{aligned} \tag{6.58}$$

if $y(t) \in \mathcal{S}$ for all $t \geq \hat{t} \in \mathbb{R}_+$. Here $\phi(\cdot)$ is an appropriate smooth function. The next abstract theorem gives a general invariance criterion.

Theorem 6.18. *A smooth manifold $\mathcal{S}$ is invariant for system* (6.58) *iff $\phi(y)$ belongs to the tangent space* $\mathrm{Tan}_{\mathcal{S}}$ *of $\mathcal{S}$ for all $y \in \mathbb{R}^n$.*

The proof of Theorem 6.57 is based on the extended Lyapunov-type technique (see [131] for details). Recall that the tangent space $\mathrm{Tan}_{\mathcal{S}}$ of $\mathcal{S}$ is determined as follows:

$$\mathrm{Tan}_{\mathcal{S}} = \{\zeta \in \mathbb{R}^n \mid Dh(y)\zeta = 0\},$$

where $Dh(y)$ denotes a derivative of $h(y)$.

In many engineering-motivated ASSs the switching times $t_i,\ i=1,\dots,r$, are determined by a specific condition that involves a smooth manifold $\mathcal{S}$, i.e.,

$$\tau^r := \{t_i,\ i=1,\dots,r \mid h(x^r(t_i))=0\}, \tag{6.59}$$

where $x(\cdot)$ is a solution of (6.55). This modeling approach corresponds to some particular mathematical models of HSs. Let us refer to [218] for some relevant real-world examples and further formal details.

Theorem 6.19. *Assume that all the conditions of* Theorem 6.16 *are satisfied and the switching times of an ASS are determined by* (6.59), *where* $\mathcal{S}$ *is a smooth manifold. Let*

$$(\{q_i\}^Z, \tau^Z, x^Z(\cdot))$$

be a Zeno *execution generated by a* Zeno *input* $\rho^Z(\cdot)$. *Assume that*

$$\rho^r(\cdot) = \chi^r(\cdot) \times \{v^r(\cdot)\}$$

is an approximating sequence from Theorem 6.16. *Then there exist projected dynamic processes* $y(\cdot)$ *given by the following initial value problem:*

$$\begin{aligned}\dot{y}(t) = \sum_{i=1}^{\infty} \beta^Z_{[t_{i-1},t_i)}(t)\mathrm{Pr}_{\mathrm{Tan}_{\mathcal{S}}}\big[a_{q_i}(y(\cdot))+ \\ b_{q_i}(y(\cdot))\hat{u}(\cdot)\big](t),\ y(0)=y_0 \in \mathcal{S},\end{aligned} \tag{6.60}$$

where $\hat{u}(\cdot) \in \mathcal{U}$ *is a strong limit of a subsequence of* $\{v^r(\cdot)\}$ *and*

$$\mathrm{Pr}_{\mathrm{Tan}_{\mathcal{S}}}[x]$$

is a projection of a vector $x \in \mathbb{R}^n$ *on the manifold* $\mathcal{S}$. *Moreover,* $\mathcal{S}$ *is an invariant manifold for* (6.60) *and*

$$\lim_{t\to t_f} ||x^Z(t) - y(t)||_{\mathbb{R}^n} = 0.$$

Proof. From the basic assumptions for the given ASS (6.55) (smoothness of the families $\mathcal{A}$, $\mathcal{B}$ and boundedness of the admissible controls) it follows that the trajectories $x(\cdot)$ of the ASS under consideration have bounded derivatives for almost all $t \in [0, t_f]$. This fact and Definition 6.13 imply

$$\lim_{t\to t_f} \mathrm{dist}[x^Z(t),\ \mathcal{S}], \tag{6.61}$$

where $\mathrm{dist}[x, \mathcal{S}]$ denotes the Euclidean distance between a vector $x \in \mathbb{R}^n$ and $\mathcal{S}$. Consider an $\mathbb{L}^1(0, t_f) \times \mathbb{L}^\infty_m(0, t_f)$-weakly convergent sequence $\{\rho^r(\cdot)\}$ of controls from Theorem 6.16 such that

$$\lim_{r\to\infty} \|x^r(\cdot) - x^Z(\cdot)\|_{\mathbb{C}_n(0,t_f)} = 0, \tag{6.62}$$

where $\{x^r(\cdot)\}$ is a sequence of solutions to (6.55) generated by $\{\rho^r(\cdot)\}$. From (6.61)–(6.62) we next deduce

$$\begin{aligned}
&\lim_{r\to\infty}\lim_{t\to t_f} ||x^r(t) - x^Z(\cdot)||_{\mathbb{R}^n} = 0,\\
&\lim_{r\to\infty}\lim_{t\to t_f} ||x^r(t) - \mathrm{Pr}_{\mathrm{Tan}_{\mathcal{S}}}(x^r(\cdot))(t)||_{\mathbb{R}^n} = 0.
\end{aligned} \tag{6.63}$$

Here

$$\mathrm{Pr}_{\mathrm{Tan}_{\mathcal{S}}}(\cdot)(\cdot)$$

is a projection on the tangent space $\mathrm{Tan}_{\mathcal{S}}$. Moreover, the above relations (6.62)–(6.63) imply

$$\lim_{r\to\infty}\lim_{t\to t_f} ||x^Z(t) - \mathrm{Pr}_{\mathrm{Tan}_{\mathcal{S}}}(x^r(\cdot))(t)||_{\mathbb{R}^n} = 0.$$

Since the mappings $||\cdot||_{\mathbb{R}^n}$ and

$$\mathrm{Pr}_{\mathrm{Tan}_{\mathcal{S}}}(\cdot)$$

are continuous, we next obtain

$$\lim_{t\to t_f} ||x^Z(t) - y(t)||_{\mathbb{R}^n} = 0,$$

where $y(\cdot)$ is a $\mathbb{C}_n(0, t_f)$-limit of

$$\mathrm{Pr}_{\mathrm{Tan}_{\mathcal{S}}}(\cdot)$$

if $r \to \infty$ and

$$y(t) := \lim_{r\to\infty} \mathrm{Pr}_{\mathrm{Tan}_{\mathcal{S}}}(x^r(\cdot))(t)\ \forall t \in [0, t_f].$$

The continuity and linearity of the above projection operator $\mathrm{Pr}_{\mathrm{Tan}_{\mathcal{S}}}(\cdot)$ on the subspace $\mathrm{Tan}_{\mathcal{S}}$ implies

$$\dot{y}(t) = \lim_{r\to\infty} \mathrm{Pr}_{\mathrm{Tan}_{\mathcal{S}}}(\dot{x}^r(\cdot))(t)$$

for almost all $t \in [0, t_f]$. The obtained relation can be rewritten as follows:

$$\dot{y}(t) = \lim_{r\to\infty} \mathrm{Pr}_{\mathrm{Tan}_{\mathcal{S}}}\Big[\sum_{i=1}^{r} \beta_{[t_{i-1},t_i)}(\cdot)\Big(a_{q_i}(x^r(\cdot))+$$
$$b_{q_i}(x^r(\cdot))u(\cdot)\Big)\Big](t) = \sum_{i=1}^{\infty} \beta^Z_{[t_{i-1},t_i)}(t) \lim_{r\to\infty} \mathrm{Pr}_{\mathrm{Tan}_{\mathcal{S}}}\Big[\Big(a_{q_i}(x^r(\cdot))+$$
$$b_{q_i}(x^r(\cdot))u(\cdot)\Big)\Big](t) = \sum_{i=1}^{\infty} \beta^Z_{[t_{i-1},t_i)}(t) \mathrm{Pr}_{\mathrm{Tan}_{\mathcal{S}}}\Big[\lim_{r\to\infty}\Big(a_{q_i}(x^r(\cdot))+$$
$$b_{q_i}(x^r(\cdot))u(\cdot)\Big)\Big](t).$$

The smoothness of the families $\mathcal{A}$, $\mathcal{B}$ from the main definition of an ASS implies the following projected differential equation:

$$\dot{y}(t) = \sum_{i=1}^{\infty} \beta^Z_{[t_{i-1},t_i)}(t) \mathrm{Pr}_{\mathrm{Tan}_{\mathcal{S}}}\big[a_{q_i}(y(\cdot))+ b_{q_i}(y(\cdot))\hat{u}(\cdot)\big](t), \tag{6.64}$$

determined on the tangent space $\mathrm{Tan}_{\mathcal{S}}$. We refer to [66] for the necessary concepts and facts from the theory of projected dynamic systems defined by ordinary differential equations. Here

$$\hat{u}(\cdot) \in \mathcal{U}$$

is an $\mathbb{L}^\infty_m(0, t_f)$-limit (strong limit) of a subsequence of $\{v^r(\cdot)\}$. Note that every $\mathbb{L}^\infty_m(0, t_f)$-weakly convergent sequence $\{v^r(\cdot)\}$ has a strongly convergent subsequence. The above inclusion $\hat{u}(\cdot) \in \mathcal{U}$ is a consequence of the closeness of the set $\mathcal{U}$ of admissible controls. Using the basic property, namely, the idempotency

$$\mathrm{Pr}^2_{\mathrm{Tan}_{\mathcal{S}}} = \mathrm{Pr}_{\mathrm{Tan}_{\mathcal{S}}}$$

of the projection operator, we finally obtain

$$\mathrm{Pr}_{\mathrm{Tan}_{\mathcal{S}}}\big[\dot{y}(\cdot)\big](t) = \dot{y}(t)$$

and, moreover, the following DI:

$$\dot{y}(t) \in \mathrm{Tan}_{\mathcal{S}}.$$

By construction, $\mathcal{S}$ is an invariant manifold for system (6.64). The proof is completed. □

The above Theorem 6.19 has a natural geometrical interpretation. The Zeno trajectory related to a Zeno execution with switching rules given by (6.59) converges to a smooth dynamic process on $\mathcal{S}$.

A specific case of the presented dynamic behavior is constituted by the celebrated sliding mode control. We refer to [30,205,267,305]. In that case we have

$$s = \dim(\mathcal{S}) \equiv m$$

(the dimension of $\mathcal{S}$ is equal to the dimension of the control vector) and the approximating process $y(\cdot)$ can directly be generated by a special control system via the so-called "equivalent control" $w_{eq}(\cdot)$. Note that we consider here the case of invertible matrices from the family $\mathcal{B}$ (see the definition of an ASS associated with the given dynamic system (6.55)), i.e.,

$$w_{eq}(y) := -(\nabla h(y) \sum_{i=1}^{\infty} \beta^{z}_{[t_{i-1},t_i)}(t) b_{q_i}(y))^{-1} \times$$
$$[\nabla h(y) \sum_{i=1}^{\infty} \beta^{z}_{[t_{i-1},t_i)}(t) a_{q_i}(y)].$$

For the Zeno-like chattering behavior in sliding mode systems see also [30,95]. We call the dynamic process $y(\cdot)$ from Theorem 6.19 the projected Zeno dynamics. This newly introduced Zeno-like dynamics constitutes a natural generalization of the conventional sliding mode control strategies from [305]. This generalization eliminates the restrictive dimension conditions of the classic sliding mode and is in fact characterized by the nonrestrictive condition

$$s = \dim(\mathcal{S}) \leq n$$

in comparison to the specific assumption $s = m$ (indispensable for the classic sliding mode control design).

6.5.2 Sliding Mode Control Approximations

The study of various types of variable structures control processes in the framework of the general discontinuous dynamic models has gained lots of interest in recent years (see, e.g., [56,74,95,254,265,267,305]). An important case of the abovementioned systems is given by a class of the conventional first-order sliding mode control systems and the corresponding mathematical models expressed in the form of differential inclusions [305]. These sophisticated mathematical models are practically motivated by many modern engineering applications and constitute a valuable part of the recent progress in systems theory. We refer to [95,265,305] for the basic theoretic facts and also for some interesting real-world applications of the sliding mode control techniques. It is necessary to stress that the control design approaches based on the traditional sliding mode technologies (see [305]) and on the high-order extensions are nowadays a mature and relatively simple methodology for the constructive synthesis of several types of robust controllers and effective observers associated with the general nonlinearly affine dynamic systems. However, the formal mathematical techniques based on the sliding mode methodology usually imply a high-frequency chattering. This chattering regime constitutes in fact a nonsuitable behavior in many real-world applications (for example, in mechanical engineering).

The above difficulties of the classic sliding mode theory imply the necessity of some "approximating sliding mode" control techniques which do not involve high-frequency chattering dynamics. In this section we propose a constructive approximation of the sliding mode. This approximation preserves some useful properties of the conventional sliding mode control and is chattering-free. Our idea is based on the β-relaxations (see Chapter 5) and also includes the variational approach. Recall that the general variational approach (the possibility to replace an original problem by an equivalent minimization problem) plays one of central roles in the applied functional and numerical analysis and constitutes the valuable contributions of many prominent mathematicians (D. Bernoulli, L. Euler, J.L. Lagrange, K.F. Gauss, D. Hilbert, and others). Our contribution pursues two main goals. Firstly, we would like to provide a strong theoretic framework for consistent approximations of the specific differential inclusions induced by the sliding mode control systems. Secondly, we elaborate an adequate variational representation of the classic sliding mode dynamics and to deduce the associated state space formalism.

Consider the nonlinearly affine control system

$$\begin{aligned} &\dot{x}(t) = a\big(x(t)\big) + b\big(x(t)\big)u(t) \text{ a.e. on } [0, t_f], \\ &x(0) = x_0, \end{aligned} \tag{6.65}$$

where $x_0 \in \mathbb{R}^n$ is a given initial state. The functions

$$a : \mathbb{R}^n \to \mathbb{R}^n, \; b : \mathbb{R}^n \to \mathbb{R}^{n \times m}$$

are assumed to be sufficiently many times differentiable. From the general existence/uniqueness theory for nonlinear ordinary differential equations (see, e.g., [185,205,265] and Chapter 5) it follows that for every admissible control function

$$u(\cdot) \in \mathbb{L}^{\infty}_{m}(0, t_f), \; u(t) \in \mathbb{R}^m,$$

the initial value problem (6.65) has a unique absolutely continuous solution $x(\cdot)$. Let us further consider the case $m < n$. We also assume that the rank of

$$\mathrm{span}\{F_1(x), ..., F_m(x)\},$$

where $F_j(x)$ are vector columns of the matrix $b(x)$, is equal to $m \in \mathbb{N}$.

The specific class of the affine dynamical models introduced above represents a wide and important family of general dynamic systems. Various control systems of the type (6.65) arise in the control of mechanical systems, electrical circuits, and aircrafts [95,205,305]. The ASSs and affine-type HSs, where different models are coupled together, are considered in [10]. Systems of the type (6.65) constitute the main mathematical object of the variable structure control methodology (see, e.g., [305]). The sliding mode character of the dynamic system (1.1) is determined by the following specific feedback control functions:

$$w(x) := \tilde{w}(\sigma(x)), \tag{6.66}$$

where $\tilde{w} : \mathbb{R}^m \to \mathbb{R}^m$ is a bounded measurable (feedback) control function and $\sigma : \mathbb{R}^n \to \mathbb{R}^m$ is a continuously differentiable m-valued function. The classic sliding mode dynamic process associated with (1.1) usually incorporates a "sliding manifold,"

$$S := \{x \in \mathbb{R}^n \mid \sigma(x) = 0\},$$

that determines the main feedback-type control strategy in the following sense: a trajectory $x(\cdot)$ generated by the closed-loop system (6.65)–(6.66) possesses the following property:

$$\sigma(x(t)) = 0, \ \forall t \geq T_{sl} \in \mathbb{R}_+, \ \ T_{sl} < t_f.$$

Let us note that the composite function $u_w(t) \equiv w(x(t))$, where $x(\cdot)$ is a trajectory of the resulting system (6.65)–(6.66), belongs to the space $\mathbb{L}^\infty_m(0, t_f)$. This is an easy consequence of the boundedness of $\tilde{w}(\cdot)$ and absolute continuity of the solution $x(\cdot)$ to the closed-loop realization of (6.65)–(6.66). Note that the main assumptions introduced above make it possible to incorporate some classes of discontinuous closed-loop systems into the considered framework. In that case the solution concept for (6.65)–(6.66) needs an additional justification. An adequate constructive definition of a solution here can be considered, for example, in the sense of the well-known Filippov approach (see [159,254,305] and Chapter 5). Let us also recall that the more general (higher-order) sliding mode control approach to systems with a relative degree $l \in \mathbb{N}$ is related to a family of smooth functions of the type

$$\bar{\sigma}(t, x)$$

and to a specific form of the feedback control law. In this case the traditional sliding mode formalism is defined by $(l-1)$ derivatives of the function $\bar{\sigma}(\cdot, \cdot)$.

Evidently, the sliding mode control systems are characterized by a specific mechanism of switchings (states transitions) and can also be considered as a specific case of general SSs. This prescribed switching mechanism is determined by a discontinuous feedback control (6.65) that is implemented for all

$$0 < t < T_{sl}.$$

The generic sliding condition

$$\sigma(x(t)) = 0, \; t \geq T_{sl},$$

evidently implies the following DI:

$$\dot{x}(t) \in \mathcal{T}_{\mathcal{S}}(x(t))$$

for all $t \geq T_{sl}$. By

$$\mathcal{T}_{\mathcal{S}}(z)$$

we denote here the tangent space of $\mathcal{S}$ at $z \in \mathcal{S}$. We next restrict our consideration to the relatively simple linear-affine sliding surfaces $\mathcal{S}$ defined by

$$\sigma(x) = Sx + c,$$

where

$$S \in \mathbb{R}^{m \times n}, \; c \in \mathbb{R}^{m}.$$

The above DI can be easily rewritten in the following form:

$$\nabla\sigma(x(t))[a\big(x(t)\big) + b\big(x(t)\big)w_{eq}(x(t))] = 0, \; \forall t \geq T_{sl}. \tag{6.67}$$

The above expression determines explicitly the celebrated equivalent control $w_{eq}(\cdot)$ in the case of an invertible matrix

$$\nabla\sigma(x)b(x).$$

The above equivalent control can be calculated as follows:

$$w_{eq}(x) := -(\nabla\sigma(x)b(x))^{-1}[\nabla\sigma(x)a(x)] = -(Sb(x))^{-1}[Sa(x)].$$

We use here the standard notation $\nabla\sigma(x)$ for a gradient of $\sigma(x)$ at $x \in \mathcal{S}$ and additionally assume the invertibility of the matrix $Sb(x)$ for all $x \in \mathbb{R}^n \setminus 0$. Let us denote

$$u_{eq}(t) := w_{eq}(x(t)).$$

The complete sliding mode control law can now be written as follows:

$$u_{sl}(t) = \left\{ \begin{array}{l} u_w(t), \text{ if } t < T_{sl}, \\ u_{eq}(t), \text{ if } t \geq T_{sl} \end{array} \right\}. \tag{6.68}$$

Note that the above form (6.68) of the designed input can be formally obtained only under the assumption of a known trajectory of the system. We call system (6.65) with the implemented control $u_{sl}(\cdot)$ the closed-loop realization of (6.65).

We now propose a new variational representation of the sliding behavior. This new theoretic development is based on the β-relaxations introduced in Chapter 5 and incorporates the classic Hamiltonian formalism. Resulting from this Hamiltonian-like representation in an extended state space the realized sliding trajectory of (6.65) can be constructively described as a solution to a specific auxiliary OCP. Moreover, the representation mentioned above makes it possible to calculate the equivalent control $u_{eq}(\cdot)$ as a time-dependent function in (6.68).

As mentioned above, the closed-loop realization (6.65)–(6.66) is in general determined by ordinary differential equations with discontinuous right-hand sides. The discontinuity effect is a natural effect in relaxation theory and is caused by the given class of the feedback control functions $w(\cdot)$ and $w_{eq}(\cdot)$. In a concrete situation a suitable sliding mode–based control design needs to possess some strong stability properties. This stability requirement is usually related to the classic Lyapunov or finite time stability concepts. We consider the corresponding closed-loop system (1.1) as a special case of a general differential equation with discontinuous right-hand side. This modeling framework makes it possible to apply the celebrated Filippov approach ([159,95,254] and Chapter 5) and to study the associated DI of the following form:

$$\dot{x}(t) \in \mathcal{K}[a,b](x(t)) \text{ a.e. on } [0,t_f], \tag{6.69}$$

where $x(0) = x_0$ and

$$\mathcal{K}[a,b](x) := \overline{\text{co}}\{\lim_{j\to\infty} [a(x_i) + b(x_i)w(x_i)] \mid x_i \to x,\ x_i \notin \mathcal{P}\}.$$

Here $\mathcal{P} \subset \mathbb{R}^n$ is a set of measure zero. Our aim now is with the development of an adequate constructive characterization of the sophisticated closed-loop realization of (6.65). Such an equivalent description of the initial mathematical model (6.69) can provide a new theoretic basis for some effective control design procedures. The expected equivalent or approximative description of (6.69) can naturally be separated into two parts, namely, for $t < T$ (Part I) and for $t \geq T$ (Part II). For the mathematical description of Part I one can use, for example, the set-valued approximation techniques proposed in [56].

Let us shortly discuss this newly elaborated approximative tool. The structure of the feedback control (6.66) motivates the consideration of the box-type admissible control set $\mathcal{U}$. We have

$$\mathcal{U} := \{v(\cdot) \in \mathbb{L}_m^{\infty}(0, t_f) \mid v(t) \in U \text{ a.e. on } [0, t_f]\},$$
$$U := \{u \in \mathbb{R}^m \ : \ v_-^j \leq u_j \leq v_+^j, \ j = 1, ..., m\},$$

where

$$v_-^j, v_+^j, j = 1, ..., m,$$

are some constants. Using the general form (6.69), the affine system (6.65)–(6.68) can now be rewritten in the equivalent form as the specific DI

$$\begin{aligned} \dot{x}(t) &\in a\big(x(t)\big) + b\big(x(t)\big)U, \\ x(0) &= x_0, \end{aligned} \tag{6.70}$$

where

$$a(t, x(t)) + b(t, x(t))U$$

is the orientor field associated with (6.65)–(6.68) for a given compact and convex set U. We denote (6.70) as an inclusion generated by the closed-loop realization (6.65)–(6.68). The original DI with an affine structure admits a conventional parametrization in the form of a relaxed control system, namely, as a Gamkrelidze system (see [173] and Chapter 5), i.e.,

$$\begin{aligned} \dot{\eta}(t) &= a\big(\eta(t)\big) + \sum_{j=1}^{n+1} \alpha^j(t) b\big(\eta(t)\big) u^j(t) \text{ a.e. on } [0, t_f], \\ \eta(0) &= x_0, \ \alpha(\cdot) \in \Lambda(n+1), \end{aligned} \tag{6.71}$$

where

$$\alpha(\cdot) := \big(\alpha^1(\cdot), \ldots, \alpha^{n+1}(\cdot)\big)^T$$

and

$$u^j(\cdot) \in \mathcal{U} \quad \forall j = 1, \ldots, n+1.$$

Moreover, we have

$$\Lambda(n+1) := \left\{ \alpha(\cdot) \;\middle|\; \alpha^j(\cdot) \in \mathbb{L}_1^1(0, t_f), \ \alpha^j(t) \geq 0, \sum_{j=1}^{n+1} \alpha^j(t) = 1, \ t \in [0, t_f] \right\},$$

as a set of admissible multipliers $\alpha(\cdot)$. Let $\eta(\cdot)$ be a solution to the Gamkrelidze system (6.71) generated by admissible controls $u(\cdot) \in \mathcal{U}$ and multipliers $\alpha(\cdot) \in \Lambda(n+1)$. We refer to

[6,32,86,154] for the exact results related to the explicit Gamkrelidze factorization of the convex-valued DI (6.70). In Chapter 5 the corresponding result is called an equivalency principle. For the specific case of DI (6.70) this result can be shortly summarized as follows.

Theorem 6.20. *A function $x(\cdot)$ is a solution of the DI* (6.70) *if and only if it is a solution of the corresponding* Gamkrelidze *system* (6.71).

Let us also recall that Theorem 6.20 is a direct consequence of the so-called Filippov selection lemma (see, e.g., [6] and Chapter 5).

Our aim now is to construct an approximative approach to the initial dynamic model (6.70) using the β-relaxations discussed in Chapter 5. Let us assume that for a given number $\epsilon_M > 0$ (the prescribed accuracy) and for every element $\omega \in U$ there exists a point

$$\omega^k \in U_M$$

(the discrete approximations of the given control set U) such that

$$||\omega - \omega^k||_{\mathbb{R}^m} < \epsilon_M.$$

Following the theoretic construction of β-relaxations, we introduce the so-called β-system

$$\begin{aligned} \dot{z}(t) &= a(z(t)) + \sum_{k=1}^{M} \beta^k(t)\big(b(z(t))\omega^k\big) \text{ a.e. on } [0, t_f], \\ z(0) &= x_0, \end{aligned} \tag{6.72}$$

where $\beta^k(\cdot)$ are elements of $\mathbb{L}^\infty_m(0, t_f)$ such that

$$\beta^k(t) \geq 0, \ \sum_{k=1}^{M} \beta^k(t) = 1 \ \ \forall t \in [0, t_f].$$

Recall that the introduced vector

$$\beta_M(\cdot) := (\beta^1(\cdot), ..., \beta^M(\cdot))^T$$

is called a β-control. The set of admissible β-controls is denoted by $\aleph(M)$. Let $\mathbb{W}^{1,1}_n(0, t_f)$ be the Sobolev space and $\mathbb{C}_n(0, t_f)$ be the space of continuous functions. Consider a sequence of approximations U_M corresponding to the given accuracy ϵ_M such that

$$\lim_{M \to \infty} \epsilon_M = 0.$$

The main result from [32] and Chapter 5 shows that the β-systems (6.72) generated by U_M are consistent approximations of the Gamkrelidze system associated with (6.70). This result for the specific case of DI (6.70) can be reformulated as follows.

Theorem 6.21. *Consider the initial system (6.70) and a solution $\eta(\cdot)$ of the corresponding* Gamkrelidze *system (6.71). Then there exists a sequence $\{\beta_M(\cdot)\}$ from $\aleph(M)$ of the β-controls and the corresponding sequence*

$$\{z_M^\beta(\cdot)\}$$

of solutions of β-systems such that $z_M^\beta(\cdot)$ approximate the solution $y(\cdot)$ in the following senses

$$\lim_{M\to\infty} ||z_M^\beta(\cdot) - y(\cdot)||_{\mathbb{C}_n(0,t_f)} = 0$$

and

$$\lim_{M\to\infty} ||z_M^\beta(\cdot) - y(\cdot)||_{\mathbb{W}_n^{1,1}(0,t_f)} = 0.$$

Moreover, every $z_M^\beta(\cdot)$ is also a solution of the DI

$$\dot{z}_M^\beta(t) \in a(z_M^\beta(t)) + b(z_M^\beta(t)) \times U \text{ a.e. on } [0, t_f], \tag{6.73}$$

where $z(0) = x_0$.

Note that the cross-operator in (6.73) indicates a simple Cartesian product of two sets, namely, the one-point set $b(z_M^\beta(t))$ and the admissible control set U. Theorem 6.21 provides an effective approximative approach to DI (6.72) and to the initial closed-loop realization (6.65)–(6.68) on the time interval $[0, T)$. Recall that this behavior of the complete sliding mode dynamics was denoted as "Part I." For the practically oriented approximative methods and for the concrete control synthesis procedures one can now eliminate the exact Gamkrelidze system (6.71) from consideration and use the generated β-systems (6.72) or the approximative DI (6.73). Note that the method based on the β-system (6.72) was effectively applied in [32] to problems of optimal control design in some particular cases of variable structure dynamics.

Let us now discuss "Part II" of the complete sliding mode dynamic behavior and concentrate on the corresponding approximative techniques for this part of the control process. Consider the time interval $[T_{sl}, t_f]$ and introduce a bounded measurable variation $\delta u(\cdot)$ of the given control $u_{eq}(t)$. The first variation of the state variable $x(\cdot)$ that corresponds to

$$u_{eq}(\cdot) + \delta u(\cdot)$$

is as usual denoted by

$$x(\cdot) + \delta x(\cdot).$$

Evidently, the original (nondisturbed) control $u_{eq}(\cdot)$ and the realized trajectory $x(\cdot)$ determine the reference sliding dynamics of the closed-loop system (6.72). Using the standard first-order argument and linearization techniques (see, e.g., [185,205,265]), we can easily derive the variational differential equation associated with the (open-loop) system (6.72) which has an input equal to the equivalent control, i.e.,

$$\begin{aligned} \dot{\delta x}(t) &= \Big[\frac{\partial a}{\partial x}(x(t)) + \sum_{j=1}^{m} \frac{\partial F_j}{\partial x}(x(t)) w_{eq}^{j}(x(t))\Big]\delta x(t) + b(x(t))\delta u(t), \\ \delta x(0) &= 0. \end{aligned} \tag{6.74}$$

Note that (6.74) is considered on the interval $[T_{sl}, t_f]$ and

$$w_{eq}^{j}(x), \; j = 1, \dots, m,$$

denotes the component of $w_{eq}(x)$.

Let now $\mathcal{U}_S$ be a subset of $\mathcal{U}$ such that

$$x(t) + \delta x(t) \in \mathcal{S}$$

for all $t \geq T_\delta$ with

$$T_\delta < t_f.$$

From the general results of the variable structure systems theory it follows that the set $\mathcal{U}_S$ is nonempty (see, e.g., [254] and the references therein). For example, $\mathcal{U}_S$ contains the equivalent control $\delta u_{eq}(\cdot)$ for the given linear system (6.74). Note that

$$\delta u_{eq}(t) \equiv 0$$

in the case of linear system (6.72). The following result is an immediate consequence of the classic equivalent control concept [305].

Theorem 6.22. *Let all conditions for the basic system (6.72) be satisfied and*

$$u_{eq}(\cdot) + \delta u(\cdot) \in \mathcal{U}_S$$

for a bounded measurable $\delta u(\cdot)$. Then

$$\delta x(t) \in \mathcal{S}$$

for all $t \geq T_\delta$.

Proof. From the definition of the set $\mathcal{U}_S$ and (6.74) we easily deduce the following:

$$S[a(x(t)+\delta x(t))+b(x(t)+\delta x(t))(w_{eq}(x(t))+\delta u(t))]=0$$

for all $t \geq T_\delta$. Since

$$u_{eq}(\cdot)=w_{eq}(x(t))$$

is an equivalent control for the original system (6.72), we have

$$S[a(x(t))+b(x(t)w_{eq}(x(t)))]=0 \ \forall t \geq T_\delta.$$

Using the conventional first-order argument, we easily obtain

$$S[(\frac{\partial a}{\partial x}(x(t))+\sum_{j=1}^{m}\frac{\partial F_j}{\partial x}(x(t))w_{eq}^j(x(t)))\delta x(t)+b(x(t))\delta u(t)]=0$$

for $t \geq T_\delta$. The last relation shows that the states $\delta x(t)$ associated with the linearized system (6.74) with $\delta u(\cdot)$ belong to the tangent space of $\mathcal{S}$ for all $t \geq T_\delta$. Since we consider the linear-affine sliding surfaces $\mathcal{S}$ defined by

$$\sigma(x)=Sx+c, \ S \in \mathbb{R}^{m \times n}, \ c \in \mathbb{R}^m,$$

we easily obtain the inclusion $\delta x(t) \in \mathcal{S}$. This completes the proof. □

Theorem 6.22 shows that the linear-affine sliding surface $\sigma(\cdot)$ also constitutes a sliding surface for the linearized trajectory $\delta x(\cdot)$. Let

$$x^{u_{eq}+\delta u}(\cdot)$$

be the solution of (6.72) generated by an admissible

$$u_{eq}(\cdot)+\delta u(\cdot).$$

Note that for the general (nonlinear) affine systems (6.72)

$$x^{u_{eq}+\delta u}(\cdot) \neq x(t)+\delta x(t).$$

By $x^{u_{eq}}(\cdot)$ we next denote a solution to (6.72) associated with $u_{eq}(\cdot)$. Assume that $\delta x(\cdot)$ solves (6.74) closed by an admissible $\delta u(\cdot)$ such that

$$u(\cdot)+\delta u(\cdot) \in \mathcal{U}_S.$$

The consistency of linear approximations given by (6.74) can be established using the general continuity result from [42].

Theorem 6.23. *Assume that the initial system (6.72) satisfies all technical assumptions. Then there exists a function* $o : \mathbb{R}_+ \to \mathbb{R}_+$ *such that*

$$s^{-1}o(s) \to 0$$

as $s \downarrow 0$ *and*

$$||x^{u_{eq}+\delta u}(\cdot) - (x^{u_{eq}}(\cdot) + \delta x(\cdot))||_{\mathbb{L}_n^\infty} \leq o(||\delta u(\cdot)||_{\mathbb{L}_m^\infty})$$

for all $u_{eq}(\cdot) + \delta u(\cdot) \in \mathcal{U}_S$.

Proof. From the celebrated Gronwall lemma and the corresponding comparison result (see [86,267]) with the comparison functions

$$\xi(t) := x^{u_{eq}}(t) + \delta x(\cdot)(t), \quad \psi(t, x) := a(x) + b(t)u + \delta u(t),$$

we obtain

$$\begin{aligned}
&||x^{u_{eq}+\delta u}(\cdot) - (x^{u_{eq}}(\cdot) + \delta x(\cdot))||_{\mathbb{L}_n^\infty} \leq e^C \int_{T_{sl}}^{t_f} ||\dot{x}^{u_{eq}}(t) + \dot{\delta x}(t) - \\
&a(x^{u_{eq}}(t) + \delta x(t)) - b(t)(u_{eq}(t) + \delta u(t))||dt = \\
&e^C \int_{T_{sl}}^{t_f} ||\langle(a_x(x^{u_{eq}}(t)), b(t)), (\delta x(t), \delta u(t))\rangle - \\
&[a(x^{u_{eq}}(t) + \delta x(t)) + b(t)\delta u(t) - a(x^{u_{eq}}(t)]||dt
\end{aligned} \tag{6.75}$$

for a constant C. Using the componentwise variant of the classic mean value theorem (see also Chapter 2), we conclude that

$$\begin{aligned}
&a_i(x^{u_{eq}}(t) + \delta x(t)) + b(t)(u_{eq}(t) + \delta u(t)) - (a_i(x^{u_{eq}}(t)) + b(t)u_{eq}(t)) = \\
&\langle(a_i)_x(x^{u_{eq}}(t) + \nu_i(t)), b(t))(\delta x(t), \delta u(t))\rangle
\end{aligned}$$

for a suitable bounded function $\nu(\cdot)$ and with $i = 1, ..., n$, $j = 1, ..., m$. The differentiability of $a(\cdot)$ implies the Lipschitz continuity of this function on a bounded set $\mathcal{R}$. The last observation causes the existence of a function $o_1 : \mathbb{R}_+ \to \mathbb{R}_+$ such that

$$s^{-1}o_1(s) \to 0$$

as $s \downarrow 0$ and

$$\begin{aligned}
&||\langle(a_x(x^{u_{eq}}(t)), b(t)), (\delta x(t), \delta u(t))\rangle - \\
&[a(x^{u_{eq}}(t) + \delta x(t)) + b(t)\delta u(t) - a(x^{u_{eq}}(t))]|| \leq o_1(||\delta u(\cdot)||_{\mathbb{L}_m^\infty})
\end{aligned}$$

for all t from the given interval $[T_{sl}, t_f]$. From (6.75) we finally deduce the expected estimation

$$||x^{u_{eq}+\delta u}(\cdot) - (x^{u_{eq}}(\cdot) + \delta x(\cdot))||_{\mathbb{L}_n^\infty} \leq o(||\delta u(\cdot)||_{\mathbb{L}_m^\infty})$$

with

$$o(s) := e^C o_1(s).$$

This completes the proof. □

We refer to [42] for the proof of a general version of Theorem 6.23. The consistency result established in Theorem 6.22 can be used for effective error estimations in linear approximations of the given sliding mode regimes. Note that in a general case the disturbed trajectory

$$x^{u_{eq}+\delta u}(\cdot)$$

does not belong to the sliding manifold $\mathcal{S}$ even if

$$x^{u_{eq}}(t) + \delta x(t) \in \mathcal{S}$$

for all

$$t \geq T := \max\{T_{sl}, T_\delta\}.$$

The sliding property of solutions $\delta x(\cdot)$ to (6.74) makes it possible to derive a Hamiltonian formalism associated with the closed-loop realization of the original nonlinearly affine system (6.72). This equivalent representation makes sense for "Part II" of the complete systems trajectory, namely, for the time interval $[T_{sl}, t_f]$. Consider a restriction of the linear system (6.74) on $[T, t_f]$, where

$$T = \max\{T_{sl}, T_\delta\}.$$

We also put $\delta x(T) = 0$. Let $\Phi(\cdot)$ be the fundamental matrix on $[T, t_f]$ solution a.e. of

$$\dot{\Phi}(t) = \Big[\frac{\partial a}{\partial x}(x(t)) + \sum_{j=1}^{m} \frac{\partial F_j}{\partial x}(x(t)) w_{eq}^j(x(t))\Big]\Phi(t),$$
$$\Phi(T) = I.$$

Here I is the unit matrix. We are now ready to formulate the main theoretic result of our contribution.

Theorem 6.24. *Consider the closed-loop realization of* (6.72)–(6.75) *on the "sliding" time interval* $[T, t_f]$. *Then there exists an absolutely continuous function* (*the adjoint vector*)

$$p(\cdot), \ p(t) \in \mathbb{R}^n$$

such that the pair

$$(u_{eq}(\cdot), x(\cdot), p(\cdot))$$

is a solution (a.e. on $[T, t_f]$*) of the following* Hamiltonian *system:*

$$\begin{aligned}
\dot{x}(t) &= \frac{\partial H}{\partial p}(u_{eq}(t), x(t), p(t)), \\
\dot{p}(t) &= -\frac{\partial H}{\partial x}(u_{eq}(t), x(t), p(t)), \\
\frac{\partial H}{\partial u}&(u_{eq}(t), x(t), p(t)) = 0,
\end{aligned} \tag{6.76}$$

where

$$H(u, x, p) := \langle p, a(x) + b(x)u \rangle$$

is the pseudo-Hamiltonian *and* $\langle \cdot, \cdot \rangle$ *is the inner product.*

Proof. Note that the first equation from (6.76) is a simple consequence of the linear structure of the Hamiltonian. Consider (6.75) for such bounded measurable functions $\delta u(\cdot)$ that guarantee

$$u_{eq}(\cdot) + \delta u(\cdot) \in \mathcal{U}_S.$$

From the results proved above it follows that the dimension of the linear space

$$\left\{ \delta x(\cdot) \in \mathbb{L}_m^{\infty}(0, t_f) \mid \Phi(t_f) \int_T^{t_f} \Phi^{-1}(t) b(x(t)) \delta u(t) dt \right\}$$

that contains the set

$$\left\{ \delta x(\cdot) \mid \Phi(t_f) \int_T^{t_f} \Phi^{-1}(t) b(x(t)) \delta u(t) dt \right\}$$

is equal to m. Therefore, there exists a vector $\rho \in \mathbb{R}^n$ such that

$$\rho^T \Phi(t_f) \Phi^{-1}(t) b(x(t)) = 0$$

for almost all $t \in [T, t_f]$. We now introduce the function $p(\cdot)$ by setting

$$p(t) := \rho^T \Phi(t_f) \Phi^{-1}(t).$$

The definition of $p(\cdot)$ implies that this function is a solution to the adjoint system

$$\dot{p}(t) = -p(t)\big[\frac{\partial a}{\partial x}(x(t)) + \sum_{j=1}^{m} \frac{\partial F_j}{\partial x}(x(t))w_{eq}^{j}(x(t))\big].$$

Moreover, it satisfies almost everywhere the equality

$$p(t)b(x(t)) = 0$$

and we get the second and the third equation in (3.1). The proof is completed. □

In the case of a nontrivial adjoint vector $p(\cdot)$, Theorem 6.24 gives a variational characterization of the sliding dynamics given by

$$(u_{eq}(\cdot), x(\cdot)).$$

Evidently, this theorem is similar to the so-called weak Pontryagin maximum principle in optimal control (see [86,96,154,173]). The restriction of the sliding trajectory $x(\cdot)$ on the interval $[T, t_f]$ satisfies the so-called weak maximum principle (see, e.g., [154]) with the Hamiltonian

$$H(u, x(t), p(t)).$$

Moreover, similar to optimal control theory (see, e.g., [86,96,154,265]) one can deduce some simple consequences of (6.76). For example, we have

$$\begin{aligned}
&\frac{dH}{dt}(u_{eq}(t), x(t), p(t)) = 0 + \frac{\partial H}{\partial x}(u_{eq}(t), x(t), p(t))[a(x(t))+ \\
&b(x(t))u_{eq}(t)] + \frac{\partial H}{\partial p}(u_{eq}(t), x(t), p(t))\dot{p}(t) = -\dot{p}(t)[a(x(t))+ \\
&b(x(t))u_{eq}(t)] + [a(x(t)) + b(x(t))u_{eq}(t)]\dot{p}(t) = 0
\end{aligned} \tag{6.77}$$

for a solution

$$(u_{eq}(\cdot), x(\cdot), p(\cdot))$$

of (6.76) on $[T, t_f]$. Evidently, (6.77) also specifies the given manifold $\mathcal{S}$. We have

$$\begin{aligned}
&\tilde{H}(t) := \langle p(t), a(x(t)) + b(x(t))w_{eq}(x(t))\rangle = \\
&\langle p(t), a(x(t)) - b(x(t))(Sb(x))^{-1}[Sa(x)]\rangle = \text{const.}
\end{aligned} \tag{6.78}$$

One can interpret the derived relations (6.77) and (6.78) as a generalized "energetic" description of the sliding dynamics $(u_{eq}(\cdot), x(\cdot))$ and the manifold $\mathcal{S}$. This interpretation can be formalized as follows:

$$\begin{aligned}\dot{x}(t) &= \frac{\partial H}{\partial p}(u_{eq}(t), x(t), p(t)),\\ \dot{p}(t) &= -\frac{\partial H}{\partial x}(u_{eq}(t), x(t), p(t)),\\ \tilde{H}(t) &= H(u_{eq}(t), x(t), p(t)) = \text{const},\ t \in [T, t_f].\end{aligned} \tag{6.79}$$

It is necessary to stress that the sliding manifold $\sigma(\cdot)$ is in general an algebraic invariant of the initial closed-loop system (6.72)–(6.75). Hence

$$\nabla\sigma(x(t))\dot{x}(t) = 0$$

holds only for $x \in \mathcal{S}$. In contrast to that case, the Hamiltonian "sliding manifold" determined by

$$H(u, x, p) - \text{const} = 0$$

is a first integral of the extended system (6.76) and the relation

$$\langle \nabla_{(x,p)} H(u_{eq}(t), x(t), p(t))(\dot{x}^T(t), \dot{p}^T(t))^T \rangle = 0$$

is true in an open subset of $\mathbb{R}^{2n}$. Let us also note that the asymptotic stability and the invariant property of the "Hamiltonian" manifold determined by

$$\tilde{H}^{-1}(\text{const})$$

can be more easily stated as the same property of $\mathcal{S}$ in (6.72)–(6.75). The corresponding stability/invariance conditions for (6.79) are given by the known Ascher–Chin–Reich theorem [129,131].

The case of a possible trivial value of the adjoint vector $p(\cdot)$ in Theorem 6.24 will be discussed below. We now consider a simple example associated with the obtained main result.

Example 6.4. *Let $n = 2,\ m = 1$, and $\mathcal{S} = (0, 1)$. Assume that*

$$\begin{aligned}&a(x) = (x_1^2 + x_2^2, 0)^T,\ b(x) = (0, 1)^T,\\ &x_1(0) = -1,\ x_2(0) = 0.\end{aligned}$$

The equivalent control can be calculated here directly, i.e.,

$$u_{eq}(t) = -(Sb(x(t)))^{-1}[Sa(x(t))] = 0.$$

The corresponding sliding trajectory is contained into S that is characterized by the simple condition $x_2 = 0$. The solutions of the one-dimensional differential equation

$$\dot{x}_1 = x_1^2$$

with $x_1(0) = -1$ are given by

$$x_1(t) = -1/(t+1).$$

Using the system Hamiltonian

$$H(u, x, p) = p_1(x_1^2 + x_2^2) + p_2 u,$$

where $p = (p_1, p_2)$, we can apply the formalism of Theorem 6.24. *From* (6.76) *we obtain*

$$\dot{p}_1 = -2p_1 x_1, \quad \dot{p}_2 = -2p_1 x_2, \quad p_2 = 0,$$

and

$$-2p_1 x_2 = 0.$$

Under assumption $p \neq 0$ we obtain the sliding surface determined by the condition $x_2 = 0$. The sliding control that guarantee $x_2 = 0$ is the same as $u_{eq}(t) = 0$. The corresponding trajectory also coincides with the dynamic behavior obtained above, i.e.,

$$x_1(t) = -1/(t+1), \; x_2(t) = 0.$$

As mentioned above the system (6.76) from Theorem 6.24 is similar to the generic conditions of a weak Pontryagin maximum principle in the classic optimal control. From the corresponding proof it follows that the pair $(u_{eq}(\cdot), x(\cdot))$ on $[T, t_f]$ can be formally considered as an optimal solution (optimal pair) to the auxiliary OCP on $[T, t_f]$ involving (6.72) with the terminal costs functional $\phi(x(t_f))$ such that

$$\frac{\partial \phi}{\partial x}(x(t_f)) = p(t_f) = \rho^T, \quad \rho^T \Phi(t_f)\Phi^{-1}(t) b(x(t)) = 0,$$

where $\rho \in \mathbb{R}^n$, $\rho \neq 0$. From the point of view of the Hamiltonian system (6.76) the information about a sliding manifold S is now presented by the vector function $p(\cdot)$. Our main variational result, namely, Theorem 6.24 does not exclude a possible zero solution for the second equation in (6.76). In that case we have a degenerate Hamiltonian system (6.76) characterized by the condition

$$p(t) \equiv 0.$$

Let us illustrate this situation by the next simple example.

Example 6.5. *Consider the following controllable linear system:*

$$\dot{x}_1 = x_2, \ \dot{x}_2 = u$$

with $x_1, x_2, u \in \mathbb{R}$, $t \in [T, t_f]$. *The sliding manifold* $\mathcal{S}$ *associated with the system is assumed to be given by the condition*

$$x_1 + x_2 = 0.$$

The differential equation for the adjoint variable from (6.76) *with*

$$H(x, p) = p_1 x_2 + p_2 u$$

implies

$$\dot{p}_1 = 0, \ \dot{p}_2 = -p_1.$$

From the Hamiltonian *stationarity condition*

$$\partial H / \partial u = 0$$

in (6.76) *we deduce that* $p_2 = 0$. *Therefore, we also have* $p_1 = 0$. *Otherwise, the equivalent control computed by the explicit formula implies*

$$w_{eq}(x) = -x_2.$$

Summarizing, we can note that in this case of the degenerate Hamiltonian *system the conditions in* (6.76) *are nonconstructive and the equivalent control cannot be found from the system* (6.76).

Recall that the case of a degenerate Hamiltonian system (6.76) is similar to the analogous situation in the classic optimal control. When solving conventional OCPs based on some necessary conditions for optimality one is often faced with two technical difficulties: the irregularity of the Lagrange multiplier associated with the state constraint [132,265] and the degeneracy phenomenon (see, e.g., [17]). Various supplementary conditions (so-called constraint qualifications) have been proposed under which it is possible to assert that the Lagrange multiplier rule holds in a "usual" constructive form (see [130]). Examples are the well-known Slater regularity condition for classic convex programming and the Mangasarian–Fromovitz regularity conditions for general nonlinear optimization problems. We refer to [36, 17,96,157,198,265] for details. Note that some regularity conditions for constrained OCPs can be formulated as controllability conditions for the linearized system [198]. Our aim now is to formulate some conditions that guarantee the existence of a nondegenerate ($p(t) \neq 0$) solution of the main Hamiltonian system in (3.1). We firstly recall the following general concept of an end point mapping (see, e.g., [63]).

Definition 6.14. *Consider system* (6.72) *and assume that all the generic technical conditions for this system are satisfied. The mapping*

$$E(x_0, t_f) : \mathcal{U} \to x(t_f)$$

is called an end point mapping associated with this system.

Note that $E(x_0, t_f)$ is defined here for fixed x_0, t_f. The differentiability properties of this mapping in the context of some applications are studied in [63]. We only mention here the following useful fact.

Theorem 6.25. *Under the basic conditions for system (6.72) the end point mapping* $E(x_0, t_f)$ *is* Fréchet *differentiable and the corresponding derivative* $E_u(x_0, t_f)$ *at* $u(\cdot) \in \mathcal{U}$ *can be computed as follows:*

$$E_u(x_0, t_f) = \delta x(t_f),$$

where $\delta x(\cdot)$ *is a solution to the linearized system* (6.74). *If the* Fréchet *derivative* $E_u(x_0, t_f)$ *of the end point mapping* $E(x_0, t_f)$ *is nonsurjective, then the Hamiltonian system in* (6.76) *is nondegenerate and there exists an absolute continuous solution* $p(\cdot)$ *to the adjoint equation from* (6.76) *with*

$$p(t) \in \mathbb{R}^n \setminus \{0\}.$$

As we can see, Theorem 6.25 expresses a sufficient condition for the nondegeneracy of the Hamiltonian system under consideration. It is easy to see that for the linear version of the basic system (6.72) with

$$a(x) = A \in \mathbb{R}^{n \times n}, \quad b(x) = B \in \mathbb{R}^{n \times m}$$

the standard controllability condition guarantees the surjectivity of the Fréchet derivative $E_u(x_0, t_f)$ of the associated end point mapping. Therefore, these systems are in the bad set and indeed they correspond to the degenerate Hamiltonian systems (6.76) (as shown in Example 6.5). Moreover, in the case of the linear system under consideration the equivalent control

$$\delta u_{eq}(\cdot) \in \mathcal{U}_S$$

in (6.74) is equal to zero and the adjoint vector $p(\cdot)$ cannot be constructively specified from the proof of the main Theorem 6.24. However, one can explicitly use the given information on the sliding surface (the algebraic invariant) S and introduce the "regularized" pseudo-Hamiltonian

$$\mathcal{H}(u, x, p) := \langle p, a(x) + b(x)u \rangle - \frac{1}{2}||Sx + c||^2$$

that corresponds to the original system (6.72) augmented by the following artificial differential equation for the new coordinate with

$$\dot{x}_{n+1} = \frac{1}{2}||Sx + c||^2.$$

It is common knowledge that $\mathcal{H}(u, x, p)$ can also be interpreted as a Hamiltonian in the following Bolza-type OCP with terminal and integral terms:

$$\begin{aligned} &\text{minimize } \phi(t_f) + \frac{1}{2}\int_0^{t_f} ||Sx(t) + c||^2 dt \\ &\text{subject to } (1.1),\ t \in [T, t_f],\ u(\cdot) \in \mathcal{U}. \end{aligned} \tag{6.80}$$

Evidently, the abovementioned augmentation of the initial system (6.72) or the equivalent consideration of the OCP (6.80) with the Bolza costs functional involves necessary "information" regarding the existent algebraic invariant in an explicit form. In that case the proof of the above Theorem 6.25 can also be realized for the extended Hamiltonian $\mathcal{H}(u, x, p)$ without any conceptual changes. Using the same orthogonality argument and the existence of the algebraic invariant for the system (6.72) one can claim the existence of a vector $\rho \in \mathbb{R}^n$ such that

$$\rho^T[\Phi(t_f)\Phi^{-1}(t)b(x(t)) - (Sx(t) + c)^2] = 0$$

for almost all $t \in [T, t_f]$. Recall that $x(t) \in \mathcal{S}$ for $t \in [T, t_f]$ and the formal augmentation above is in fact an augmentation by an effective zero. The last condition implies the modified adjoint equation similar to the second equation in (3.1) with respect to the augmented Hamiltonian $\mathcal{H}(u, x, p)$. Let us apply the proposed technique to the system from Example 6.4 with the initially degenerate Hamiltonian system.

Example 6.6. *Consider the dynamic model from* Example 6.5 *and extend this original system by the auxiliary equation*

$$\dot{x}_3 = \frac{1}{2}(x_1 + x_2)^2.$$

The augmented Hamiltonian *has the form*

$$\mathcal{H} = p_1 x_2 + p_2 u + p_3 \frac{1}{2}(x_1 + x_2)^2$$

and evidently corresponds to minimization of the auxiliary functional

$$\frac{1}{2}\int_0^{t_f} (x_1(t) + x_2(t))^2 dt$$

subject to the dynamic system of this example. From (6.76) *we deduce*

$$\dot{p}_1 = -p_3(x_1 + x_2), \ \dot{p}_2 = -p_1(x_1 + x_2), \ \dot{p}_3 = 0,$$

and

$$p_2 = 0.$$

The obtained adjoint system has a nonzero solution $(p_1, 0, p_3)$ *with some constant* p_1 *and* p_3 *if*

$$x_1 + x_2 = 0.$$

That leads to the correct variational characterization of the equivalent control

$$u_{eq}(t) = x_1(t) = -x_2(t), \ t \in [T, t_f].$$

Note that the linearized system (6.74) for the given augmented system in Example 6.6 is a controllable system. On the other hand, the surjectivity condition in Theorem 6.25 is only a sufficient condition for the nondegeneracy of the Hamiltonian system in (6.76).

We now discuss a computational scheme for a concrete evaluation of the extremal controls determined by (6.76). For a nonlinearly affine system of the type (6.72) the extremal condition for the Hamiltonian in (6.76) implies the following:

$$\langle p(t), F_j(x(t))\rangle = 0 \ \forall j = 1, ..., m \text{ a.e. on } [T, t_f]. \tag{6.81}$$

Differentiating (6.81) with respect to t, we get

$$q(x(t), p(t)) + Q(x(t), p(t))u_{eq}(t) = 0, \tag{6.82}$$

where $q(x, p)$ is the m-dimensional vector with components

$$q_j(x, p) := \langle p, [F_j, a](x)\rangle$$

and $Q(x, p)$ is the $m \times m$ skew-symmetric matrix with

$$Q_{i,j} := \langle p, [F_i, F_j](x)\rangle.$$

By $[\cdot, \cdot]$ we denote here the Lie brackets defined by the usual convention, i.e.,

$$[Z_1, Z_2](\xi) := \frac{\partial Z_1}{\partial \xi}(\xi)Z_2(\xi) - \frac{\partial Z_2}{\partial \xi}(\xi)Z_1(\xi).$$

Clearly, the last relation has a constructive character only in the case $[F_j, a](x) \neq 0$ and

$$[F_i, F_j](x) \neq 0$$

for all $x \in \mathbb{R}^n$ and all $i, j = 1, ..., m$. This relation represents a constructive basis for the constructive numerical treatment of Theorem 6.24.

Consider system (6.72) and suppose that $m \in \mathbb{N}$ is even. Moreover, assume that

$$\max_{(x,p)} \operatorname{rank} Q(x, p) = m$$

and $det\{Q(x, p)\} \neq 0$ for the all pairs (x, p) that satisfy (6.82). Under these additional assumptions relation (6.82) implies

$$u_{eq}(t) = -Q^{-1}(x(t), p(t))q(x(t), p(t)). \tag{6.83}$$

Here $x(\cdot)$ and $p(\cdot)$ are calculated by (6.76). For an odd number $m \in \mathbb{N}$ one can obtain a similar formula, the only complication being the existence of a kernel for $Q(x, p)$. In that case one can extend the original system (6.72) by an additional differential equation of the type

$$\dot{x}_{n+1} = u_{m+1}$$

such that the rank conditions for the newly defined

$$\operatorname{span}\{F_1(x), ..., F_m(x)\}$$

are also satisfied. In such a way we evidently come back to the case of an even control dimension. Let us now apply the numerical idea discussed above to an illustrative computational example.

Example 6.7. *Consider the nonlinear system*

$$\begin{cases} \dot{x}_1 = x_1^2 + x_2^2 + x_1 u_1 \\ \dot{x}_2 = x_1^2 + x_2 u_2 \end{cases}$$

with some suitable initial conditions $x_1(0)$ and $x_2(0)$. Here

$$x_1(t),\ x_2(t) \in \mathbb{R}$$

for all $t \in [0, t_f]$. Assume that the sliding manifolds $\mathcal{S}_1$ and $\mathcal{S}_2$ are expressed by

$$x_1 + x_2 = 0$$

and $x_1 - x_2$, respectively. The corresponding initial conditions for these two cases are assumed to be nontrivial and, moreover,

$$x_1(0) + x_2(0) = 0$$

or $x_1(0) - x_2(0) = 0$. *It is easy to see that the given control strategy*

$$(\hat{u}_1(\cdot), \hat{u}_2(\cdot)), \ \hat{u}_1(t) := -3x_1(t), \ \hat{u}_2(t) \equiv 0$$

implies the sliding condition $x(t) \in S_1$ *for all* $t \geq 0$. *For* $x(t) \in S_2$ *we get the control strategy*

$$(\tilde{u}_1(\cdot), \tilde{u}_2(\cdot)), \ \tilde{u}_1(t) := -x_1(t), \ \hat{u}_2(t) \equiv 0.$$

Note that control pairs $(u_1(\cdot), u_2(\cdot))$ *that guarantee the above two sliding behaviors are nonuniquely defined. For example, the condition* $x(t) \in S_2$ *can also be guaranteed for the given control system closed by*

$$u_1(t) = 0, \ u_2(t) = x_2(t).$$

Here we use the natural notation

$$x(t) := (x_1(t), x_2(t))^T.$$

The first components of the obtained control pairs are calculated using the explicit equivalent control formalism (6.76).

We now characterize the same sliding dynamic behavior by the Hamiltonian *system* (6.76). *Evidently,*

$$H(u, x, p) = p_1(x_1^2 + x_2^2 + x_1 u_1) + p_2 x_1^2 + p_2 x_2 u_2, \ p := (p_1 p_2)^T$$

and we easily obtain the corresponding adjoint system

$$\begin{cases} \dot{p}_1 = -(2p_1 x_1 + p_1 u_1 + 2p_2 x_1) \\ \dot{p}_2 = -(2p_1 x_2 + p_2 u_2). \end{cases} \tag{6.84}$$

The last condition in (6.76) *implies the algebraic condition*

$$(p_1 x_1, p_2 x_2)^T = 0.$$

Since the assumption $x_2 = 0$ *implies* $x_1 = 0$, *we get the contradiction with the nontriviality of the initial conditions. Therefore,* $p_2 = 0$. *Differentiating the expression* $p_1 x_1 = 0$ *with respect to the time and replacing the derivatives of the adjoint variable using* (6.84), *we get the simple orthogonality condition*

$$p_1(x_1 - x_2)(x_1 + x_2) = 0.$$

From the assumption $p \neq 0$ *the characterization of the possible sliding manifolds* S_1 *and* S_2 *follows. Note that the formalism* (6.83) *can be obtained applying the initially given system.*

For example, in the case $x_1 = x_2$ we use the given information about the derivatives (from the dynamic system) and derive the expression

$$u_1 - u_2 = -x_1.$$

Clearly, the specific control pair $(\tilde{u}_1(\cdot), \tilde{u}_2(\cdot))$, where $\tilde{u}_1(\cdot)$ is an equivalent control, satisfies the obtained general relation between the control components for the case $x(t) \in \mathcal{S}_2$.

Let us finally note that theoretic results from this section as well as the corresponding numerical tool do not use the dimensionality restriction $\dim \mathcal{S} = m < n$ that is typical for the sliding mode control design. Therefore, theory presented in this section can be considered as a generalization of the equivalent control concept for the general case

$$\dim \mathcal{S} < n.$$

The OCP-based approximation methodology discussed in this chapter can be developed not only to the classical sliding mode dynamics but also to higher-order realizations. In that case the generic variational Hamiltonian-based description needs to be extended by additional state/derivatives constraints related to the admissible higher-order sliding mode feedback control law. These restrictions are usually given in the following form:

$$\omega(x) := \tilde{\omega}(\sigma(x), \dot{\sigma}(x), ..., \sigma^{(l-1)}(x)).$$

Here $l \in N$ is a relative degree of the system under consideration. The resulting higher-order design procedure constitutes a sophisticated problem that not only incorporates the usual state constraint $\sigma(x(t)) = 0$ but is also a set of additional algebraical constraints, i.e.,

$$\dot{\sigma}(x(t)) = 0, \dots, \sigma^{(l-1)}(x(t)) = 0.$$

Under some concrete restrictions the specific structure of the abovementioned feedback control law can guarantee so-called "finite time" stability of the corresponding closed-loop system. A formal consequence of this dynamic property is a singularity of the end point mapping. This fact implies a possibility to rewrite the closed-loop model in an equivalent Hamiltonian form.

6.6 Notes

We have completed here a list of useful works devoted to the OCPs for hybrid, switched dynamic systems and also for the sliding mode control approach [2,3,7,10,19,33–68,74,80,81, 103–105,108,122–125,140–143,150–152,166,169–172,186,189,199,212,218,219,222–224, 231,232,240,245] and [255,257,270,294,298,299,305,310–313,316–320,323–325,333,334].

CHAPTER 7

Numerically Tractable Relaxation Schemes for Optimal Control of Hybrid and Switched Systems

7.1 The Gamkrelidze–Tikhomirov Generalization for HOCPs

7.1.1 Relaxation of the General HOCPs

Similar to the main optimal control problem (OCP) from Chapter 6 we now study the following Bolza-type hybrid OCP (HOCP):

$$\begin{aligned} &\text{minimize } J(x(\cdot), u(\cdot)) = \int_0^{t_f} f_0(t, x(t), u(t))dt \\ &\text{subject to } \dot{x}(t) = f_{q_i}(t, x(t), u(t)) \text{ a.e. on } [t_{i-1}, t_i] \\ &q_i \in Q, \ i = 1, ..., r+1, \ x(0) = x_0 \in M_{q_1}, \\ &u(\cdot) \in \mathcal{U}, \end{aligned} \tag{7.1}$$

where $f_0 : [0, t_f] \times \mathbb{R}^n \times \mathbb{R}^m \to \mathbb{R}$ is a continuously differentiable function and $x_0 \in \mathbb{R}^n$ is a given initial state. Let us assume that the functions f_q, $q \in Q$, are continuously differentiable. The set $U \subseteq \mathbb{R}^m$ of admissible controls in (7.1) is assumed to be compact and convex. We will restrict here our consideration to the "box-type" admissible control sets of the type

$$U := \{u \in \mathbb{R}^m \ : \ b_-^i \leq u_i \leq b_+^i, \ i = 1, ..., m\},$$

where

$$b_-^i, b_+^i, i = 1, ..., m,$$

are some constants. The admissible control functions $u : [0, t_f] \to \mathbb{R}^m$ are square integrable functions of time and the set of admissible control inputs is defined as follows:

$$\mathcal{U} := \{v(\cdot) \in \mathbb{L}_m^2([0, t_f]) \ : \ v(t) \in U \text{ a.e. on } [0, t_f]\}.$$

We consider the HOCP in the absence of additional state restrictions. Note that it is possible to extend the relaxation theory approach to classes of HOCPs with several state conditions.

A Relaxation-Based Approach to Optimal Control of Hybrid and Switched Systems
https://doi.org/10.1016/B978-0-12-814788-7.00013-8

We next consider the compact form (1.3) for the hybrid dynamic system in (7.1) and assume that for each admissible $u(\cdot) \in \mathcal{U}$ the initial value problem

$$\begin{aligned} \dot{x}(t) &= \tilde{f}(t, x(t), u(t)) \text{ a.e. on } [0, t_f], \\ \tilde{f}(t, x, u) &:= \sum_{i=1}^{r+1} \beta_{[t_{i-1}, t_i)}(t) f_{q_i}(t, x, u), \\ x(0) &= x_0 \end{aligned} \tag{7.2}$$

has a unique absolutely continuous solution $x^u(\cdot)$. Some constructive uniqueness conditions for solutions of the Carathéodory-type ordinary differential equations can be found in [86, 154,192,269]. Given an admissible control function $u(\cdot)$ a solution to the initial value problem (7.1) is an absolutely continuous function $x : [0, t_f] \to \mathbb{R}^n$. Let us note that the class of HOCPs (7.1) is broadly representative (see [35–38,103–105,231,232] and references therein).

Similar to the abstract constructions of Chapter 5 we use here the following notation:

$$\aleph(n+1) := \Big\{ (\alpha^1(\cdot), ..., \alpha^{n+1}(\cdot))^T \; : \; \alpha^j(\cdot) \in \mathbb{L}_1^1([0, t_f]), \; \alpha^j(t) \geq 0,$$

$$\sum_{j=1}^{n+1} \alpha^j(t) = 1 \; \forall t \in [0, t_f] \Big\}, \; \alpha(\cdot) := (\alpha^1(\cdot), ..., \alpha^{n+1}(\cdot))^T.$$

We next assume that the number $r \in \mathbb{N}$ of switchings in (7.2) is finite such that we avoid a possible Zeno behavior in the hybrid system (HS) under consideration. We also refer to Chapter 6 for an effective Zeno-free approximation of the existing Zeno dynamics.

For the given hybrid control system (7.2) we introduce the relaxed system in the form of a Gamkrelidze system (the Gamkrelidze chattering, see Chapter 5). We have

$$\begin{aligned} \dot{y}(t) &= \sum_{j=1}^{n+1} \alpha^j(t) \tilde{f}(t, y(t), u^j(t)) \text{ a.e. on } [0, t_f], \\ y(0) &= x_0, \end{aligned} \tag{7.3}$$

where

$$\alpha(\cdot) \in \aleph(n+1), \; u^j(\cdot) \in \mathcal{U},$$

for $j = 1, ..., n+1$. Under some generic assumptions there exists an absolutely continuous solution $y^\nu(\cdot)$ of the specific initial value problem (i.v.p.) (7.3) generated by the (admissible) generalized control

$$\nu(\cdot) \in \aleph(n+1) \times \mathcal{U}^{n+1},$$

where

$$\nu(t) := (\alpha^1(t), ..., \alpha^{n+1}(t), u^1(t), ..., u^{n+1}(t))^T.$$

Evidently, the relaxed system (7.3) can be rewritten as follows:

$$\begin{aligned} \dot{y}(t) &= \sum_{j=1}^{n+1} \alpha^j(t) \sum_{i=1}^{r+1} \beta_{[t_{i-1},t_i)}(t) f_{q_i}(t, y(t), u^j(t)) \text{ a.e. on } [0, t_f], \\ y(0) &= x_0. \end{aligned} \tag{7.4}$$

Since functions $\beta_{[t_{i-1},t_i)}(\cdot)$ associated with an HS depend on the "continuous" trajectory $x^u(\cdot)$, the generalized control $\nu(\cdot)$ does not include these functions and the Gamkrelidze systems (7.3) and (7.4) are similar to the corresponding relaxations of classical OCPs (see Chapter 5).

Taking into consideration the last remark, consider the Radon probability measure ς on the Borel sets of U (a regular positive measure such that $\varsigma(U) = 1$). Recall that by $\mathcal{M}^1_+(U)$ we denote the space of all probability measures on the Borel sets of U. A relaxed control $\mu(\cdot)$ associated with the Gamkrelidze system (7.3) is a measurable function

$$\mu : [0, t_f] \to \mathcal{M}^1_+(U).$$

The set of relaxed controls $\mu(\cdot)$ is denoted $R_c([0, t_f], U)$. The Young relaxation of the initially given HS (7.2) can now be written similarly to the corresponding relaxed schemes from Chapter 5, i.e.,

$$\begin{aligned} \dot{\eta}(t) &= \int_U \tilde{f}(t, \eta(t), u)\mu(t)(du) \text{ a.e on } [0, t_f], \\ \eta(0) &= x_0. \end{aligned} \tag{7.5}$$

Using the concept of an orientor field, the initially given hybrid control system can be rewritten in the form of a differential inclusion (DI), i.e.,

$$\begin{aligned} \dot{x}(t) &\in \tilde{f}(t, x(t), U) \text{ a.e. on } [0, t_f], \\ x(0) &= x_0, \end{aligned}$$

with

$$\tilde{f}(t, x(t), U) := \{\tilde{f}(t, x(t), u) \ : \ u \in U\}.$$

A function $x^u(\cdot)$ is a solution of the initial control system (7.1) with $u(\cdot) \in \mathcal{U}$ if and only if it is a solution of the DI given above (see [154] and Chapter 5). The relaxed DI (RDI) associated with the initially given DI (7.5) has a generic form, i.e.,

$$\begin{aligned} \dot{x}(t) &\in \operatorname{conv} \tilde{f}(t, x(t), U) \text{ a.e. on } [0, t_f], \\ x(0) &= x_0. \end{aligned} \tag{7.6}$$

We are now ready to reformulate the fundamental equivalence theorem (see [127,321,154] and Chapter 5).

Theorem 7.1. *Let the initially given control system (7.2) from HOCP (7.1) satisfy the above technical assumptions. A function $y(\cdot)$ is a solution of the* Gamkrelidze *system (7.3) with*

$$\nu(\cdot) \in \aleph(n+1) \times \mathcal{U}^{n+1}$$

if and only if it is a solution of the RDI (7.6). *Moreover, a function $x(\cdot)$ is a solution of the* RDI (7.6) *if and only if it is a solution of the Young relaxed control system (7.5) with a generalized control*

$$\mu(\cdot)(\cdot) \in R_c([0, t_f], U).$$

Let now

$$\nu(\cdot) = (\alpha^1(\cdot), ..., \alpha^{n+1}(\cdot), u^1(\cdot), ..., u^{n+1}(\cdot))^T,$$
$$\bar{\nu}(\cdot) = (\bar{\alpha}^1(\cdot), ..., \bar{\alpha}^{n+1}(\cdot), \bar{u}^1(\cdot), ..., \bar{u}^{n+1}(\cdot))^T.$$

Similar to the useful estimations presented in Chapter 5 one can obtain the corresponding inequalities for HSs of the type (7.2).

Theorem 7.2. *Let the initial control system* (7.2) *satisfy the basic technical assumptions. There exist finite constants c_1, c_2, c_3, and c_4 such that*

$$\begin{aligned}
&||x^u(\cdot)||_{\mathbb{L}_n^\infty([0,t_f])} \leq c_1, \\
&||x^u(\cdot) - x^{\tilde{u}}(\cdot)||_{\mathbb{L}_n^\infty([0,t_f])} \leq \\
&\leq c_2 ||u(\cdot) - \tilde{u}(\cdot)||_{\mathbb{L}_m^2([0,t_f])}, \\
&||y^\nu(\cdot)||_{\mathbb{L}_n^\infty([0,t_f])} \leq c_3, \\
&||y^\nu(\cdot) - y^{\bar{\nu}}(\cdot)||_{\mathbb{L}_n^\infty([0,t_f])} \leq \\
&\leq c_4 \sum_{j=1}^{n+1} ||u^j(\cdot) - \bar{u}^j(\cdot)||_{\mathbb{L}_m^2([0,t_f])},
\end{aligned}$$

for all

$$u(\cdot), \tilde{u}(\cdot) \in \mathcal{U},\ \nu(\cdot), \bar{\nu}(\cdot) \in \aleph(n+1) \times \mathcal{U}^{n+1},$$

where $x^u(\cdot)$, $x^{\tilde{u}}(\cdot)$ are the solutions of (7.2) *and $y^\nu(\cdot)$, $y^{\bar{\nu}}(\cdot)$ are the solutions of* (7.3) *associated with the admissible controls $u(\cdot)$, $\tilde{u}(\cdot)$ and $\nu(\cdot)$, $\bar{\nu}(\cdot)$, respectively.*

The Gamkrelidze relaxation of the initially given HOCP (7.1) is now determined as follows:

$$\begin{aligned}&\text{minimize } \bar{\mathcal{J}}(\nu(\cdot))\\&\text{subject to (7.3) } \nu(\cdot) \in \aleph(n+1) \times \mathcal{U}^{n+1},\\&q_i \in Q,\ i = 1, ..., r+1,\end{aligned} \tag{7.7}$$

where

$$\bar{\mathcal{J}}(\nu(\cdot)) := \int_0^{t_f} \int_U f_0(t, x(t), u)\mu(t)(du)dt.$$

Under some mild assumptions the relaxed HOCP (7.7) has an optimal solution $\nu^{opt} \in \aleph(n+1) \times \mathcal{U}^{n+1}$. We prove the corresponding result in Chapter 8. An optimal solution of (7.7) is hereafter denoted by $\nu^{opt}(\cdot)$. The following result uses the equivalence theorem (Theorem 7.1) and establishes the relation between the initially given HOCP (7.1) and the corresponding Gamkrelidze relaxation (7.7).

Theorem 7.3. *Let $\nu(\cdot) \in \aleph(n+1) \times \mathcal{U}^{n+1}$ be an optimal solution of the relaxed HOCP (7.7) and $u^{opt}(\cdot) \in \mathcal{U}$ be an optimal solution of the initially given HOCP (7.1). Then*

$$\bar{\mathcal{J}}(\nu^{opt}(\cdot)) \leq \mathcal{J}(u^{opt}(\cdot)).$$

Theorem 7.3 can be proved similarly to the corresponding result from Chapter 5. Similar to the conventional Gamkrelidze relaxation we obtain a natural relaxation gap

$$Gap := \mathcal{J}(u^{opt}(\cdot)) - \bar{\mathcal{J}}(\nu^{opt}(\cdot))$$

associated with the HOCPs (7.1) and (7.7).

Let $\mathbb{L}^1([0, t_f], \mathbb{C}(U))$ be the space of absolutely integrable functions from $[0, t_f]$ to $\mathbb{C}(U)$ (the space of all continuous function on U). Recall (see Chapter 5) that the topology imposed on $R_c([0, t_f], U)$ is the weakest topology such that the mapping

$$\mu \to \int_0^{t_f} \int_U \psi(t, u)\mu(t)(du)dt$$

is continuous for all $\psi(\cdot, \cdot) \in \mathbb{L}^1([0, t_f], \mathbb{C}(U))$. Finally we recall ([321], p. 287) that the space of the Young measures $R_c([0, t_f], U)$ is a compact and convex space of the dual space to $\mathbb{L}^1([0, t_f], \mathbb{C}(U))^*$. The topology of the "weak" norm associated with the space

$$\mathbb{L}^1([0, t_f], \mathbb{C}(U))^*$$

and restricted to the space $R_c([0, t_f], U)$ coincides with the weak star topology [321].

We now observe that the Gamkrelidze relaxation (7.4) is equivalent to the following i.v.p.:

$$\dot{y}(t) = \sum_{i=1}^{r+1} \beta_{[t_{i-1},t_i)}(t) \sum_{j=1}^{n+1} \alpha^j(t) f_{q_i}(t, y(t), u^j(t)) \text{ a.e. on } [0, t_f],$$
$$y(0) = x_0. \tag{7.8}$$

The above form implies the following simple fact.

Theorem 7.4. *A Gamkrelidze relaxation of the HS (7.2) is equivalent to the weighted (by functions $\beta_{[t_{i-1},t_i)}(\cdot)$) sum of the Gamkrelidze relaxed subsystems f_{q_i}. The characteristic functions $\beta_{[t_{i-1},t_i)}(\cdot)$ for the above sum are determined by the sequence τ of the switching times associated with the trajectory $x^u(\cdot)$ of the initially given HS (7.2).*

The proof of Theorem 7.4 immediately follows from (7.8).

7.1.2 Full Relaxation of the HOCPs Associated With the Switched Mode Dynamics

We next consider a switched mode system of the type (6.6) and the associated HOCP (6.8) (Chapter 6). As noted in Chapter 5 the celebrated chattering lemma constitutes an important analytic tool for the general theory of DIs (see [10,15]). Moreover, it also provides a necessary theoretic fundament for the proof of the Pontryagin maximum principle (see, e.g., [10]). We next indicate the approach based on the chattering lemma applied to the HOCP (6.8) as a "fully relaxed" approach. A practical application of this methodology can cause a significant relaxation gap (see, e.g., [2,6,26,32]). This is specifically true in the case of the discrete-like part of the complete control input in the above HOCP. A novel theoretical consideration of a specific subclass of HSs, namely, the switched mode models, makes it possible to apply a new relaxation concept that involves an infimal prox convolution and causes a lower gap (see Section 7.4). Moreover, the particular approach we propose in the next sections can also be applied to a specific class of the general OCPs, namely, to singular OCPs associated with the switched mode dynamics.

Recall that the phrase "relaxing an initial problem" has various meanings in applied mathematics, depending on the area where it is defined, depending also on what one relaxes (a functional, the underlying space, etc.). In the context of HOCPs of the type (6.8), when dealing with the minimization of $J(u(\cdot))$, the most general way of looking at relaxation is to consider the lower semicontinuous hull of $J(u(\cdot))$ determined on a convexified set of admissible controls. We also refer to [279] for some classic relaxation procedures in optimal control. Applying the theoretic construction of a closed convex hull $\overline{co}\{J(u(\cdot))\}$ (described in detail in

Chapter 5) for the objective $J(u(\cdot))$ and considering the convexification of the set of admissible control functions, we formulate the relaxed variant of the sophisticated HOCP (6.8). Then we have

$$\begin{aligned}&\text{minimize } \bar{co}\{J(u(\cdot))\}\\&\text{subject to } u(\cdot) \in \text{conv}\{\mathcal{B}_I \bigotimes (\mathcal{Q}_K)^I\}.\end{aligned} \tag{7.9}$$

Problem (7.9) constitutes a simple relaxation of the initially given "switched mode" HOCP (6.8). In accordance with the main results from Chapter 5 the optimization procedure in (7.9) is determined over a convex hull of the set of admissible controls, i.e.,

$$\text{conv}\{\mathcal{B}_I \bigotimes (\mathcal{Q}_K)^I\} = \text{conv}\{\mathcal{B}_I\} \bigotimes \text{conv}\{(\mathcal{Q}_K)\}^I.$$

We refer to [26,31] for some additional mathematical details. For the numerical stage we assume that problem (7.9) possesses an optimal solution (an optimal pair)

$$(u^{relaxed}(\cdot), x^{relaxed}(\cdot)).$$

Since the image of the elements of the control vector, namely, of $\beta_{[t_{i-1},t_i)}(t)$ and $q_{1,[t_{i-1},t_i)}(t)$, takes values from the discrete set $\{0,\ 1\}$, the image of the corresponding elements of

$$\text{conv}\{\mathcal{B}_I\},\ \text{conv}\{(\mathcal{Q}_K)\}$$

is the interval (the convexification) $[0, 1]$. The possible well-approximability condition (so-called "zero gap property") of the above fully relaxed HOCP (7.9) can formally be expressed as follows:

$$\inf_{u(\cdot)} J(u(\cdot)) = \inf_{u(\cdot)} \bar{co}\{J(u(\cdot))\}.$$

This relation is a simple consequence of the following facts:

$$\inf_{u(\cdot)} J(u(\cdot)) = -J^*(0), \quad J^*(u(\cdot)) = \bar{co}\{J(u(\cdot))\},$$

where $J^*(u(\cdot))$ is a conjugate of $J(u(\cdot))$.

Let us observe that for a given objective functional $J(\cdot)$ getting its closed convex hull is a complicated, but at the same time fascinating, operation. Note that the objective functional $J(u(\cdot))$ in (7.9) is a composite functional in a real Hilbert space. This situation is a typical case for the general optimal control processes governed by ordinary differential equations. The numerical calculation of $\bar{co}\{J(u(\cdot))\}$ is not broached in our book. Let us note that some novel effective computational methods for a constructive computation of $\bar{co}\{J(u(\cdot))\}$ are proposed in [32]. These methods are in fact a sophisticated generalization of the celebrated

McCormic relaxation scheme (see Chapter 5) in the specific case of a composite cost functional in a general OCP. The closed convexification of such a functional is realized by solving an "auxiliary control system" (see [26,32] for details).

We now introduce the extended state vector for the initial switched mode systems (6.6), i.e.,

$$\tilde{x} := (x^T, x_{n+1})^T,$$

where

$$\dot{x}_{n+1}(t) = f_0(x(t))$$

a.e. on $[0, t_f]$, $x_{n+1}(t_0) := 0$, and the cost functional has a "terminal" (Mayer) form

$$J(u(\cdot)) = \phi(\tilde{x}(t_f)) := x_{n+1}(t_f).$$

The projected gradient method discussed in Chapter 3 applied to the fully relaxed problem (7.9) can now be expressed as follows:

$$\begin{aligned} u_{(l+1)}(\cdot) = \gamma_l \mathcal{P}_{\text{conv}\{\mathcal{B}_I\} \bigotimes \text{conv}\{(\mathcal{Q}_K)\}^I}[u_{(l)}(\cdot) - \\ \alpha_l \nabla \bar{co}\{J(u_{(l)}(\cdot))\}] + (1 - \gamma_l) u_{(l)}(\cdot), \; l \in \mathbb{N}, \end{aligned} \tag{7.10}$$

where

$$\mathcal{P}_{\text{conv}\{\mathcal{B}_I\} \bigotimes \text{conv}\{(\mathcal{Q}_K)\}^I}$$

is the conventional projection operator on the convex set

$$\text{conv}\{\mathcal{B}_I\} \bigotimes \text{conv}\{(\mathcal{Q}_K)\}^I$$

and $\{\alpha\}$ and $\{\gamma_l\}$ are sequences of some suitable step sizes.

By ∇ we denote here the Fréchet derivative of the convexified functional $\bar{co}\{J(u_{(l)}(\cdot))\}$. Since $J(u(\cdot))$ is a continuously (Fréchet) differentiable functional, the closed convex hull

$$\bar{co}\{J(u(\cdot))\}$$

is a Fréchet differentiable functional (see [22] for the formal proof). Note that the projection here is defined in the real Hilbert space

$$\mathbb{L}^2\{[0, t_f]; \mathbb{R}^{I+K \times I}\}.$$

Applying the analytic results from [5,7] related to the explicit evaluation of the reduced gradient in OCPs involving HSs, we can now calculate

$$\nabla \bar{co}\{J(u_{(l)}(\cdot))\}$$

in a closed form, i.e.,

$$\nabla \bar{co}\{J(u_{(l)}(\cdot))\}(t) = -\frac{\partial H(x(t), u_{(l)}(t), p(t), p_{n+1}(t))}{\partial u},$$
$$H(t, x, u, p, p_{n+1}) = \langle p, \langle \beta(t), \left(\langle q^i(t), F(x(t))\rangle_K\right)_{i=1,\dots,I} \rangle_I \rangle_n + p_{n+1} f_0(x). \tag{7.11}$$

Note that $H(x, u, p, p_{n+1})$ in (7.11) is the formal Hamiltonian associated with the fully convexified OCP (7.9) (extended by the additional differential equation introduced above). Recall that $p \in R^n$, $p_{n+1} \in R$ are adjoint variables and

$$\tilde{p} := (p^T, p_{n+1})^T.$$

Adjoint and state variables are determined as solutions of the generic Hamilton-type boundary value problem

$$\frac{d\tilde{p}(t)}{dt} = -\frac{\partial H(x(t), u_{(l)}(t), p(t), p_{n+1}(t))}{\partial \tilde{x}} = -((\frac{\partial H(x(t), u_{(l)}(t), p(t), p_{n+1}(t))}{\partial x})^T, 0)^T,$$
$$\tilde{p}(t_f) = -\frac{\partial(\bar{co}\{\phi(\tilde{x}(t_f))\})}{\partial \tilde{x}},$$
$$\frac{d\tilde{x}(t)}{dt} = \frac{\partial H(x(t), u_{(l)}(t), p(t), p_{n+1}(t))}{\partial \tilde{p}},$$
$$\tilde{x}(t_0) = (x_0^T, 0)^T.$$

Several choices are possible for the step sizes α_l and γ_l in (7.10). We describe here shortly the application of the Armijo line search methods from the abstract convex programming (see Chapter 3) to the concrete case of the sophisticated "switched mode"–type HOCP (7.9).

- Armijo line search along the boundary of the admissible control set

  $$\text{conv}\{\mathcal{B}_I \bigotimes (\mathcal{Q}_K)^I\}$$

  with $\gamma_l = 1$ for all $l \in \mathbb{N}$ and α_l is determined by

  $$\alpha_l := \bar{\alpha}\theta^{\chi(l)}$$

  for some $\bar{\alpha} > 0, \theta, \ \delta \in (0, 1)$, where

  $$\chi(l) := \min\{\chi \in \mathbb{N} \mid \bar{co}\{J(\mathcal{P}_{\text{conv}(\mathcal{S})}\left[u_{(l,s)}(\cdot)\right])\} \le \bar{co}\{J(u_{(l)}(\cdot))\} - \delta\langle \nabla \bar{co}\{J(u_{(l)}(\cdot))\}, u_{(l)}(\cdot) - \mathcal{P}_{\text{conv}\{\mathcal{B}_I \otimes (\mathcal{Q}_K)^I\}}[u_{(l,s)}(\cdot)]\rangle\}$$

  and

  $$u_{(l,s)}(\cdot) := u_{(l)}(\cdot) - \bar{\alpha}\theta^{\chi} \nabla \bar{co}\{J(u_{(l)}(\cdot))\}.$$

- Armijo line search along the feasible direction:

$$\{\alpha_l\} \subset [\bar{\alpha}, \hat{\alpha}], \quad \bar{\alpha} < \hat{\alpha} < \infty$$

  and γ_l is determined by the Armijo rule

$$\gamma_l := \theta^{\chi(l)},$$

  for some $\theta, \ \delta \in (0,1)$, where

$$\chi(l) := \min\{\chi \in \mathbb{N} \mid \bar{co}\{J(u_{(l,s)}(\cdot))\} \leq \\ \bar{co}\{J(u_{(l)}(\cdot))\} - \theta^{\chi}\delta\langle \nabla \bar{co}\{J(u_{(l)}(\cdot))\}, u_{(l)}(\cdot) - \mathcal{P}_{\text{conv}\{\mathcal{B}_I \bigotimes (\mathcal{Q}_K)^I\}}[w_l(\cdot)]\rangle\}$$

  and

$$u_{(l,s)}(\cdot) := \theta^{\chi}\mathcal{P}_{\text{conv}\{\mathcal{B}_I \bigotimes (\mathcal{Q}_K)^I\}}[w_l(\cdot)] + (1-\theta^{\chi})u_{(l)}(\cdot).$$

- Exogenous step size before projecting: $\gamma_l = 1$ for all $l \in \mathbb{N}$ and α_l given by

$$\alpha_l := \frac{\delta_l}{||\nabla \bar{co}\{J(u_{(l)}(\cdot))\}||_{\mathbb{L}^2\{[t_0,t_f];\mathbb{R}^m\}}}, \quad \sum_{l=0}^{\infty} \delta_l = \infty, \ \sum_{l=0}^{\infty} \delta_l^2 < \infty.$$

Let us also mention the possible constant step size selection in (7.10):

$$\gamma_l = 1, \ \alpha_l = \alpha > 0$$

for all $l \in \mathbb{N}$. Recall that this simple strategy was analyzed in [19] and its weak convergence was proved under Lipschitz continuity of

$$\nabla \bar{co}\{J(\cdot)\}.$$

The main difficulty here is the necessity of taking $\alpha \in (0, 2/L)$, where L is the Lipschitz constant for

$$\nabla \bar{co}\{J(\cdot)\}.$$

Under some nonrestrictive assumptions (see Chapter 2) the projected gradient iterations (7.10) generate a minimizing sequence for the fully relaxed optimization problem (7.9). Many mathematically exact convergence theorems for the gradient iterations (7.10) can be found in [178]. A comprehensive discussion of the weakly and strongly convergent variants of the basic gradient method can be found in [11,19]. We also refer to [3,8,9,36–39] for some specific convergence results associated with the gradient-based solution schemes for various types of "switched" OCPs. Note that the first Armijo gradient-based strategy requires one projection onto the convexified control set

$$\mathrm{conv}\{\mathcal{B}_I \bigotimes (\mathcal{Q}_K)^I\}$$

for each step of the inner loop resulting from the Armijo line search. Therefore, many projections might be performed for each iteration l, making the first strategy inefficient when the projection onto the set

$$\mathrm{conv}\{\mathcal{B}_I \bigotimes (\mathcal{Q}_K)^I\}$$

cannot be computed explicitly. On the other hand, the second Armijo strategy demands only one projection for each outer step, i.e., for each iteration l. The Armijo optimization strategies presented above are the constrained versions of the line search proposed in the initial work [1] for solving unconstrained optimization problems. Under existence of minimizers and convexity assumptions for a convex minimization problem, it is possible to prove, for the presented Armijo strategies, convergence of the whole (minimizing) sequence to a solution of the above optimization problem in finite-dimensional spaces.

Last strategy from the approaches presented above, as its counterpart in the unconstrained case, fails to be a decent method. Furthermore, it is easy to show that this approach implies

$$||u_{(l+1)}(\cdot) - u_{(l)}(\cdot)|| \leq \delta_l$$

for all l, with δ_l given above. This reveals that convergence of the sequence of points generated by this exogenous approach can be very slow (step sizes are small). Note that the second and the third strategy allow for occasionally long step sizes because both strategies employ all information available at each l-iteration. Moreover, the last strategy does not take into account the values of the objective functional for determining the "best" step sizes. These characteristics, in general, entail poor computational performance.

The basic convergence results for the approximating sequence $\{u_{(l)}(\cdot)\}$ generated by (7.10) in combination with the Armijo line search can be stated as follows.

Theorem 7.5. *Assume that all hypotheses for the switched mode system (6.6) and for the HOCP (6.8) are satisfied and*

$$p_{n+1} \neq 0.$$

Consider a sequence $\{u_{(l)}(\cdot)\}$ generated by method (7.10) *with an* Armijo *step size α. Then for an admissible initial point*

$$u_{(0)}(\cdot) \in conv\{\mathcal{B}_I \bigotimes (\mathcal{Q}_K)^I\}$$

the resulting sequence $\{u_{(l)}(\cdot)\}$ *is a minimizing sequence for the relaxed* HOCP (7.9), *i.e.*,

$$\lim_{l\to\infty} \bar{co}\{J(u_{(l)}(\cdot))\} = \bar{co}\{J(u^{relaxed}(\cdot))\} = \min_{Problem(3.1)} J(u(\cdot)).$$

Additionally assume that

$$\partial f_k(x)/\partial u, \; k = 1, ..., K,$$

are Lipschitz *continuous. Then* $\{u_{(l)}(\cdot)\}$ *converges weakly to a solution* $u^{relaxed}(\cdot)$ *of* (7.9).

Proof. As mentioned above the convexified cost functional $J(\cdot)$ in (7.9) is Fréchet differentiable. The property of $\{u_{(l)}(\cdot)\}$ to be a minimizing sequence for (7.9) is an immediate consequence of [178]. Consider (7.11) and the associated Hamiltonian system. The differentiability of $x^u(t)$ (see, e.g., [22]) implies the Lipschitz continuity of this control state mapping defined on the bounded set

$$\text{conv}\{\mathcal{B}_I \bigotimes (\mathcal{Q}_K)^I\}.$$

Since $\partial f_k(x)/\partial u$ are also assumed to be Lipschitz continuous and the composition of two Lipschitz continuous mappings also possesses the same property, we derive the Lipschitz continuity of the derivative

$$\nabla \bar{co} J(u(\cdot))(t)$$

uniformly in $t \in [t_0, t_f]$. The weak convergence to $u^{relaxed}(\cdot)$ of the sequence $\{u_l(\cdot)\}$ generated by (7.10) with an Armijo step size follows now from the results of Chapter 3. The proof is completed. □

Let us also note that Theorem 7.5 justifies the convergence result obtained in [40] and [43] for the class of HOCPs related to the switched mode systems.

Note that from the mathematical point of view the full relaxation (7.9) is a formal consequence of the celebrated chattering lemma (see, e.g., [2,9,10]). In this case the abovementioned zero gap property is in fact an impossible request; we have

$$\bar{co}\{J(u^{relaxed}(\cdot))\} \leq J(u^{opt}(\cdot)).$$

A challenging problem related to the HOCPs of the type (6.8) (and also to the general HOCPs) is to determine some alternative relaxation schemes that involve a smaller relaxation gap.

7.2 The Bengea–DeCarlo Approach

Let us consider shortly a relaxation approach proposed in [81]. This technique can be applied to a simple class of switched OCPs (SOCPs) of the following type (we use here the improved original notation from [81]):

$$\begin{aligned}
&\text{minimize } J(x_0, u(\cdot), v(\cdot)) = g(x_0, t_f, x_f)+ \\
&\int_0^{t_f} f^0_{v(t)}(t, x(t), u(t))dt, \\
&\text{subject to} \\
&\dot{x}(t) = f_{v(t)}(t, x(t), u(t)), \\
&x(0) = x_0.
\end{aligned} \tag{7.12}$$

Here $v(t) \in \{0, 1\}$ is the switching control and $u(\cdot)$ is a conventional measurable control function. The authors consider the dynamic (switched) system in (7.12) in the absence of possible state jumps. Moreover, some terminal conditions for the initial and final states are assumed. The right-hand sides f_1 and f_2 of the switched dynamic system in (7.12) are assumed sufficiently smooth. The same assumptions are made for the objective functions f_1^0 and f_2^0. The admissible control region $U \subset \mathbb{R}^m$ in (7.12) is assumed to be compact and convex.

The switched-type dynamic system from the SOCP (7.12) constitutes a simple variant of the general switched system (SS) (see Chapter 1 and Chapter 6). This system contains only two possible locations.

Using the natural parametrization idea, one can introduce an "embedding" system associated with the initially given switched-type dynamics in (7.12). Then we have

$$\begin{aligned}
&\dot{x}(t) = [1 - v(t)] f_0(t, x(t), u_0(t)) + v(t) f_1(t, x(t), u_1(t)), \\
&x(0) = x_0, \;\; v(t) \in [0, 1].
\end{aligned} \tag{7.13}$$

The associated objective functional can also be correspondingly relaxed, i.e.,

$$\begin{aligned}
&J_R(x_0, v(\cdot), u_0(\cdot), u_1(\cdot)) = g(x_0, t_f, x_f)+ \\
&\int_0^{t_f} \big([1 - v(t)] f^0_{v(t)}(t, x(t), u_0(t)) + v(t) f^1(t, x(t), u_1(t))\big)dt.
\end{aligned} \tag{7.14}$$

The authors in [81] develop a relaxation based on the chattering lemma (see Chapter 5) but in fact the obtained relaxation (7.13) constitutes a simple particular case of the general Gamkrelidze relaxation scheme.

Under some condition of openness of the terminal condition set, an optimal solution of the initially given SOCP is also the solution of the "embedded" OCP: minimize the objective functional J_R from (7.14) subject to the embedded dynamic system (7.13). The inverse statement is also true (see [81]).

In [302] the (Gamkrelidze) relaxed OCP associated with the above "embedded" OCP is next used for a constructive approximation of the initially given SOCP (7.12). Taking into consideration the hybrid β-relaxation theory (see Section 7.3) and some general results from Section 7.6 the obtained approximability results in [302] are not surprising. Let us finally note that the authors of [81] do not consider the conceptually (and numerically) important question of a relaxation gap.

7.3 The β-Relaxations Applied to Hybrid and Switched OCPs

Similar to the OCPs from Chapter 5 we now study the possible application of the β-relaxation technique to the following constrained HOCP (or SOCP):

$$\begin{aligned}
&\text{minimize } \mathcal{J}(u(\cdot))\\
&\text{subject to } \dot{x}(t) = \tilde{f}(t, x(t), u(t)) \text{ a.e. on } [0, t_f],\\
&x(0) = x_0,\\
&u(t) \in U \text{ a.e. on } [0, t_f],\\
&h_j(x(t_f)) \leq 0 \; \forall j \in I,\\
&g(t, x(t)) \leq 0 \; \forall t \in [0, t_f],\\
&\int_0^{t_f} s(t, x(t), u(t))dt \leq 0,\\
&q_i \in Q, \; i = 1, ..., r+1, \; x(0) = x_0 \in M_{q_1}.
\end{aligned} \tag{7.15}$$

Here

$$\mathcal{J}(u(\cdot)) := \phi(x^u(1)),$$

where $x^u(\cdot)$ is an absolutely continuous trajectory of the dynamic system in (7.15) generated by an admissible control input $u(\cdot)$ and $\phi(\cdot)$ is assumed to be continuous. We also assume that all the basic hypotheses from Section 5.2 are satisfied. Additionally we suppose that every function

$$f_q(t, \cdot, \cdot), \; q \in Q,$$

is differentiable and that f_q, $\partial f_q/\partial x$, $\partial f_q/\partial u$ are continuous for all $q \in Q$. Moreover, there exists a constant $C < \infty$ such that

$$||\partial f_q/\partial x(t, x, u)|| \leq C$$

for all $(t, x, u) \in [0, t_f] \times \mathbb{R}^n \times U$ and $q \in Q$. We also assume that functions g and h_j, $j \in I$, in (7.15) are continuous. Function s in (7.15) is assumed to be integrable.

Recall that $\tilde{f}$ is defined in (7.2) and constitutes a right-hand side of the given HS or SS. Similar to the conventional OCPs the constructive approximations of the relaxed control system and the corresponding HOCPs are based on the idea of the possible approximation of the generic Gamkrelidze system (7.3). We now introduce the auxiliary β-type control system (see also Chapter 5). For the aims of visibility the β-multipliers used in the relaxation procedure will be denoted by γ (in order to have a different notation for the characteristic functions associated with HSs and SSs).

As in the classic case the first step in generating the relaxed β-controls consists in approximation of the admissible control set U by a finite set U_M of points (grid)

$$v^k \in U, \; k = 1, ..., M,$$

where

$$M = (n+1)\tilde{M}, \; \tilde{M} \in \mathbb{N}.$$

We assume that for a given number $\epsilon > 0$ (accuracy) there exists a natural number $\tilde{M}_\epsilon$ such that for every $v \in U$ one can find a point $v^k \in U_M$ with

$$||v - v^k||_{\mathbb{R}^m} < \epsilon,$$

where

$$M = (n+1)\tilde{M}, \; \tilde{M} > \tilde{M}_\epsilon.$$

Consider a sequence of the accuracies $\{\epsilon_M\}$ mentioned above such that

$$\lim_{M \to \infty} \epsilon_M = 0.$$

The following auxiliary system is next denoted as a β-system associated with the HS (or SS) from OCP (7.15):

$$\begin{aligned} \dot{z}(t) &= \sum_{k=1}^{M} \gamma^k(t) \tilde{f}(t, z(t), v^k) \text{ a.e. on } [0, t_f], \\ z(0) &= x_0, \; q \in Q, \end{aligned} \tag{7.16}$$

where

$$\gamma^k(\cdot) \in \mathbb{L}_1^1([0, t_f]), \; \gamma^k(t) \geq 0,$$

and

$$\sum_{k=1}^{M} \gamma^k(t) = 1 \; \forall t \in [0, t_f].$$

We consider the set of admissible β-controls

$$\aleph(M) := \{(\beta^1(\cdot), ..., \beta^M(\cdot))^T \; : \; \gamma^k(\cdot) \in \mathbb{L}_1^1([0, t_f]), \; \gamma^k(t) \geq 0,$$
$$\sum_{k=1}^{M} \gamma^k(t) = 1 \; \forall t \in [0, t_f]\}, \; \gamma_M(\cdot) := (\beta^1(\cdot), ..., \beta^M(\cdot))^T.$$

We now call the introduced control system (7.16) (with the β-controls) the hybrid β-control system. Note that under the basic assumptions for a fixed U_M the given hybrid β-control system (7.16) has an absolutely continuous solution $z_M^\gamma(\cdot)$ for every admissible β-control

$$\gamma_M(\cdot) \in \aleph(M).$$

Let us note that the set $\aleph(M)$ from Definition 5.6 is similar to the set $\aleph(n+1)$ of admissible controls in the Gamkrelidze relaxed dynamic system (5.3) from Section 5.2.

We now are interested to establish some basic properties of the introduced hybrid β-control systems.

Theorem 7.6. *Let the right-hand side of the dynamic system from* (7.15) *satisfy the basic technical assumptions. For every*

$$(\nu(\cdot), y^\nu(\cdot)), \; \nu(\cdot) \in \aleph(n+1) \times \mathcal{U}^{n+1}$$

there exist a sequence of β-controls

$$\{\gamma_M(\cdot)\} \subset \aleph(M)$$

and the corresponding sequence $\{z_M^\beta(\cdot)\}$ of solutions of an appropriate hybrid β-control system such that $z_M^\gamma(\cdot)$ approximate the solution $y^\nu(\cdot)$ of the (constrained) Gamkrelidze system (7.3) in the following sense:

$$\lim_{M\to\infty} ||z_M^\gamma(\cdot) - y^\nu(\cdot)||_{\mathbb{C}_n([0,t_f])} = 0$$

and

$$\lim_{M\to\infty} ||z_M^\gamma(\cdot) - y^\nu(\cdot)||_{\mathbb{W}_n^{1,1}([0,t_f])} = 0.$$

Proof. For a fixed number M such that $M/(n+1) \in \mathbb{N}$ we put

$$\tilde{M} := M/(n+1)$$

and introduce an equidistant partition of [0, 1], i.e.,

$$G_{\tilde{M}} := \{t_1, t_2, ..., t_{\tilde{M}+1}\},\ t_1 = 0,\ t_{\tilde{M}+1} = 1,\ t_R = t_1 + \frac{R-1}{\tilde{M}},$$

where $R = 1, ..., \tilde{M}+1$. By $\mathbb{L}_m^{2,\tilde{M}}(G_{\tilde{M}})$ we next denote the Euclidean space of the piecewise constant control functions $u_{\tilde{M}}(\cdot)$, where

$$u_{\tilde{M}}(t) := \sum_{R=1}^{\tilde{M}} \chi^R(t) u^R,\ u^R = u(t_R),\ u(\cdot) \in \mathbb{L}_m^2([0,1]),$$

$$\chi^R(t) := \begin{cases} 1 & \text{if } t \in [t_R, t_{R+1}), \\ 0 & \text{otherwise.} \end{cases}$$

The scalar product and the norm in this space $\mathbb{L}_m^{2,\tilde{M}}(G_{\tilde{M}})$ are defined as follows:

$$\langle u_{\tilde{M}}(\cdot), p_{\tilde{M}}(\cdot)\rangle_{\mathbb{L}_m^{2,\tilde{M}}(G_{\tilde{M}})} := \sum_{R=1}^{\tilde{M}} \langle u^R, p^R\rangle_{\mathbb{R}^m},$$

$$||u_{\tilde{M}}(\cdot)||_{\mathbb{L}_m^{2,\tilde{M}}(G_{\tilde{M}})} :=$$

$$(\sum_{R=1}^{\tilde{M}} ||u^R||^2_{\mathbb{R}^m})^{1/2}$$

for all

$$u_{\tilde{M}}(\cdot), p_{\tilde{M}}(\cdot) \in \mathbb{L}_m^{2,\tilde{M}}(G_{\tilde{M}}).$$

We next consider the constrained Gamkrelidze system (7.3). The set of all step functions $u_{\tilde{M}}(\cdot)$ for all possible partitions

$$G_{\tilde{M}},\ \tilde{M} \in \mathbb{N},$$

is dense in $\mathbb{L}_m^2([0,1])$ with respect to the topology generated by the sup-norm

$$||\theta(\cdot)||_\infty := \sup_{t \in [0,1]} ||\theta(t)||_{\mathbb{R}^m},$$

where $\theta(\cdot) \in \mathbb{L}_m^2([0,1])$. Therefore, for every

$$u^j(\cdot) \in \mathcal{U}, \ j = 1, ..., n+1,$$

and for a given accuracy $\epsilon_M > 0$ we can choose an approximate step function

$$u^j_{\tilde{M}}(\cdot) \in \mathbb{L}_m^{2,\tilde{M}}(G_{\tilde{M}})$$

such that

$$u^j_{\tilde{M}}(t) := \sum_{R=1}^{\tilde{M}} \chi^R(t) v^{j,R}, \ v^{j,R} \in U_M, \text{ and}$$
$$||u^j(\cdot) - u^j_{\tilde{M}}(\cdot)||_\infty < \epsilon_M.$$

Consequently,

$$||u^j(\cdot) - u^j_{\tilde{M}}(\cdot)||_{\mathbb{L}_m^2([0,1])} < \epsilon_M.$$

Since $U_M \subset U$, the control function $u^j_{\tilde{M}}(\cdot)$ belongs to the set of admissible controls $\mathcal{U}$. Let

$$\tilde{y}^\nu_M(\cdot)$$

be an absolutely continuous solution of the following (quasi-Gamkrelidze) system

$$\dot{\tilde{y}}(t) = \sum_{j=1}^{n+1} \alpha^j(t) \tilde{f}(t, \tilde{y}(t), u^j_{\tilde{M}}(t)) \text{ a.e. on } [0,1],$$

for $\tilde{y}(0) = x_0$. The solutions $\tilde{y}^\nu_M(\cdot)$ and $y^\nu(\cdot)$ of the corresponding i.v.p. depend continuously on parameters [5] and the function f is continuous. Moreover,

$$\lim_{M \to \infty} \epsilon_M = 0.$$

Using these facts and the hybrid version of the equivalence theorem (Theorem 7.1), we obtain

$$\lim_{M \to \infty} ||\tilde{f}(t, y^\nu(t), u^j(t)) - \tilde{f}(t, \tilde{y}^\nu_M(t), u^j_{\tilde{M}}(t))||_{\mathbb{R}^n} = 0$$

uniformly in $t \in [0,1]$. The last fact implies the following relation:

$$\lim_{M \to \infty} ||\dot{y}^\nu(\cdot) - \dot{\tilde{y}}^\nu_M(\cdot)||_{\mathbb{R}^n} = 0 \tag{7.17}$$

uniformly in $t \in [0, 1]$. Moreover, we also get

$$||y^{\nu}(t) - \tilde{y}^{\nu}_{M}(t)||_{\mathbb{R}^n} =$$

$$= || \int_0^t \sum_{j=1}^{n+1} \alpha^j(\tau)(\tilde{f}(\tau, y^{\nu}(\tau), u^j(\tau)) - \tilde{f}(\tau, \tilde{y}^{\nu}_M(\tau), u^j_{\tilde{M}}(\tau)))d\tau ||_{\mathbb{R}^n} \leq$$

$$\leq \sum_{j=1}^{n+1} \int_0^t \alpha^j(\tau) ||\tilde{f}(\tau, y^{\nu}(\tau), u^j(\tau)) - \tilde{f}(\tau, \tilde{y}^{\nu}_M(\tau), u^j_{\tilde{M}}(\tau))||_{\mathbb{R}^n} d\tau \leq$$

$$\leq \sum_{j=1}^{n+1} \int_0^1 \alpha^j(\tau) ||\tilde{f}(\tau, y^{\nu}(\tau), u^j(\tau)) - \tilde{f}(\tau, \tilde{y}^{\nu}_M(\tau), u^j_{\tilde{M}}(\tau))||_{\mathbb{R}^n} d\tau \leq$$

$$\leq (n+1) \max_{t \in [0,1]} \alpha^j(t) ||\tilde{f}(t, y^{\nu}(t), u^j(t)) - \tilde{f}(t, \tilde{y}^{\nu}_M(t), u^j_{\tilde{M}}(t))||_{\mathbb{R}^n}.$$

Since

$$\lim_{M \to \infty} ||\tilde{f}(t, y^{\nu}(t), u^j(t)) - \tilde{f}(t, \tilde{y}^{\nu}_M(t), u^j_{\tilde{M}}(t))||_{\mathbb{R}^n} = 0$$

uniformly in $t \in [0, 1]$, we obtain

$$\lim_{M \to \infty} ||y^{\nu}(t) - \tilde{y}^{\nu}_M(t)||_{\mathbb{R}^n} = 0 \tag{7.18}$$

uniformly in $t \in [0, 1]$. That is,

$$\lim_{M \to \infty} ||z^{\gamma}_M(\cdot) - y^{\nu}(\cdot)||_{\mathbb{C}_n([0,1])} = 0. \tag{7.19}$$

From (7.17) and (7.18) it follows that

$$\lim_{M \to \infty} ||y^{\nu}(\cdot) - \tilde{y}^{\nu}_M(\cdot)||_{\mathbb{W}^{1,1}_n([0,1])} = 0. \tag{7.20}$$

We claim that $\tilde{y}^{\nu}_M(\cdot)$ coincides with $z^{\gamma}_M(\cdot)$ for an admissible β-control $\gamma_M(\cdot)$. To see this, note that

$$\tilde{f}(t, \tilde{y}^{\nu}_M(t), u^j_{\tilde{M}}(t)) = \sum_{R=1}^{\tilde{M}} \chi^R(t) \tilde{f}(t, \tilde{y}^{\nu}_M(t), v^{j,R})$$

and

$$\sum_{j=1}^{n+1} \alpha^j(t) \tilde{f}(t, \tilde{y}^{\nu}_M(t), u^j_{\tilde{M}}(t)) = \sum_{j=1}^{n+1} \sum_{R=1}^{\tilde{M}} \alpha^j(t) \chi^R(t) \tilde{f}(t, \tilde{y}^{\nu}_M(t), v^{j,R}).$$

Put now

$$\gamma^k(t) := \alpha^j(t)\chi^k(t)$$

and $v^k := v^{j,r}$, where $j = 1, ..., n+1$, $R = 1, ...\tilde{M}$, and $k = 1, ..., M$. Clearly, $\tilde{y}_M^{\nu}(\cdot)$ satisfies the differential equation

$$\dot{\tilde{y}}_M^{\nu}(t) = \sum_{k=1}^{M} \beta^k(t)\tilde{f}(t, \tilde{y}_M^{\nu}(t), v^k),$$

with $\tilde{y}(0) = x_0$. Moreover,

$$v^k \in U_M, \; k = 1, ..., M,$$

and $\beta_M(\cdot) \in \aleph(M)$. We finally obtain

$$\tilde{y}_M^{\nu}(\cdot) \equiv z_M^{\beta}(\cdot).$$

The proof is completed. □

Similar to the theory of β-relaxation for the classic OCPs (see Chapter 5) we derive the next property of the hybrid β-system (7.16).

Theorem 7.7. *Let the right-hand side of the system from OCP* (7.15) *satisfy the hypotheses of Theorem 7.6. Additionally assume that*

$$\gamma_M(\cdot) \in \aleph(M)$$

and $z_M^{\gamma}(\cdot)$ is the corresponding solution of (7.16). *Then*

$$\dot{z}_M^{\gamma}(t) \in \text{conv}\{\tilde{f}(t, z_M^{\gamma}(t), U)\} \text{ a.e. on } [0, t_f].$$

Proof. Let δ_{v_k} be a Dirac measure at $v_k \in U_M$. For a given $\gamma_M(\cdot) \in \aleph(M)$, we next define the admissible relaxed control

$$\mu(t) = \sum_{k=1}^{M} \gamma_k(t)\delta_{v_k} \in \mathcal{M}_+^1(U).$$

Evidently, $z_M^{\gamma}(\cdot)$ is a solution of the Young relaxed HS (7.5) corresponding to the relaxed control $\mu(\cdot)$ determined above. From the hybrid version of the equivalence theorem (Theorem 7.1) we now deduce

$$\dot{z}_M^{\gamma}(t) \in \text{conv}\{\tilde{f}(t, z_M^{\gamma}(t), U)\} \text{ a.e. on } [0, t_f].$$

The proof is completed. □

Note that the set-valued function in the right-hand side of the DI in Theorem 7.7 is compact and convex-valued. The existence of a solution to this DI follows from the theoretical results of Chapter 4. Let us give here the explicit form of the orientor field $\tilde{f}(t, z(t), U)$ from Theorem 7.7:

$$\tilde{f}(t, z(t), U) := \sum_{i=1}^{r+1} \beta_{[t_{i-1},t_i)}(t) f_{q_i}(t, z(t), U).$$

Theorem 7.6 and Theorem 7.7 imply that $z_M^{\gamma}(\cdot)$ is a consistent approximation to the solution of the (convexified) RDI (7.6) from Section 7.1. This approximability property is proved with respect to the convergence in the Sobolev space $\mathbb{W}_n^{1,1}([0, t_f])$. It constitutes a fundamental approximability property for the introduced hybrid β-system (7.16).

Taking into consideration the celebrated Tikhomirov relaxation idea (see [302]), we now generate the approximating controls from the space $\mathcal{U}$ of the nonrelaxed control function $u(\cdot)$. Similar to the classic OCPs we consider the approximative solutions.

Theorem 7.8. *Let the right-hand side of the system from HOCP* (7.15) *satisfy the assumptions of Theorem 7.6. Let*

$$\gamma_M(\cdot) \in \aleph(M)$$

and $z_M^{\gamma}(\cdot)$ be the corresponding solution of the β-system (7.16). *Then there exists a piecewise constant control function $\tilde{u}_M(\cdot) \in \mathcal{U}$ such that the solution $x^{\tilde{u}_M}(\cdot)$ of the initial system from the HOCP* (7.15) *exists on the time interval* $[0, t_f]$ *and*

$$\lim_{M \to \infty} ||z_M^{\gamma}(t) - x^{\tilde{u}_M}(t)||_{\mathbb{R}^n} = 0 \tag{7.21}$$

uniformly in $t \in [0, t_f]$.

Proof. By construction of the solution $z_M^{\gamma}(\cdot)$ this function is a solution of the relaxed HS (7.6) with the specific admissible relaxed control (measure) $\mu(\cdot)$ defined as

$$\mu(t) = \sum_{k=1} \beta_k(t) \delta_{v_k}.$$

Then, there exists an admissible piecewise constant control $u_M(\cdot) \in \mathcal{U}$ such that (7.21) holds (see [32], Theorem 13.2.1, p. 677 and Theorem 13.2.2, p. 678). □

The presented analytic results make it possible to approximate (in a strong sense) the initially given constrained HOCP (7.15).

We now introduce an additional notation

$$\bar{\mathcal{J}}_M, \bar{h}_M, \bar{s}_M : \aleph(M) \to \mathbb{R}$$

and

$$\bar{g}_M : \aleph(M) \to \mathbb{C}([0, t_f])$$

defined by

$$\begin{aligned}
&\bar{\mathcal{J}}_M(\gamma_M(\cdot)) := \phi(z_M^{\gamma}(t_f)),\ \bar{h}_M(\gamma_M(\cdot)) := h(z_M^{\gamma}(t_f)),\\
&\bar{g}_M(\gamma_M(\cdot))(t) := q(t, z_M^{\gamma}(t))\ \forall t \in [0, t_f],\\
&\bar{s}_M(\gamma_M(\cdot)) := \int_0^{t_f} \sum_{k=1}^{M} \gamma^k(t) s(t, z_M^{\gamma}(t), v^k) dt.
\end{aligned}$$

The following main approximability result gives a possibility to approximate the constrained Gamkrelidze relaxed HOCP associated with the initially given problem (7.15) using the proposed β-relaxations.

Theorem 7.9. *For every $\delta > 0$ there exists a number $M_\delta \in \mathbb{N}$ such that for all natural numbers $M > M_\delta$ the β-relaxed OCP*

$$\begin{aligned}
&\text{minimize } \bar{\mathcal{J}}_M(\gamma_M(\cdot)) \text{ subject to } \gamma_M(\cdot) \in \aleph(M),\\
&\bar{h}_M(\gamma_M(\cdot)) \le \delta,\\
&\bar{q}_M(\gamma_M(\cdot))(t) \le \delta\ \forall t \in [0, t_f],\\
&\bar{s}_M(\gamma_M(\cdot)) \le \delta,\\
&q_i \in Q,\ i = 1, ..., r+1,\ x(0) = x_0 \in M_{q_1}
\end{aligned} \tag{7.22}$$

has an optimal solution

$$\gamma_M^{opt}(\cdot) \in \aleph(M)$$

under the assumption that it has an admissible solution. Moreover,

$$|\bar{\mathcal{J}}_M(\gamma_M^{opt}(\cdot)) - \bar{\mathcal{J}}(\nu^{opt}(\cdot))| \le \delta,$$

where

$$\nu^{opt}(\cdot) \in \aleph(n+1) \times \mathcal{U}^{n+1}$$

is an optimal solution of the constrained Gamkrelidze system (7.3).

Proof. From Theorem 7.6 the existence of a sequence of β-controls $\{\gamma_M(\cdot)\}$ such that

$$\lim_{M\to\infty} ||z^{\gamma}_M(t_f) - y^{opt}(t_f)||_{\mathbb{R}^n} = 0,$$

where $y^{opt}(\cdot)$ is the optimal trajectory of the Gamkrelidze system (7.3) for $\nu^{opt}(\cdot)$, follows. The continuity property of the constraint functions

$$h(\cdot),\ g(\cdot),\ s(\cdot)$$

implies that, for every $\delta > 0$, there exists a number $M_{1,\delta} \in \mathbb{N}$ such that

$$\bar{h}_M(\gamma_M(\cdot)) \leq \delta,\ \bar{g}_M(\gamma_M(\cdot))(t) \leq \delta,\ \forall t \in [0, t_f], \bar{s}_M(\gamma_M(\cdot)) \leq \delta,$$

for all $M > M_{1,\delta}$.

The structure of the right-hand side of the β-system (7.16) implies convexity of the extended orientor field. Additionally, the control set

$$\mathcal{F}(M) := \{(\gamma^1, ..., \gamma^M)^T : \gamma^k \geq 0, \sum_{k=1}^{M} \gamma^k = 1\}$$

is compact. The solutions of the β-system are uniformly bounded. The state, target, and integral constraints in (7.22) also define a bounded and closed set. Since the set of admissible solutions to (7.22) is nonempty, the existence of an optimal solution

$$\gamma_M^{opt}(\cdot) \in \aleph(M)$$

for this problem follows from the Filippov–Himmelberg theorem (see Chapter 4).

Recall that the function $\phi(\cdot)$ is assumed to be continuous. Then, for every $\delta > 0$, there exists a number $M_{2,\delta} \in \mathbb{N}$ such that

$$|\tilde{\mathcal{J}}_M(\gamma_M(\cdot)) - \tilde{\mathcal{J}}(\nu^{opt}(\cdot))| \leq \delta/3,$$

for all

$$M > M_{2,\delta}.$$

Let now $z_M^{opt}(\cdot)$ be the solution of β-system (7.16) associated with $\gamma_M^{opt}(\cdot)$. Evidently, $z_M^{opt}(\cdot)$ is also a solution of the Young relaxed HS (7.5) associated with the relaxed control

$$\hat{\mu}(t) = \sum_{k=1}^{M} \gamma^{k,opt}(t)\delta_{v_k},$$

where $\gamma^{k,opt}(t)$ is the kth component of $\gamma_M^{opt}(t)$.

From the hybrid version of the equivalence theorem (Theorem 7.1) we now deduce that $z^{\gamma_M(\cdot)}$ is also a solution of the Gamkrelidze system (7.3) for an admissible generalized control $\hat{\nu}(\cdot)$. The continuity properties of the functions

$$h_j, \ g, \ s(\cdot)$$

imply that, for every $\delta > 0$, there exist a number $M_{3,\delta} \in \mathbb{N}$ and an admissible generalized control $\nu(\cdot)$ such that

$$|\bar{\mathcal{J}}_M(\gamma_M^{opt}(\cdot)) - \bar{\mathcal{J}}(\nu(\cdot))| \leq \delta/3,$$

for all natural numbers $M > M_{3,\delta}$. Hence,

$$\bar{\mathcal{J}}_M(\gamma_M^{opt}(\cdot)) = \bar{\mathcal{J}}(\nu(\cdot)) + \delta, \ -\delta/3 \leq \delta \leq \delta/3.$$

Now let us observe that for all natural numbers

$$M > M_\delta := \max(M_{1,\delta}, \ M_{2,\delta}, \ M_{3,\delta})$$

we have

$$\begin{aligned} &\delta/3 \geq |\bar{\mathcal{J}}_M(\gamma_M(\cdot)) - \bar{\mathcal{J}}(\nu^{opt}(\cdot))| \geq \\ &|\bar{\mathcal{J}}_M(\gamma_M^{opt}(\cdot)) - \bar{\mathcal{J}}(\nu^{opt}(\cdot))| = \\ &\bar{\mathcal{J}}(\nu(\cdot)) - \bar{\mathcal{J}}(\nu^{opt}(\cdot)) + \delta. \end{aligned}$$

Since

$$\bar{\mathcal{J}}(\nu(\cdot)) \geq \bar{\mathcal{J}}(\nu^{opt}(\cdot)),$$

we have

$$\delta/3 - \delta \geq |\bar{\mathcal{J}}(\nu(\cdot)) - \bar{\mathcal{J}}(\nu^{opt}(\cdot))|.$$

Therefore,

$$|\bar{\mathcal{J}}_M(\gamma_M^{opt}(\cdot)) - \bar{\mathcal{J}}(\nu^{opt}(\cdot)) - \delta| \geq \delta/3 - \delta$$

and finally

$$|\bar{\mathcal{J}}_M(\gamma_M^{opt}(\cdot)) - \bar{\mathcal{J}}(\nu^{opt}(\cdot)) \leq \delta.$$

The proof is now completed. □

From Theorem 7.9 we can deduce that the hybrid β-system (7.16) introduced above and the corresponding β-relaxed HOCP (7.22) constitute a theoretical basis for the consistent numerical treatment of the corresponding constrained Gamkrelidze relaxed HOCP.

The above theory of the hybrid β-relaxations is designed for the HOCPs. As mentioned in Chapter 1, the general HSs and SSs have the same symbolical representation in a form of (1.3), i.e.,

$$\begin{aligned} \dot{x}(t) &= \sum_{i=1}^{r+1} \beta_{[t_{i-1},t_i)}(t) f_{q_i}(t, x(t), u(t)), \\ x(0) &= x_0. \end{aligned} \tag{7.23}$$

The above "compact" dynamic system incorporates a huge conceptual difference in the case of an HS or an SS. For the HSs the characteristic functions $\beta_{[t_{i-1},t_i)}(\cdot)$ in (7.23) depend on the trajectory (solution) $x(\cdot)$ of the dynamic system (7.23) closed by an admissible control. We have in fact the dependence of the switching times

$$t_i = t_i(x(\cdot)), \ i = 1, ..., r+1.$$

In the case of an SS the above functions $\beta_{[t_{i-1},t_i)}(\cdot)$ constitute an additional (to the conventional control $u(\cdot)$) control input such that the full control vector is a pair

$$(\beta(\cdot), u(\cdot)).$$

The classical relaxation techniques we considered, namely, the Gamkrelidze method and the approximating β-relaxations, provide adequate approaches for the conventional "continuous" control input $u(\cdot)$. Since the first component of the combined control input $(\beta(\cdot), u(\cdot))$ in an SS is in fact a "discrete" control ("switching control") the possible β-relaxation applied to the SOCPs will generate a big relaxation gap. The same conclusion is also true with respect to the Gamkrelidze relaxation approach. The above conclusion implies a necessity of some problem-oriented relaxation schemes for the general SOCPs. In the next sections we will propose some specific problem-related relaxation techniques for the classes of SOCPs.

7.4 Weak Approximation Techniques for Hybrid Systems

In this section we study a switched mode system of the type (6.6) (see Chapter 6). The corresponding OCP associated with this model, namely, HOCP (6.8), is also presented in Chapter 6. Recall that the set

$$\mathcal{B}_I \bigotimes (\mathcal{Q}_K)^I$$

of admissible control functions in problem (6.8) is a subset of the real Hilbert space (see Chapter 6)

$$\mathbb{L}^2\{[0, t_f]; \mathbb{R}^{I+K\times I}\}.$$

By $\mathcal{B}_I$ we have denoted the set of all vector functions $\beta(\cdot)$ such that the time intervals $[t_{i-1}, t_i)$ are disjunct and, moreover,

$$I \leq I_{max}.$$

An admissible vector function (the sequencing control vector) $q^i(\cdot) \in \mathcal{Q}_K$ in (6.6) is an element of the set

$$\mathcal{Q}_K := \{\theta : [0, t_f] \to \{0,\ 1\}^K \mid \sum_{k=1}^{K} q_{k,[t_{i-1},t_i)}(t) = 1\}$$

which represents the schedule of admissible modes for every time instant $t \in [0, t_f]$.

Due to the highly restrictive nonlinear constraint for an admissible input $u(\cdot)$, where

$$u(t) := (\beta^T(t), q^T(t))^T \in \mathcal{B}_I \bigotimes (\mathcal{Q}_K)^I,$$

and because $q := \{q^1, ..., q^I\}$ is a family of sequencing controls for every time interval $[t_{i-1}, t_i)$, an application of the conventional relaxation theory for classic OCPs (see Chapter 5) to the HOCP (6.8) implies numerical difficulties. Moreover, an application of the conventional (continuous) relaxations, namely, of the Gamkrelidze or Young (or Rubio) relaxation schemes, will imply a very large relaxation gap. This fact is a simple consequence of the discrete nature of the combined control input $u(\cdot)$ in a switched mode system (6.6). Recall that a "sequencing control vector" $q_{[t_{i-1},t_i)}(t)$ (for the time interval $[t_{i-1}, t_i)$) has the discrete components

$$q_{k,[t_{i-1},t_i)}(t) \in \{0,\ 1\}$$

for all $k = 1, ..., K$, $t \in [0, t_f]$ such that

$$\sum_{k=1}^{K} q_{k,[t_{i-1},t_i)}(t) = 1$$

for all $k = 1, ..., K$. A "continuous-type" relaxation from the schemes mentioned above will replace $\mathcal{Q}_K$ by the corresponding convex hull such that the relaxed version $\tilde{q}_{k,[t_{i-1},t_i)}(\cdot)$ of the second component of the combined control input $u(\cdot)$ will have a "fuzzy" structure and belong to the interval [0, 1].

The fully (globally) convexified HOCP (7.9) (see Section 7.1) evidently constitutes a mathematically and computationally sophisticated technique. It is associated with a heavy calculation of the convex envelope of a composite functional in a real Hilbert space. Motivated by this fact we next consider an alternative relaxation procedure and call it a "weak relaxation" scheme associated with the initial HOCP (6.8). Our idea originated from the "local convexification" procedure involving the so-called "infimal (prox) convolution"

$$J_\lambda(u(\cdot)) := J(u(\cdot)) + \frac{\lambda}{2}||u(\cdot)||^2_{\mathbb{L}^2\{[0,t_f];\mathbb{R}^{I+K\times I}\}}, \quad \lambda > 0,$$

of the given objective functional $J(u(\cdot))$. Let us note that the above infimal convolution makes it possible to obtain a minimal-gap relaxation procedure for the initial optimization problem. We use the abstract concepts from [19,31] and apply them to our case, namely, to the HOCP (6.8).

Definition 7.1. *The cost functional $J(u(\cdot))$ is called locally paraconvex around*

$$u(\cdot) \in \mathbb{L}^2\{[0,t_f];\mathbb{R}^{I+K\times I}\}$$

if the infimal convolution $J_\lambda(u(\cdot))$ is convex and continuous on a δ-ball $\mathcal{W}_\delta(u(\cdot))$ around $u(\cdot)$ for some

$$\delta > 0, \ \lambda > 0.$$

Note that the infimal convolution $J_\lambda(u(\cdot))$ from the above Definition 7.1 is a locally convex functional.

Definition 7.2. *We say that $J(u(\cdot))$ is prox-regular at*

$$\hat{u}(\cdot) \in \mathbb{L}^2\{[0,t_f];\mathbb{R}^{I+K\times I}\}$$

if there exist $\epsilon > 0$ and $r > 0$ such that

$$\begin{aligned} &J(u_1(\cdot)) > J(u_2(\cdot)) + \\ &\langle \nabla J(\hat{u}(\cdot)), u_1(\cdot) - u_2(\cdot)\rangle_{\mathbb{L}^2\{[0,t_f];\mathbb{R}^{I+K\times I}\}} - \frac{r}{2}||u_1(\cdot) - u_2(\cdot)||^2_{\mathbb{L}^2\{[0,t_f];\mathbb{R}^{I+K\times I}\}} \end{aligned}$$

for all $u_1(\cdot)$ from an ϵ-ball $\mathcal{W}_\epsilon(\hat{u}(\cdot))$ around $\hat{u}(\cdot)$ whenever

$$u_2(\cdot) \in \mathcal{W}_\epsilon(\hat{u}(\cdot))$$

and

$$|J(u_1(\cdot)) - J(\hat{u}(\cdot))| < \epsilon.$$

For HOCP (6.8) we can define

$$J_\lambda(u(\cdot)) := \int_0^{t_f} f_0(x(t))dt + \frac{\lambda}{2}\int_0^{t_f} \langle u(t), u(t)\rangle_{I+K\times I}.$$

Consider now the infimal convolution-based (auxiliary) HOCP

$$\begin{aligned} &\text{minimize } J_\lambda(u(\cdot)) \\ &\text{subject to } u(\cdot) \in \text{conv}\{\mathcal{B}_I \bigotimes (\mathcal{Q}_K)^I\} \end{aligned} \tag{7.24}$$

and assume that it possesses an optimal solution $u_\lambda^{opt}(\cdot)$. The corresponding optimal trajectory is denoted by $x_\lambda^{opt}(\cdot)$. Similar to the classic Mayer-type problem we introduce the auxiliary variable x_{n+1} as follows:

$$\dot{x}_{n+1}(t) = f_0(x(t)) + \frac{\lambda}{2}\langle u(t), u(t)\rangle_{I+K\times I} \text{ a.e. on } [0, t_f],$$

for $x_{n+1}(t_0) := 0$. The infimal convolution-based OCP (7.24) is assumed to be equipped with the objective functional of the following Mayer-type:

$$J(u(\cdot)) = \phi(\tilde{x}(t_f)) := x_{n+1}(t_f).$$

We introduce a specific λ-Hamiltonian associated with the HOCP (6.8), i.e.,

$$H_\lambda(x, u, p, p_{n+1}) = H(x, u, p, p_{n+1}) + \frac{p_{n+1}\lambda}{2}\langle u(t), u(t)\rangle_{I+K\times I},$$

where the original Hamiltonian $H(x, u, p, p_{n+1})$ for HOCP (6.8) is defined as follows:

$$H(t, x, u, p, p_{n+1}) = \langle p, \langle\, \beta(t),\ \left(\langle q^i(t), F(x(t))\rangle_K\right)_{i=1,\ldots,I}\ \rangle_I\rangle_n + p_{n+1} f_0(x).$$

Note that $H(x, u, p, p_{n+1})$ defined above is also the formal Hamiltonian associated with the fully relaxed HOCP (7.9) from Section 7.1.

We next assume (a technical assumption)

$$p_{n+1} \neq 0.$$

Evidently, (7.24) constitutes a partial relaxation (convexification) of the initially given "switched mode"–type HOCP (6.8). Note that in the case $\lambda = 0$ the HOCP (7.24) represents a partially relaxed variant of the initial HOCP (6.8) with an original functional $J(u(\cdot))$ and a convexified control set. The optimal pair for this specific value of λ is next denoted by

$$(u_0^{opt}(\cdot), x_0^{opt}(\cdot)).$$

Under some weak assumptions the prox-regularity of a functional in a Hilbert space implies paraconvexity of the corresponding infimal convolution (see [31]). Using this fact, we can finally prove a local convergence result for the sequence $\{u_l(\cdot)\}$ generated by the projected gradient method applied to the infimal convolution-based OCP (7.24). We have

$$u_{(l+1)}(\cdot) = \gamma_l \mathcal{P}_{\text{conv}\{\mathcal{B}_I \bigotimes (\mathcal{Q}_K)^I\}} \left[u_{(l)}(\cdot) - \alpha_l \nabla J_\lambda(u_{(l)}(\cdot)) \right] + (1-\gamma_l) u_{(l)}(\cdot), \; l \in \mathbb{N}. \tag{7.25}$$

Note that $\nabla J_\lambda(u_{(l)}(\cdot))$ can be determined using the λ-Hamiltonian $H_\lambda(x, u, p, p_{n+1})$ introduced above.

Theorem 7.10. *Assume that all hypotheses for the switched mode system (6.6) and for the HOCP (6.8) are satisfied and* $p_{n+1} \neq 0$. *Let*

$$u_0^{opt}(\cdot) \in \text{int}\{conv\{\mathcal{B}_I \bigotimes (\mathcal{Q}_K)^I\}\}.$$

Consider a sequence $\{u_{(l)}(\cdot)\}$ *generated by* (7.25) *with an Armijo step size (described in Section 7.1). Then there exists an initial point*

$$u_{(0)}(\cdot) \in conv\{\mathcal{B}_I \bigotimes (\mathcal{Q}_K)^I\}$$

such that

$$\lim_{\lambda \to 0} \lim_{l \to \infty} J_\lambda(u_{(l)}(\cdot)) = \min_{conv\{\mathcal{B}_I \bigotimes (\mathcal{Q}_K)^I\}} J(u(\cdot)) = J(u_0^{opt}(\cdot)).$$

Additionally assume that

$$\partial f_k(x)/\partial u, \; k = 1, ..., K,$$

are Lipschitz *continuous. Then for a fixed* $\lambda > 0$ *sequence* $\{u_{(l)}(\cdot)\}$ *converges weakly to a solution* $u_\lambda^{opt}(\cdot)$ *of the infimal convolution-based HOCP* (7.24).

Proof. Consider an ϵ-ball $\mathcal{W}_\epsilon(u_0^{opt}(\cdot))$ around the optimal control function $u_0^{opt}(\cdot)$. Since

$$u_0^{opt}(\cdot) \in \text{conv}\{\mathcal{B}_I \bigotimes (\mathcal{Q}_K)^I\} \subset \mathbb{L}^2\{[0, t_f]; \mathbb{R}^{I+K \times I}\},$$

the ϵ-neighborhood $\mathcal{W}_\epsilon(u_0^{opt}(\cdot))$ is determined by the norm

$$|| \cdot ||_{\mathbb{L}^2\{[0,t_f];\mathbb{R}^{I+K\times I}\}}$$

of the real Hilbert space $\mathbb{L}^2\{[0, t_f]; \mathbb{R}^{I+K \times I}\}$. Let us now estimate the difference

$$J(u_2(\cdot)) - J(u_1(\cdot))$$

for some

$$u_1(\cdot),\ u_2(\cdot) \in \mathcal{W}_\epsilon(u_0^{opt}(\cdot)).$$

First note that the gradient $\nabla J(u(\cdot))(\cdot)$ in (7.24) with a $\lambda \geq 0$ can be calculated similarly to (7.11), i.e.,

$$\nabla J_\lambda(u(\cdot))(t) = -\frac{\partial H_\lambda(x(t), u(t), p(t), p_{n+1}(t))}{\partial u},$$

where functions $x(\cdot),\ u(\cdot),\ p(\cdot),\ p_{n+1}(\cdot)$ can be determined from the boundary value problem for the Hamiltonian system (see Section 7.1) replacing $H(x(t), u(t), p(t), p_{n+1}(t))$ by

$$H_\lambda(x(t), u(t), p(t), p_{n+1}(t)).$$

From the weak version of the Pontryagin maximum principle we have

$$\partial H_\lambda(x_0^{opt}(t), u_0^{opt}(t), p(t), p_{n+1}(t))/\partial u = 0$$

(see [29]). Therefore,

$$\langle \nabla J(u_0^{opt}(\cdot)), u_1(\cdot) - u_2(\cdot)\rangle_{\mathbb{L}^2\{[0,t_f];\mathbb{R}^{I+K\times I}\}} = 0$$

for all admissible $u_1(\cdot),\ u_2(\cdot)$. From the Lipschitz continuity of the control state mapping $x^u(\cdot)$ and taking into consideration the boundedness of the admissible control set, we deduce the inequality

$$J(u_2(\cdot)) - J(u_1(\cdot)) < \frac{r}{2}||u_1(\cdot) - u_2(\cdot)||^2_{\mathbb{L}^2\{[0,t_f];\mathbb{R}^{I+K\times I}\}}$$

for a suitable constant $r > 0$. The combination of the two last relations implies the prox-regularity property of the functional $J(u(\cdot))$ at $u_0^{opt}(\cdot)$ (see Definition 7.2).

Since $J(u(\cdot))$ is continuous, the prox-regularity property implies the paraconvexity of the infimal convolution $J_\lambda(u(\cdot))$ in a neighborhood of the optimal solution $u_0^{opt}(\cdot)$. Using the convergence results of the gradient method for locally convex functions defined on a convex set (see [11,31]), we conclude that

$$\lim_{l\to\infty} J_\lambda(u_{(l)}(\cdot)) = J_\lambda(u_\lambda^{opt}(\cdot)). \tag{7.26}$$

Using the continuity property of the infimal convolution $J_\lambda(u(\cdot))$ and (7.26) we obtain the final relation. The weak convergence of $\{u_{(l)}(\cdot)\}$ can be established by the same argument as in Theorem 7.5. The proof is completed. □

As one can see, Theorem 7.10 establishes the consistency property of the infimal convolution-based relaxation (7.24) with $\lambda = 0$. The last one constitutes a "weak" convexification in comparison with the fully relaxed problem (7.9). The relaxation gap relations for three optimization problems under consideration, namely, for the original HOCP (6.8), for the fully relaxed problem (7.9), and for the weakly relaxed version (7.24), can be expressed as follows:

$$\bar{co}\{J(u^{relaxed}(\cdot))\} \leq J(u_0^{opt}(\cdot)) \leq J(u^{opt}(\cdot)).$$

Note that the above fundamental inequality is an immediate consequence of the absence of the operator $\bar{co}\{J(u(\cdot))\}$ in the weakly relaxed problem (7.24).

7.5 A Remark on the Rubio Generalization

Consider an HS determined by Definitions 1.1–1.3 (see Chapter 1) in the context of the Rubio relaxation procedure from [281] and [92,93]. The basic HOCP (6.38) is rewritten as

$$\begin{aligned}
&\text{minimize } \int_0^{t_f} f_0(\tau, u(\tau), x(\tau))d\tau \\
&\text{subject to } \dot{x}(t) = \sum_{i=1}^{r+1} \beta_{[t_{i-1},t_i)}(t) f_{q_i}(t, x(t), u(t)) \text{ a.e. on } [0, t_f] \\
&q_i \in \mathcal{Q},\ i = 1, \ldots, r+1, \\
&x(0) = x_0 \in M_{q_1},\ x(t_f) = x_f \in M_{q_{r+1}} \\
&u(\cdot) \in \mathcal{U},
\end{aligned} \tag{7.27}$$

where $f_0(\cdot,\cdot,\cdot)$ is assumed to be continuous. Note that we consider here an HOCP with a final point constraint (compare with (5.10)). The above HS is evidently interpreted as an ordinary differential equation with the Carathéodory-type right-hand sides

$$\tilde{f}(t, x, u) = \sum_{i=1}^{r+1} \beta_{[t_{i-1},t_i)}(t) f_{q_i}(t, x, u),\ i = 1, \ldots, r+1.$$

Moreover, $u(t) \in U$ and the admissible control set $U \subset \mathbb{R}^m$ is assumed to be compact.

Let us give a formal description of the switching conditions in an HS in the case of smooth switching manifolds $M_q,\ q \in Q$, determined by the corresponding algebraic equations

$$M_q := \{z \in \mathbb{R}^n : g_q(z) = 0\},$$

where $g_q : \mathbb{R}^n \to \mathbb{R}^{n-1}$ are continuous functions. For an admissible trajectory associated with the HS from problem (7.27) we evidently have

$$g_q(x(t_i)) = 0,\ t_i \in \tau,$$

where τ is a sequence of the realized switching times in the above HS (see Definition 1.1). For every admissible τ associated with the above HS the Rubio relaxation of the constrained OCP (7.27) can now be expressed as follows (see also Chapter 5): find a positive linear bounded functional

$$\Lambda \in \mathbb{C}^*([0, t_f] \times X \times U)$$

which solves the generalized optimization problem

$$\begin{aligned} &\text{minimize } \Lambda(f_0(\cdot, \cdot, \cdot)) \\ &\text{subject to} \\ &\Lambda(\varphi^{\tilde{f}}) = \Delta\varphi \text{ f.a. } \varphi \in \mathbb{C}^1(B_{[0,t_f]\times X}), \\ &\Lambda(\delta(t - t_i)\tilde{g}_q) = 0 \text{ f.a } [0, t] \subseteq [0, t_f]. \end{aligned} \tag{7.28}$$

We formally defined here

$$\tilde{g}_q(t, x, u) = g_q(x) \ \forall (t, x, u) \in \mathbb{R}_+ \times X \times U$$

for all $q \in Q$. A "Dirac delta" $\delta(\cdot)$ in (7.28) is understood as a substitution of the sequence of zero-centered normal distributions. Recall that for a given continuous trajectory of the HS from HOCP (7.27) we have

$$\int_0^{t_f} g_q(x(t))\delta(t - t_i)dt = g_q(x(t - i)).$$

As in Chapter 5 $\varphi^{\tilde{f}}$ denotes the formal derivative $\varphi_t + \varphi_x \tilde{f}$, and

$$\Delta\varphi := \varphi(t_f, x_f) - \varphi(0, x_0).$$

Moreover, $B_{[0,t_f]\times X}$ in (7.28) is an open neighborhood of the space $[0, t_f] \times X$. Let us recall that a (positive) linear functional Λ in (7.28) satisfies

$$\Lambda(F(\cdot, \cdot, \cdot)) \geq 0$$

for a continuous function $F(\cdot, \cdot, \cdot) \geq 0$.

Note that the existence of an optimal solution

$$\Lambda^{opt} \in \mathbb{C}^*([0, t_f] \times X \times U)$$

to the abstract HOCP (7.28) is a simple consequence of the linear program result in the dual space

$$\mathbb{C}^*([0, t_f] \times X \times U).$$

Note that the minimum in (7.28) coincides with the infimum, i.e.,

$$\Lambda^{opt}(f_0(\cdot,\cdot,\cdot)) \equiv \inf_{Problem (7.28)} \Lambda(f_0(\cdot,\cdot,\cdot)). \tag{7.29}$$

We refer to [61] for the necessary theoretical details.

The compact form of the hybrid dynamic system from HOCP (7.27) has the same symbolic form as an SS. In the HS case the characteristic functions $\beta_{[t_{i-1},t_i)}(t)$ incorporate the dependence of the switching mechanism τ on the "continuous" trajectory (see Definition 1.3) $x(\cdot)$. In fact, the switching time instants are functions of $x(\cdot)$: $t_i(x(\cdot))$ for $i = 1, ...r + 1$. Note that this difference between SSs and HSs does not affect essentially the form of the Rubio relaxation (7.28). The corresponding Rubio relaxation associated with an SOCP of the type (7.27) (in which the characteristic functions $\beta_{[t_{i-1},t_i)}(t)$ do not depend on $x(\cdot)$) is given as follows:

$$\begin{aligned}
&\text{minimize } \Lambda(f_0(\cdot,\cdot,\cdot)) \\
&\text{subject to} \\
&\Lambda(\varphi^{\tilde{f}}) = \Delta\varphi \text{ f.a. } \varphi \in \mathbb{C}^1(B_{[0,t_f]\times X}), \\
&\Lambda(\delta(t - t_i)) = 0 \text{ f.a } [0,t] \subseteq [0,t_f].
\end{aligned} \tag{7.30}$$

As in the case of the conventional relaxation theory the Rubio relaxations (7.27) and (7.30) make it possible to establish the lower bound for the Gamkrelidze relaxations. Let $u^{opt}(\cdot)$ be an optimal solution to the original OCP (7.27). We present here the corresponding result for the HOCPs of the type (7.27) (associated with the HSs).

Theorem 7.11. *Consider an HOCP of the type (7.27) and assume that all the above technical conditions are satisfied. Then*

$$\inf_{Problem (7.28)} \Lambda(f_0(\cdot,\cdot,\cdot)) \le \bar{\mathcal{J}}(\nu^{opt}(\cdot)) \le \mathcal{J}(u^{opt}(\cdot)).$$

The formal proof of Theorem 7.11 is similar to the proof of Theorem 5.9 (see [281]). As in the case of classic OCPs the above result seems to be very natural due to the natural consideration of the powers of sets

$$\mathbb{C}^*([0,t_f] \times X \times U)$$

and $R_c([0,t_f],U)$. The corresponding Rubio relaxation gap for HOCPs can now be expressed as follows (similar to the corresponding result from Chapter 5):

$$Gap := \mathcal{J}(u^{opt}(\cdot)) - \Lambda^{opt}(f_0(\cdot,\cdot,\cdot)).$$

Using the Rubio relaxation (7.30) of an SOCP of the type (7.27), one can also prove a theoretic result for SOCPs similar to Theorem 7.11.

Finally note that similarly to the classic case discussed in Chapter 5 the optimal Rubio generalization presented above is mathematically equivalent to the abstract optimization problem from the classic Ky Fan theory [261].

7.6 Special Topics

7.6.1 A Constrained LQ-Type Optimal Control

The aim of this section is to elaborate a consistent computational algorithm for a linear quadratic (LQ)-type OCP in systems with piecewise constant control inputs. Evidently, the switched control structure of the above OCP involves a general SS approach. On the other side the switched structure constitutes a specific (hard) restriction in the corresponding LQ dynamic optimization. We consider here a specific case of a dynamic system discussed in Chapter 5. The given restrictive structure of the admissible control function under consideration is motivated by some important engineering applications (see [73,89,169,179,206,207, 270,295,300]) as well as by the application of common quantization procedures to the original dynamics (see, e.g., [108,231,265]). Note that quadratic optimal control of piecewise linear systems was addressed earlier in [207,270]. The treatment there was based on the backward solutions of Riccati differential equations, and the optimum had to be recomputed for each new final state. Computation of nonlinear gain using the Hamilton–Jacobi–Bellman equation and the convex optimization techniques has also been done in [270]. Let us also refer to a sophisticated solution technique for the general nonlinear OCPs proposed in [61]. This approach is based on a newly developed relaxation procedure.

On the other hand, the common optimization approaches to linear constrained and switched-type systems are not sufficiently advanced to LQ-type problems for linear systems with fixed level controls. We consider here a new numerical method that includes a specific relaxation scheme in combination with the classic projection approach. Recall that the general SSs constitute a formal framework of systems where two types of dynamics are present, continuous and discrete event behavior (see, e.g., [103,231]). Evidently, a dynamic model with fixed level control inputs constitutes a simple example of an SS. The nonstationary linear systems we study include a particular family of SSs with the time-driven location transitions. We refer to [97,124,140,150,289,298,310,323] for some relevant examples and abstract concepts.

Consider the following linear nonstationary system with a switched control structure:

$$\dot{x}(t) = A(t)x(t) + B(t)u(t), \ t \in [t_0, t_f], \ x(t_0) = x_0, \tag{7.31}$$

where

$$A(\cdot) \in \mathbb{L}^{\infty}[t_0, t_f; \mathbb{R}^{n \times n}], \quad B(\cdot) \in \mathbb{L}^{\infty}[t_0, t_f; \mathbb{R}^{n \times m}].$$

Here $\mathbb{L}^{\infty}[t_0, t_f; R^{n\times n}]$ and $\mathbb{L}^{\infty}[t_0, t_f; \mathbb{R}^{n\times m}]$ are the standard Lebesgue spaces of the essentially bounded matrix functions defined on a bounded time interval $[t_0, t_f]$. Similarly to the classic LQ regulator (LQR) theory it is desired to minimize the following quadratic cost functional associated with (7.31):

$$J(u(\cdot)) = \frac{1}{2}\int_{t_0}^{t_f} (\langle Q(t)x(t), x(t)\rangle + \langle R(t)u(t), u(t)\rangle)dt + \frac{1}{2}\langle Gx(t_f), x(t_f)\rangle, \qquad (7.32)$$

where $G \in \mathbb{R}^{n\times n}$ and the matrix functions $Q(\cdot)$ and $R(\cdot)$ are assumed to be integrable.

Following the conventional LQR theory we next introduce the standard regularity/positivity hypothesis:

$$G \geq 0, \quad Q(t) \geq 0, \; R(t) \geq \delta I, \; \delta > 0 \; \forall t \in [t_0, t_f].$$

It is well known that the classic LQ optimal control strategy $u^{opt}(\cdot)$ does not incorporate any additional (state or control) restrictions into the resulting design procedure. Let us recall here the explicit formula for $u^{opt}(\cdot)$ (see, e.g., [154,206]):

$$u^{opt}(t) = -R^{-1}(t)\left[B^T(t)P(t)\right]x^{opt}(t), \qquad (7.33)$$

where $P(\cdot)$ is the matrix function, namely, the solution to the classic differential matrix Riccati equation associated with the LQ problem (7.31)–(7.32). In the abovementioned conventional case the above optimization problem is formally studied in the full space $\mathbb{L}^2[t_0, t_f; \mathbb{R}^m]$ of square integrable control functions. In contrast to the classic case, we consider system (7.31) in combination with the specific piecewise constant admissible inputs $u(\cdot)$. Let us also refer to Chapter 5 for a similar problem statement. Resulting from the above admissibility assumption the main minimization problem for the linear system (7.31) can be interpreted as a restricted LQ (classic) optimization problem.

Let us now specify formally the set of admissible piecewise constant control functions for system (7.31) in a general case. For each component $u^{(k)}(\cdot)$ of the feasible control input

$$u(\cdot) = [u^{(1)}(\cdot), \ldots, u^{(m)}(\cdot)]^T$$

we introduce the following finite set of feasible value levels:

$$\mathbb{Q}^k := \{q_j^{(k)} \in \mathbb{R}, \; j = 1, \ldots, M_k\}, \; M_k \in \mathbb{N}, \quad k = 1, \ldots, m.$$

In general, all the sets $\mathbb{Q}^k$ are different (contain different levels) and have various numbers of elements. In addition, each $\mathbb{Q}^k$ possesses a strict order property

$$q_1^{(k)} < q_2^{(k)} < \ldots < q_{M_k}^{(k)}.$$

We now introduce the set of switching times associated with an admissible control function

$$T^k := \{t_i^{(k)} \in \mathbb{R}_+, i = 1, ..., N_k\}, \quad N_k \in \mathbb{N}, \qquad k = 1, \ldots, m.$$

The sets T^k are assumed to be defined for each control component

$$u^{(k)}(\cdot), \; k = 1, ..., m,$$

where $\mathbb{R}_+$ denotes a nonnegative semiaxis. Let us consider an ordered sequence of time instants

$$t_0 < t_1^{(k)} < ... < t_{N_k}^{(k)}.$$

For the final time instants associated with each set T^k we put

$$t_{N_1}^{(1)} = \ldots = t_{N_m}^{(m)} = t_f.$$

Using the notation of the level sets $\mathbb{Q}^k$ and the fixed switching times T^k introduced above, the set of admissible controls $\mathcal{S}$ can now be easily specified by the Cartesian product

$$\mathcal{S} := \mathcal{S}_1 \times \ldots \times \mathcal{S}_m, \tag{7.34}$$

where each set $\mathcal{S}_k$, $k = 1, ..., m$, is defined as follows:

$$\mathcal{S}_k := \{v(\cdot) \mid v(t) = \sum_{i=1}^{N_k} I_{[t_{i-1}^{(k)}, t_i^{(k)})}(t) q_{j_i}^{(k)}; \; q_{j_i}^{(k)} \in \mathbb{Q}^k; \; j_i \in \mathbb{Z}[1, M_k]; \; t_i^{(k)} \in T^k\}.$$

By $\mathbb{Z}[1, M_k]$ we denote here the set of all integers into the interval $[1, M_k]$ and

$$I_{[t_{i-1}^k, t_i^k)}(t)$$

is the characteristic function of the interval $[t_{i-1}^k, t_i^k)$. Evidently, the set of admissible control inputs $\mathcal{S}$ can be qualitatively interpreted as the set of all the possible functions $u : [t_0, t_f] \to \mathbb{R}^m$, such that each component $u^{(k)}(\cdot)$ of $u(\cdot)$ attains a constant level value

$$q_{j_i}^{(k)} \in \mathbb{Q}^k$$

for $t \in [t_{i-1}^k, t_i^k)$. Moreover, the component level changes occur only at the prescribed times

$$t_i^k \in T^k, \; i = 1, ..., N_k - 1.$$

Motivated by various engineering applications, we can now formulate the following specific constrained LQ-type OCP:

$$\begin{aligned} &\text{minimize } J(u(\cdot)) \\ &\text{subject to } u(\cdot) \in \mathcal{S}, \end{aligned} \tag{7.35}$$

where $J(\cdot)$ is the costs functional defined in (7.32). Note that $\mathcal{S}$ constitutes a nonempty subset of the space $\mathbb{L}^2[t_0, t_f; \mathbb{R}^m]$. However, the classically LQ-optimal control input $u^{opt}(\cdot)$ in (7.33) does not belong to the introduced specific set $\mathcal{S}$. Due to the highly restrictive condition $u(\cdot) \in \mathcal{S}$, the main optimization problem (7.35) cannot be generally solved by a direct application of the classic Pontryagin maximum principle. A possible application of a suitable hybrid version of the conventional maximum principle from [39,103,174,289,298,323,324] is also complicated by a nonstandard structure of the simple control inputs under consideration. Let us additionally note that the value of an exponentially growing problem cardinality $|\mathcal{S}|$ exacerbates crucially a possible application of some combinatorial and various state/control discretization-based numerical algorithms for OCPs.

We will propose here a relatively simple implementable computational procedure for a consistent numerical treatment of the constrained OCP (7.35). We use a simple (problem-oriented) relaxation technique associated with the main OCP. Next we use these relaxations in a constructive solution procedure for the original problem (7.35).

Consider now a solution $x^u(\cdot)$ to the initial value problem (7.31) generated by an admissible control $u(\cdot) \in \mathcal{S}$. Evidently, every component of $x^u(\cdot)$ is an affine function (functional) of $u(\cdot)$, i.e.,

$$x(t,u) = \Phi(t,t_0)x_0 + \int_{t_0}^{t} \Phi(t,\tau)B(\tau)u(\tau)d\tau.$$

Here $\Phi(\cdot,\tau)$ is the fundamental solution matrix associated with (7.31). Let us note that the set of admissible controls $\mathcal{S}$ constitutes a nonconvex set. This fact is due to the originally combinatorial structure of $\mathcal{S}$ determined in (7.34).

Example 7.1. *Consider $\mathcal{S} := \mathcal{S}_1 \times \mathcal{S}_2$ and, moreover,*

$$\begin{aligned} &\mathcal{S}_1 = \{(0 \times I_{[0,0.5)}(t) + 1 \times I_{[0.5,1)}(t));\ (1 \times I_{[0,0.5)}(t) + 0 \times I_{[0.5,1)}(t));\\ &(0 \times I_{[0,0.5)}(t) + 2 \times I_{[0.5,1)}(t));\ (2 \times I_{[0,0.5)}(t) + 0 \times I_{[0.5,1)}(t));\\ &(1 \times I_{[0,0.5)}(t) + 2 \times I_{[0.5,1)}(t));\ (2 \times I_{[0,0.5)}(t) + 1 \times I_{[0.5,1)}(t));\ (0);\ (1);\ (2)\}\\ &\mathcal{S}_2 = \{(0 \times I_{[0,0.33)}(t) + (-1) \times I_{[0.33,0.66)}(t) + (-1) \times I_{[0.66,1)}(t));\\ &(0 \times I_{[0,0.33)}(t) + (-1) \times I_{[0.33,0.66)}(t) + 0 \times I_{[0.66,1)}(t));\\ &(0 \times I_{[0,0.33)}(t) + 0 \times I_{[0.33,0.66)}(t) + (-1) \times I_{[0.66,1)}(t)); \end{aligned}$$

$$((-1) \times I_{[0,0.33)}(t) + (-1) \times I_{[0.33,0.66)}(t) + 0 \times I_{[0.66,1)}(t));$$
$$((-1) \times I_{[0,0.33)}(t) + 0 \times I_{[0.33,0.66)}(t) + 0 \times I_{[0.66,1)}(t));$$
$$((-1) \times I_{[0,0.33)}(t) + 0 \times I_{[0.33,0.66)}(t) + (-1) \times I_{[0.66,1)}(t)); \ (0); \ (-1)\}.$$

The combinatorial structure of $\mathcal{S}$ *is evident. Recall that a combinatorial set is a nonconvex set. The convex hull* $\text{conv}(\mathcal{S})$ *of the original set* $\mathcal{S}$ *has here a simple expression, i.e.,*

$$\text{conv}(\mathcal{S}) = \{(C_1 \times I_{[0,0.5)}(t), C_2 \times I_{[0.5,1)}(t))\} \times$$
$$\{(D_1 \times I_{[0,0.33)}(t) + D_2 \times I_{[0.33,0.66)}(t) + D_3 \times I_{[0.66,1)}(t))\},$$

where

$$C_1, C_2 \in [0, 2], \ D_1, D_2, D_3 \in [0, -1].$$

Motivated by the above example let us consider the convex hull $\text{conv}(\mathcal{S})$ associated with the general set $\mathcal{S}$, i.e.,

$$\text{conv}(\mathcal{S}) := \{v(\cdot) \mid v(t) = \sum_{s=1}^{|\mathcal{S}|} \lambda_s u_s(t), \ \sum_{s=1}^{|\mathcal{S}|} \lambda_s = 1,$$

where

$$\lambda_s \geq 0, \ u_s(\cdot) \in \mathcal{S}, \ s = 1, \ldots, |\mathcal{S}|\}.$$

From the definition of $\mathcal{S}$ we conclude that the convex set $\text{conv}(\mathcal{S})$ is closed and bounded. Using (7.34), we also can give the alternative characterization, i.e.,

$$\text{conv}(\mathcal{S}) = \text{conv}(\mathcal{S}_1) \times \ldots \times \text{conv}(\mathcal{S}_m),$$

where $\text{conv}(\mathcal{S}_k)$ is a convex hull of $\mathcal{S}_k$ $k = 1, \ldots, m$. Since

$$\text{conv}(\mathbb{Q}^k) \equiv [q_1^{(k)}, q_{M_k}^{(k)}],$$

we have

$$\text{conv}(\mathcal{S}_k) := \{v(\cdot) \mid v(t) =$$
$$\sum_{i=1}^{N_k} I_{[t_{i-1}^{(k)}, t_i^{(k)})}(t) q_{j_i}^{(k)}; \ q_{j_i}^{(k)} \in [q_1^{(k)}, q_{M_k}^{(k)}]; \ j_i \in \mathbb{Z}[1, M_k]; \ t_i^{(k)} \in T^k\}.$$

Roughly speaking $\text{conv}(\mathcal{S})$ contains all the piecewise constant functions $u(\cdot)$ such that the constant value $u^{(k)}(t)$ belongs to the interval

$$[q_1^{(k)}, q_{M_k}^{(k)}]$$

for all $t \in [t_{i-1}^{(k)}, t_i^{(k)})$. Let us note that in contrast to the initial set $\mathcal{S}$, the corresponding convex hull conv$(\mathcal{S})$ is an infinite-dimensional space. Using the above convex construction, we can formulate the following auxiliary OCP:

$$\begin{aligned} &\text{minimize } J(u(\cdot)) \\ &\text{subject to } u(\cdot) \in \text{conv}(\mathcal{S}). \end{aligned} \tag{7.36}$$

The problem (7.36) formulated above is in fact a simple convex relaxation of the initial OCP (7.35). We will study this problem and use it for a constructive numerical treatment of the initial switched-type OCP (7.35).

Let us firstly formulate the following key property of the auxiliary OCP (7.36).

Theorem 7.12. *The cost functional*

$$J : \text{conv}(\mathcal{S}) \to \mathbb{R}$$

determined as

$$J(u(\cdot)) = \frac{1}{2}\int_{t_0}^{t_f} [\langle Q(t)x^u(t), x^u(t)\rangle + \langle R(t)u(t), u(t)\rangle]dt + \frac{1}{2}\langle Gx^u(t_f), x(t_f)\rangle$$

is convex and the auxiliary OCP (7.36) *constitutes a convex optimization problem in the* Hilbert *space* $\mathbb{L}^2[t_0, t_f; \mathbb{R}^m]$.

Proof. Evidently, conv$(\mathcal{S})$ is a bounded closed and convex subset of $\mathbb{L}^2[t_0, t_f; \mathbb{R}^m]$. The cost functional $J(\cdot)$ is in fact a sum of two functionals:

$$J(u(\cdot)) = J_1(u(\cdot)) + J_2(u(\cdot)), \; J_1(u(\cdot)) := \frac{1}{2}\int_{t_0}^{t_f} [\langle R(t)u(t), u(t)\rangle]dt,$$

$$J_2(u(\cdot)) := \frac{1}{2}\int_{t_0}^{t_f} [\langle Q(t)x^u(t), x^u(t)\rangle]dt + \frac{1}{2}\langle Gx^u(t_f), x^u(t_f)\rangle.$$

The first one, namely, the functional $J_1(\cdot)$, is convex (recall that its Hessian is a positive definite matrix).

Moreover, $J_2(\cdot)$ is a composition of a convex (quadratic) functional of $x^u(\cdot)$, where $x^u(\cdot)$ is an affine mapping with respect to $u(\cdot)$. We now easily deduce the convexity property of $J_2(\cdot)$. Since the sum of two convex functions is convex, we obtain the desired convexity result for $J(\cdot)$. The proof is completed. □

As we can see (7.36) is a convex relaxation of the initial OCP (7.35). The proved convexity of the auxiliary (relaxed) OCP (7.36) makes it possible to apply the powerful numerical convex programming approaches to this auxiliary optimization problem (see Chapter 3). We next use a variant of the projected gradient method for a concrete numerical treatment of (7.36).

Note that under the basic assumptions introduced in Section 7.1 the mapping

$$x^u(t) : \mathbb{L}^2[t_0, t_f; \mathbb{R}^m] \to \mathbb{R}^n$$

is Fréchet differentiable for every $t \in [t_0, t_f]$ (see [17,23]). Therefore, the quadratic costs functional $J(\cdot)$ in (7.36) is also Fréchet differentiable. We refer to [23,27] for the corresponding differentiability concept. Assume $u^*(\cdot) \in \text{conv}(\mathcal{S})$ is an optimal solution of (7.36). The existence of an optimal input $u^*(\cdot)$ is guaranteed in the convex problem (7.36) (see Chapter 3). By $x^*(\cdot)$ we next denote the corresponding optimal trajectory (solution) of (7.35) generated by $u^*(\cdot)$. The projected gradient method for the relaxed problem (7.36) can now be easily expressed, i.e.,

$$u_{l+1}(\cdot) = \mathcal{P}_{\text{conv}(\mathcal{S})}\left[u_l(\cdot) - \alpha_l \nabla J(u_l(\cdot))\right], \tag{7.37}$$

where $\mathcal{P}_{\text{conv}(\mathcal{S})}$ is the operator of projection onto convex set $\text{conv}(\mathcal{S})$ and $\{\alpha_l\}$ is a sequence of step sizes. The conventional projection operator $\mathcal{P}_{\text{conv}(\mathcal{S})}$ is defined as follows:

$$\mathcal{P}_{\text{conv}(\mathcal{S})}(u(\cdot)) := \text{Argmin}_{v(\cdot) \in \text{conv}(\mathcal{S})} \left(||v(\cdot) - u(\cdot)||_{\mathbb{L}^2[t_0, t_f; \mathbb{R}^m]} \right).$$

Recall that the projected gradient iterations (7.37) generate a minimizing sequence for the convex optimization problem (7.36) (see Chapter 3). In the context of OCP (7.36) and the applied computational optimization method (7.37) the basic convergence result from [259, 276] can now be reformulated as follows.

Theorem 7.13. *Assume that all the technicalities are satisfied. Consider a sequence of control functions generated by* (7.37). *Then there exist admissible initial data* $(u^0(\cdot), x^0(\cdot))$ *and a sequence of the step sizes* $\{\alpha_l\}$ *such that* $\{u_l(\cdot)\}$ *is a minimizing sequence for the relaxed OCP* (7.36), *i.e.,*

$$\lim_{l \to \infty} J(u_l(\cdot)) = J(u^*(\cdot)).$$

The proposed gradient-type method (3.4) provides a basis for the computational approach to (7.36). Using an optimal solution $u^*(\cdot) \in \text{conv}(\mathcal{S})$ we next need to determine a suitable approximation for a solution to the original OCP (7.35).

Note that in contrast to the relaxed optimization problem (7.36) the original switched-type OCP (7.35) does not possess any convexity property. However, the simple relaxed OCP (7.36) can be effectively used for an approximative numerical treatment of the original problem (7.35).

Let us introduce the formal Hamiltonian associated with the given OCPs

$$H(t, x, u, p) = \langle p, A(t)x + B(t)u \rangle - \frac{1}{2} \left(\langle Q(t)xt, x \rangle + \langle R(t)u, u \rangle \right),$$

where $p \in \mathbb{R}^n$ is the adjoint variable. By $\hat{u}(\cdot) \in \mathcal{S}$ we now denote an optimal solution to the initial OCP (7.35). Using the explicit representation of the gradient $\nabla J(u_l(\cdot))$ in OCPs with ordinary differential equations (see, e.g., [151,261,289,300]), we can propose an implementable conceptual computational scheme for (7.35) using the relaxation (7.36).

Algorithm 7.1.

(0) *Set the initial condition for the iterative scheme*

$$u_{(0)}(\cdot) := \mathcal{P}_{\text{conv}(\mathcal{S})}(u^{opt}(\cdot)),$$

where $u^{opt}(\cdot)$ is the optimal control input (7.33) *from the classic* LQ *problem. Calculate the corresponding trajectory $x_{(l)}(\cdot)$ of* (7.31) *and put the iterations index $l := 0$.*

(1) *Calculate $\nabla J(u_{(l)})(\cdot)$ as*

$$\nabla J(u_{(l)})(t) = -\frac{\partial H(t, x_{(l)}(t), u_{(l)}(t), p(t))}{\partial u} = -B^T(t)p(t) + R(t)u_{(l)},$$

where the adjoint variable $p(\cdot)$ is a solution to the usual boundary value problem

$$\dot{p}(t) = -\frac{\partial H(t, x_{(l)}(t), u_{(l)}(t), p)}{\partial x} = -A^T(t)p(t) + Q(t)x_{(l)}(t),$$
$$p(t_f) = -Gx_{(l)}(t_f).$$

(2) *Calculate the projection of*

$$u_{(l)}(\cdot) - \alpha_{(l)} \nabla J(u_{(l)}(\cdot))$$

on the convex (relaxed) restriction set $\text{conv}(\mathcal{S})$ *and determine*

$$\bar{u}_{(l+1)}(\cdot) := \mathcal{P}_{\text{conv}(\mathcal{S})}(\bar{u}_{(l)}(\cdot)).$$

(3) *Evaluate the $(l+1)$-iteration of the control function given by components*

$$u^{(k)}_{(l+1)}(t) = \sum_{i=1}^{N_k} I_{[t^{(k)}_{i-1}, t^{(k)}_i)}(t) \bar{q}^{(k)}_{i,n} \quad \forall k = 1, \ldots, m,$$

where

$$\bar{q}_{i,l}^{(k)} := \begin{cases} q_1^{(k)}, & \bar{\bar{q}}_{i,l}^{(k)} < q_1^{(k)} \\ \bar{\bar{q}}_{i,l}^{(k)}, & q_1^{(k)} \leq \bar{\bar{q}}_{i,l}^{(k)} \leq q_{M_k}^{(k)} \\ q_{M_k}^{(k)}, & q_{M_k}^{(k)} \leq \bar{\bar{q}}_{i,l}^{(k)} \end{cases}, \; i = 1, \ldots, N_k,$$

and

$$q_j^{(k)} \in \mathbb{Q}^k, \; \forall j = 1, \ldots, M_k,$$

$$\bar{\bar{q}}_{i,l}^{(k)} := \frac{1}{\Delta_i} \int_{t_{i-1}}^{t_i} \bar{u}_{(l)}^{(k)}(t)dt, \; \Delta_i := t_i - t_{i-1}.$$

(4) *Calculate the difference*

$$|J(u_{(l+1)}(\cdot)) - J(u_{(l)}(\cdot))|.$$

If it is less than a prescribed accuracy $\varepsilon > 0$*, then we put*

$$u^*(\cdot) \equiv u_{(l+1)}(\cdot)$$

(an approximating optimal solution to (7.35)*) and* Stop. *Else, update the iteration register and go to* Step (1).

(5) *Using the evaluated function* $u^*(\cdot)$ *the approximating optimal control*

$$\hat{u}(\cdot) \in \mathcal{S}$$

can finally be calculated by components

$$\hat{u}^{(k)}(\cdot) = \sum_{i=1}^{N_k} I_{[t_{i-1}^{(k)}, t_i^{(k)})}(t)\hat{q}_i^{(k)} \quad \forall k = 1, ..., m,$$

where

$$\hat{q}_i^k := \text{Arg} \min_{v \in \mathbb{Q}^k} |v - \bar{q}_{i,l+1}^{(k)}|.$$

Solve (7.31) *with the obtained control input* $\hat{u}(\cdot) \in \mathcal{S}$ *and obtain the approximating optimal trajectory* $\hat{x}(\cdot)$. Stop.

Using the above theorems and the continuity property of the objective functional one can establish the convergence of the proposed conceptual Algorithm 7.1. Note that this type of convergence is determined as a “convergence in functional.” To put it another way, the control sequence $\{u_l(\cdot)\}$ generated by Steps (0)–(4) of the above algorithm is a minimizing sequence.

By implementation and taking into consideration the continuity of the objective functional, we can finally establish the convergence property ("in functional") of the resulting sequences, i.e.,

$$\{\hat{u}^{(k)}(\cdot)\},\ k = 1, \ldots, m,$$

obtained in Step (5) of the above algorithm.

We have developed a problem-oriented implementable numerical approach to a constrained LQ-type OCP. The convex structure of the auxiliary OCP makes it possible to take into consideration diverse powerful computational algorithms from the classic convex programming. Let us note that various variants of the basic gradient method, namely, Armijo-type gradient schemes, can be applied to the obtained relaxed OCP (see [16,44,140,152,300] and Chapter 6).

It is common knowledge that modern numerical algorithms mainly use specific nonequidistant discretizations with the aim to increase the effectiveness of the resulting algorithm. The specific type of the control functions discussed in this section is motivated by the initially given physical nature of the class of controlled processes under consideration. Note that there are various formal models that involve a nonequally spaced inputs grid. The necessary investigation of these types of models and the corresponding engineering applications (mainly from modern communication science) constitute an interesting theoretic and numerical subject.

7.6.2 A Simple Switched System and the Corresponding SOCP

In recent years there has been a revival of efficient optimization techniques developed for various types of nonlinear HOCPs. This fact is due to the valuable progress in the area of computational engineering and computer sciences. Nowadays the most powerful numerical approaches to switched OCPs (SOCPs) and HOCPs are based on so-called "optimality zone" algorithms developed in [13,27] and on the first-order (gradient-based) techniques.

We now study a specific class of SSs and apply the classic proximal point method (see Chapter 3) in the corresponding optimal control design procedure. The elaborated approach involves a specific (problem-oriented) relaxation procedure associated with the initially given dynamic model. After the basic convexity properties of the OCP under consideration are established, one can apply a combination of the proximal point regularization algorithm and the usual convex programming techniques.

In this section we consider a simple SS and the corresponding SOCP. Recall that the abstract compact form for an SS is the same as for the HS, namely, the specific ordinary differential equation in (7.2). The "hybrid nature" of this mathematical model is characterized by an explicit dependence

$$t_i,\ i = 1, ..., r + 1$$

(switching times) on the trajectory $x(\cdot)$ of the given HS. In this case the corresponding characteristic functions $\beta_{[t_{i-1},t_i)}(\cdot)$ in (7.2) can be interpreted as an additional system output (a causal variable) and the classic relaxation techniques (for example, the Gamkrelidze or Rubio relaxations) can be directly applied to the initial HS (7.2).

In the case of an SS the characteristic functions $\beta_{[t_{i-1},t_i)}(\cdot)$ in (7.2) are in fact the additional time-dependent system inputs. The full control vector of an SS is a pair $(\beta(\cdot), u(\cdot))$, where $\beta(\cdot)$ is an $(r+1)$-dimensional vector function containing the above functions, i.e.,

$$\beta_{[t_{i-1},t_i)}(\cdot),\ i = 1, ..., r+1.$$

A formal implementation of a classic relaxation procedure in that case, for example, application of the Gamkrelidze (or Rubio) relaxation to the SS, will imply a big relaxation gap. This conclusion is a simple consequence of the specific "jump"-type character of the first component of the SS control pair $(\beta(\cdot), u(\cdot))$. For example, the relaxed control $\mu(\cdot)$ (see Section 7.1) is a measure and the initial control component $\beta(\cdot)$ is a simple discrete-valued function.

Motivated by the above observation, we now consider a simple problem-oriented relaxation procedure for a simplified switched control system introduced in [2,3,17,18].

Definition 7.3. *A simple SS with controllable location transitions is a collection*

$$\{\mathcal{Q}, \mathcal{F}, \tau, \mathcal{S}\},$$

where:

- *$\mathcal{Q}$ is a finite set of indices called (locations);*
- *$\mathcal{F} = \{f_q\}_{q\in\mathcal{Q}}$ is a family of vector fields $f_q : \mathbb{R}^n \to \mathbb{R}^n$;*
- *$\tau = \{t_i\},\ i = 1, ..., r$, is an admissible sequence of switching times such that*

$$0 = t_0 < t_1 < ... < t_{r-1} < t_r = t_f;$$

- *and*

$$\mathcal{S} \subset \Xi := \{(q, q') \ : \ q, q' \in \mathcal{Q}\}$$

is a reset set.

We next assume that all functions f_q, $q \in \mathcal{Q}$, are continuously differentiable and the corresponding derivatives are bounded. Elements of the collection of vector fields $\mathcal{F}$ do not contain any conventional control parameter. An input of this SS is given as a sequence τ of switching times such that the length of this sequence is equal to $r \in \mathbb{N}$. In fact an admissible sequence of switchings τ determines a partitioning of the given time interval $[0, t_f]$ by the adjoint subintervals $[t_{i-1}, t_i)$ associated with every location

$$q_i \in \mathcal{Q}, \; i = 1, ..., r.$$

A switched control system from Definition 7.3 remains in a location $q_i \in \mathcal{Q}$ for all $t \in [t_{i-1}, t_i)$. The dynamic behavior of the SS under consideration is given by the following differential equation associated with every location:

$$\dot{x}_i(t) = f_{q_i}(x_i(t)).$$

The above dynamics is considered for almost all times $t \in [t_{i-1}, t_i]$, where

$$x_i(\cdot) = x(\cdot)|_{(t_{i-1}, t_i)}$$

is an absolutely continuous function on (t_{i-1}, t_i) continuously prolongable to $[t_{i-1}, t_i]$, $i = 1, ..., r$. By $x(\cdot)$ we denote here a complete admissible trajectory of an SS such that

$$x(0) = x_0 \in \mathbb{R}^n.$$

This trajectory is determined by a selection of an admissible sequence τ and constitutes an absolutely continuous solution of the resulting i.v.p. We have

$$\begin{aligned} \dot{x}(t) &= \sum_{i=1}^{r} \beta_{[t_{i-1}, t_i)}(t) f_{q_i}(x(t)), \text{ a.e. on } [0, t_f], \\ x(0) &= x_0. \end{aligned} \tag{7.38}$$

As usual $\beta_{[t_{i-1}, t_i)}(\cdot)$ is a characteristic function of the time interval $[t_{i-1}, t_i)$ for $i = 1, ..., r$. The complete input for system (7.38) can now be written as follows:

$$\beta(\cdot) := (\beta_{[t_0, t_1)}(\cdot), ..., \beta_{[t_{r-1}, t_r)}(\cdot))^T.$$

A set $\mathcal{B}_r$ of all admissible vectors $\beta(\cdot)$ introduced above contains specific control inputs characterized by the following conditions:

$$\beta(t) \in \{0, 1\}^r, \; \sum_{i=1}^{r} \beta_{[t_{i-1}, t_i)}(t) = 1.$$

As mentioned above $\beta(\cdot)$ from $\mathcal{B}_r$ does not depend on the SS trajectory $x(\cdot)$. Evidently, $\mathcal{B}_r$ is in one-to-one correspondence to the set of all admissible sequences τ from Definition 7.3. Since trajectory $x(\cdot)$ of an SS is a continuous function, the above system concept describes a class of dynamical systems without impulsive components such that

$$x(t_i) = x(t_{i+1})$$

for all $i = 1, ..., r - 1$.

Let us also note that the set $\mathcal{B}_r$ of admissible control inputs depends on a fixed number r. This condition restricts the number of possible switchings of input $\beta(\cdot)$. Dynamic models (7.38) are widely used in analysis and prototyping of many real-world electronic systems with high-frequency switching. In particular, (3.1) provides a useful theoretic framework for the control design of various power converters.

Given the SS defined above we now consider the following Bolza-type SOCP:

$$\begin{aligned} &\text{minimize } J(\beta(\cdot)) := \int_0^{t_f} f_0(x(t))dt \\ &\text{subject to (3.1)}, \;\; \beta(\cdot) \in \mathcal{B}_r. \end{aligned} \tag{7.39}$$

Here $f_0 : \mathbb{R}^n \to \mathbb{R}$ is assumed to be a continuously differentiable function. Evidently, (7.39) can be interpreted as a specific problem of dynamic switchings optimization. Throughout this section we assume that (7.39) has an optimal solution

$$(\beta^{opt}(\cdot)) \in \mathcal{B}_r.$$

The corresponding optimal trajectory is next denoted by $x^{opt}(\cdot)$. Since $\mathcal{B}_r$ is in one-to-one correspondence to the set of all τ from Definition 7.3, the optimization procedure (7.39) is equivalent to the system optimization over all possible switching times from τ.

Let us note that $\mathcal{B}_r \subset \mathbb{L}^2_r(0, t_f)$, where $\mathbb{L}^2_r(0, t_f)$ is the standard Lebesgue space of r-dimensional square integrable functions. Evidently, the celebrated Pontryagin maximum principle (see [19]) cannot be applied directly to problem (7.39). Recall that the necessary optimality condition in the form of the classic Pontryagin maximum principle provides an adequate (also in a numerical sense) solution procedure in the full space of measurable control functions (the space $\mathbb{L}^2_r(0, t_f)$). A possible direct application of the conventional maximum principle to (7.39) does not imply admissibility of the obtained optimal solution, namely, the condition

$$\beta^{opt}(\cdot) \in \mathcal{B}_r.$$

The same theoretic observation is also true with respect to a hybrid version of the maximum principle (see [5,20,27,30,31] and Chapter 6). One of the possible computational approaches to (7.39) is based on a suitable problem-oriented relaxation procedure.

Let us reformulate the initial SOCP (7.39) in the full space, namely, as a dynamic optimization problem in $\mathbb{L}^2_r(0, t_f)$. We have

$$\begin{aligned} &\text{minimize } J(v(\cdot)) := \int_0^{t_f} f_0(y(t))dt \\ &\text{subject to } \dot{y}(t) = \langle v(t), \Psi(y(t))\rangle_{\mathbb{R}^r}, \; y(0) = x_0, \\ &v(\cdot) \in \mathbb{L}_r^2(0, t_f), \; \sum_{i=1}^{r} v_i(t) = 1, \\ &0 \leq v_i(t)(1 - v_i(t)) \leq \epsilon \; \forall i = 1, ..., r, \end{aligned} \tag{7.40}$$

where

$$v(\cdot) := (v_1(\cdot), ..., v_r(\cdot))^T$$

is a new (auxiliary) square integrable control variable,

$$\Psi(\cdot) := (f_{q_1}(\cdot), ..., f_{q_r}(\cdot))^T,$$

and ϵ is a small positive number. Note that conditions

$$v_i(t)(1 - v_i(t)) = 0, \; i = 1, ..., r,$$

in fact select the characteristic functions ($v_i = 0$ or $v_i = 1$). The inequalities in (7.40) involve a relaxed solution procedure associated with the original OCP (7.39). In that case $\epsilon > 0$ can be interpreted as a required numerical accuracy of the proposed approximation. The SOCP (7.40) is in fact a relaxation of the initial problem (7.39). This relaxed problem provides a simple computational basis for the optimal control design in the original SOCP (7.39).

We now consider the linear variant of the vector field $\mathcal{F}$ from Definition 7.3,

$$f_q(x) = A_q x,$$

where $A_q \in \mathbb{R}^{n \times n}$ for all $q \in \mathcal{Q}$. Assume that every component of all matrices A_q is positive. Recall that such matrices are called Metzler matrices (see, e.g., [241]). The ensuing analysis is also restricted to a monotonically nondecreasing, convex function $f_0(\cdot)$. In that specific case the control system in (7.40) has the simple linear form

$$\dot{y}(t) = \langle v(t), \Psi(y(t))\rangle_{\mathbb{R}^r}, \; y(0) = x_0, \tag{7.41}$$

where

$$\Psi(y) := (A_{q_1} y, ..., A_{q_r} y)^T$$

is an ordered set of given matrices. We next consider the initial and relaxed SOCPs in the framework of the convexity concept (see [4,10] and Chapter 6). Similar to the concept presented in Chapter 6, the relaxed SOCP (7.40) which is equivalent to a convex optimization problem (2.1) is called a convex OCP (see Chapter 6).

Definition 7.4. *Let $y^v(\cdot)$ be an absolutely continuous solution to the relaxed control system (7.41) generated by an admissible control input $v(\cdot)$. Then (7.41) is called a convex control system if every functional*

$$V^k(v(\cdot)) := y^v_k(t),\ k = 1, ..., n,\ t \in [0, t_f],$$

is convex.

Let us present a useful convexity criterion (in the sense of Definition 7.4).

Theorem 7.14. *Consider a classical control system*

$$\dot{z}(t) = h(t, z(t), u(t)) \text{ a.e. on } [0, t_f],\ \ z(0) = x_0$$

and assume that function h is continuous and satisfies the Lipschitz condition

$$||h(t, z_1, u) - h(t, z_2, u)|| \leq L||z_1 - z_2||$$

for all $z_1, z_2 \in \mathbb{R}^n,\ u \in U \subseteq \mathbb{R}^m$. Let

$$h_k(t, \omega),\ k = 1, ..., n,$$

be convex functionals with respect to the variable $\omega := (z, u)$ for every $t \in [0, t_f]$. Moreover, let

$$h_k(t, \cdot, u),\ k = 1, ..., n,$$

be monotonically nondecreasing functionals for every $t \in [0, t_f],\ u \in U$. Then the above control system is convex in the sense of Definition 7.4.

We now show that under the technical assumptions given above the relaxed SOCP (7.40) is equivalent to a convex problem (2.1).

Theorem 7.15. *Assume that the above technical assumptions are satisfied. Then SOCP (7.40) associated with the given linear SS is equivalent to the following convex OCP:*

$$\begin{aligned} &\text{minimize } J(v(\cdot)) := \int_0^{t_f} f_0(y(t))dt \\ &\text{subject to (4.1)},\ v(\cdot) \in \mathbb{L}^2_r(0, t_f),\ \sum_{i=1}^{r} v_i(t) = 1, \\ &0 \leq v_i(t) \leq \delta_1,\ 1 \geq v_j(t) \geq \delta_2,\ i \neq j, \end{aligned} \tag{7.42}$$

where $i,\ j = 1, ..., r$ and $\delta_1 < \delta_2$ are solutions of the simple quadratic equation

$$v^2 - v + \epsilon = 0$$

for a small enough $\epsilon > 0$.

Proof. The right-hand side of the linear switched system (7.41) is linear with respect to $\omega := (v, y)$. From Theorem 7.14 it follows that (7.41) is a convex control system. It means that every $V^k(v(\cdot))$ is a convex functional for every $k = 1, ..., n$. This fact implies convexity of $f_0(y^v(t))$ for all $t \in [0, t_f]$ and also the convexity property of the functional $J(v(\cdot))$. Let

$$v^1(\cdot),\ v^2(\cdot)$$

be admissible controls for OCP (7.42). Then for a convex combination

$$v^3(\cdot) := \lambda v^1(\cdot) + (1-\lambda)v^2,$$

where $\lambda \in (0, 1)$, we obtain

$$\sum_{i=1}^{r} v_i^3(t) = \sum_{i=1}^{r} (\lambda v_i^1(t) + (1-\lambda)v_i^2(t)) = \lambda + 1 - \lambda = 1.$$

Thus the equality constraint in (7.42) determines a convex subset of the generic Hilbert space $\mathbb{L}_r^2(0, t_f)$.

The linear inequality constraints from (7.42) also determine a convex subset of $\mathbb{L}_r^2(0, t_f)$. Since an intersection of a finite number of convex sets constitutes a convex set, problem (7.42) is a convex OCP. Observe that the system of the original inequality/equality constraints

$$0 \le v_i(t)(1 - v_i(t)) \le \epsilon,\ i = 1, ..., r,\ \sum_{i=1}^{r} v_i(t) = 1$$

in (7.40) is equivalent to the system of inequalities from (7.42) for $i = 1, ..., r$. We can conclude that the relaxed SOCP (7.40) is equivalent to a convex OCP of the type (7.42). The proof is completed. □

Problem (7.42) from Theorem 7.15 is equivalent to the abstract convex program (2.1) discussed in Chapter 2. As we have seen the relaxed SOCP (7.40) can finally be replaced by a simple convex program. This fact illustrates an applicability of the problem-oriented relaxation procedures. We now can effectively apply some numerical methods from convex programming for a computational treatment of the sophisticated OCP. Let us consider here the proximal point algorithm (see Chapter 3) and define the regularized variant of (7.42), i.e.,

$$\begin{aligned} &\text{minimize } J^l(v(\cdot)) := J(v(\cdot)) + \frac{\chi_l}{2}||v(\cdot) - v^l(\cdot)||^2_{\mathbb{L}_r^2(0,t_f)} \\ &\text{subject to (4.1)},\ v(\cdot) \in \mathbb{L}_r^2(0, t_f), \\ &\sum_{i=1}^{r} v_i(t) = 1,\ 0 \le v_i(t) \le \delta_1,\ 1 \ge v_j(t) \ge \delta_2,\ i \ne j, \end{aligned} \tag{7.43}$$

where $i, \ j = 1, ..., r$. Here $v^l(\cdot)$ is an admissible l-iteration of the proximal point method and $v^l(\cdot)$ is an admissible initial control. A sequence $\{v^l(\cdot)\}$ of optimal solutions to (7.43) is a weakly convergent minimizing sequence (see Chapter 3 for technical details). We have derived the final representation of the approximating OCP, namely, problem (7.43) that provides a necessary basis for the computational control design for the relaxed SOCP (7.40). The regularized OCP (7.43) can finally be solved by a standard numerical convex optimization technique. Note that (7.43) is a convex optimization problem with a strictly convex functional $J^l(\cdot)$ and can effectively be solved by a simple first-order numerical procedure, for example, by a suitable gradient-type procedure.

The gradient $\bigtriangledown J^l(v(\cdot))$ of $J^l(v(\cdot))$ can be computed as follows:

$$\bigtriangledown J^l(v(\cdot)) = \bigtriangledown J(v(\cdot)) + \chi_l ||v(\cdot) - v^l(\cdot)||_{\mathbb{L}^2_r(0,t_f)},$$

$$\bigtriangledown J(v(\cdot))(t) = -\frac{\partial H}{\partial v}(v(t), y(t), p(t)) = -\langle p(t), \sum_{i=1}^{r} A_{q_i} y(t)\rangle_{\mathbb{R}^n},$$

where

$$H(v, y, p) := \langle p, \langle v(t), \Psi(y(t))\rangle_{\mathbb{R}^r}\rangle_{\mathbb{R}^n} + f_0(y)$$

is the Hamiltonian of (7.42) and $p(\cdot)$ is a solution of the corresponding system of the adjoint equation

$$\dot{p}(t) = -\frac{\partial H}{\partial y}(v(t), y(t), p(t)) = \sum_{i=1}^{r} v_i(t) A^T_{q_i} p(t), \quad p(t_f) = 0.$$

Let us now discuss the quality of the approximating problem (7.43), namely, the convergence property of the sequence $\{v^l(\cdot)\}$ generated by the proposed regularization (7.43). Since problem (7.43) approximates the relaxed SOCP (7.40), we also discuss the relationship to the original optimal solution $\beta^{opt}(\cdot)$ of the SOCP (7.39). From the qualitative point of view, the numerically stable (for example, the proximal-based) optimal solutions to (7.42) with a "small enough" parameter $\epsilon > 0$ represent a well-determined approximation of the optimal vector of characteristic functions in (7.39).

Observe that the measurable function $\beta(\cdot)$ takes its values in the set ver(T_r) of vertices of the following r-dimensional simplex:

$$T_r := \{\beta \in \mathbb{R}^r \mid \beta_i \geq 0, \sum_{i=1}^{r} \beta_i = 1\}.$$

We now give a constructive geometric characterization of the weak convergence of $\{\chi^s(\cdot)\}$, $s \in \mathbb{N}$, from T_r.

Theorem 7.16. *Let $\{\chi^s(\cdot)\}$ be a sequence of $\mathbb{L}_r^2(0,t_f)$ functions such that $\chi^s(t) \in T_r$ for almost all $t \in [0,t_f]$. Assume that $\{\chi^s(\cdot)\}$ converges weakly to a measurable function $\beta(\cdot)$ and*

$$\beta(t) \in \text{ver}(T_r).$$

Then this sequence converges strongly to $\beta(\cdot)$.

We refer to [10] for the complete proof of this auxiliary geometrical result. Theorem 7.16 can be interpreted as follows: the given $\mathbb{L}_r^2(0,t_f)$-weak convergence to a characteristic function coincides with the strong convergence. Note that this analytic fact is true due to a specific type of functions under consideration, namely, for the concrete type of the characteristic functions. Using this abstract result, we are now able to formulate an "improved" convergence result for the proposed proximal point–based algorithm.

Theorem 7.17. *Let all the additional technical assumptions be satisfied. Consider a sequence $\{v^l(\cdot)\}$ of solutions to the OCP (7.43). Then this sequence converges strongly (in the norm topology of $\mathbb{L}_r^2(0,t_f)$) to a solution $\beta^{opt}(\cdot)$ of the initial OCP (7.39).*

Proof. From the standard properties of the classical proximal point method in real Hilbert spaces we immediately deduce the weak convergence of $\{v^l(\cdot)\}$ to an optimal solution of (7.39). Using the equivalency of problems (7.39) and (7.40) established in Theorem 7.15, we conclude that $\{v^l(\cdot)\}$ converges weakly to an optimal solution of the relaxed SOCP (7.40), namely, to an admissible function $v^{opt}(\cdot)$. Evidently,

$$\lim_{\epsilon \to 0} ||v^{opt}(\cdot) - \beta^{opt}(\cdot)||_{\mathbb{L}_r^2(0,t_f)} = 0. \tag{7.44}$$

Recall that $\beta^{opt}(\cdot)$ is an optimal solution to the original OCP (7.39).

It is easy to see that for the decreasing ϵ the solution δ_1 from this theorem also decreases. The solution δ_2 increases for the decreasing ϵ. This means that the diameters of the convex sets that are determined by the corresponding inequality constraints in Theorem 7.15 decrease. This observation and the equivalency relation between problems (7.43) and (7.40) implies the above relation (7.44). A combination of the weak and strong (determined by (7.44)) convergent sequences generates a weakly convergent sequence. Therefore, the control sequence $\{v^l(\cdot)\}$ converges weakly to $\beta^{opt}(\cdot)$.

Finally, from the abstract Theorem 7.16 we deduce the strong convergence of the proximal point sequence $\{v^l(\cdot)\}$ to an optimal solution $\beta^{opt}(\cdot)$ of the initially given SOCP (7.39). The proof is finished. □

Theorem 7.17 gives an answer to the question of a possible relaxation gap. The sequence $\{v^l(\cdot)\}$ generated by the proximal point algorithm for the regularized problem (7.43) converges in the strong sense (in the sense of the usual $\mathbb{L}_r^2(0, t_f)$-norm) to an optimal solution of the initially given SOCP (7.39). This result indicates an absence of the relaxation gap in that specific case.

Let us note that we have additionally shown that the proximal point approach can be used for numerically tractable approximations of some relaxed OCPs associated with switched mode dynamics. An abstract proximal regularization scheme in combination with a specific problem-oriented relaxation approach, extended by an effective convex programming algorithm, makes it possible to establish numerically stable properties of the resulting control design algorithm. The iterative updates of the approximation level we obtained in this section are based on an increasing sequence of "embedded" consistent approximations of the relaxed problem.

7.6.3 On the Hybrid Systems in Mechanics and the Corresponding HOCPs

Let us close this section by consideration of a class of specific types of HSs, namely, HSs of a mechanical nature. We also consider the corresponding HOCPs and possible relaxations. In many real-world applications a controlled mechanical system presents the main modeling framework and is a strongly nonlinear dynamical system of high order [9,10,22]. Let us note that the majority of applied OCPs governed by sophisticated mechanical systems are problems of hybrid nature. The most real-world mechanical control problems are becoming too complex to allow for an analytical solution. Thus, computational algorithms are inevitable in solving these problems.

Let us consider the following variational problem:

$$\begin{aligned} &\text{minimize } \int_0^1 \sum_{i=1}^{r} \beta_{[t_{i-1},t_i)}(t)\tilde{L}_{p_i}(t, q(t), \dot{q}(t))dt \\ &\text{subject to } q(0) = c_0, \;\; q(1) = c_1, \end{aligned} \tag{7.45}$$

where $\tilde{L}_{p_i}$ are Lagrangian functions associated with a family of noncontrolled mechanical systems, $p_i \in \mathcal{P}$ (a finite set of indices), and

$$q(\cdot), \;\; q(t) \in \mathbb{R}^n$$

is a continuously differentiable function; $\mathcal{P}$ is called the set of possible locations associated with a given hybrid system. Moreover, $\beta_{[t_{i-1},t_i)}(\cdot)$ are characteristic functions of the time intervals $[t_{i-1}, t_i)$, $i = 1, ...r$.

Note that a full time interval [0, 1] is assumed to be separated into disjunct subintervals of the above type for a sequence of switching times, i.e.,

$$\tau := \{t_0 = 0, t_1, ..., t_r = 1\}.$$

We consider the generic hybrid mechanical systems which can be represented by n generalized configuration coordinates $q_1, ..., q_n$. The components

$$\dot{q}_\lambda(t), \lambda = 1, ..., n,$$

of $\dot{q}(t)$ are the so-called generalized velocities. We next assume that $\tilde{L}_{p_i}(t, \cdot, \cdot)$ are twice continuously differentiable convex functions. The necessary optimality conditions for the variational problem (7.31) describe the dynamics of a mechanical system with variable structure. In this case the system is free from some possible external influences or forces. These optimality conditions for (7.45) can be written in the form of the second-order Euler–Lagrange equations (see [265]), i.e.,

$$\begin{aligned} &\frac{d}{dt}\frac{\partial \tilde{L}_{p_i}(t,q,\dot{q})}{\partial \dot{q}} - \frac{\partial \tilde{L}_{p_i}(t,q,\dot{q})}{\partial q} = 0, \\ &q(0) = c_0, \ q(1) = c_1, \end{aligned} \tag{7.46}$$

for all $p_i \in \mathcal{P}$. The Hamilton principle (see, e.g., [1]) gives a variational description of the solution of the two-point boundary value problem (7.46).

For hybrid mechanical systems determined by a family of Lagrangians

$$L_{p_i}(t, q, \dot{q}, u), \ \ p_i \in \mathcal{P},$$

one usually considers the equations of motion

$$\begin{aligned} &\frac{d}{dt}\frac{\partial L_{p_i}(t,q,\dot{q},u)}{\partial \dot{q}} - \frac{\partial L_{p_i}(t,q,\dot{q},u)}{\partial q} = 0, \\ &q(0) = c_0, \ q(1) = c_1, \end{aligned} \tag{7.47}$$

where $u(\cdot) \in \mathcal{U}$ is a control input from the set of admissible controls $\mathcal{U}$. Let

$$\begin{aligned} &U := \{u \in \mathbb{R}^m \ : \ b_{1,\nu} \leq u_\nu \leq b_{2,\nu}, \ \nu = 1, ..., m\}, \\ &\mathcal{U} := \{v(\cdot) \in \mathbb{L}^2_m([0,1]) \ : \ v(t) \in U \text{ a.e. on } [0,1]\}, \end{aligned}$$

where

$$b_{1,\nu}, b_{2,\nu}, \nu = 1, ..., m$$

are constants. The introduced set $\mathcal{U}$ provides a standard example of an admissible control set for problems in mechanics. In this specific case we deal with the following set of admissible controls $\mathcal{U} \bigcap \mathbb{C}^1_m(0, 1)$.

Note that L_{p_i} depends directly on the control function $u(\cdot)$. Let us assume that functions $L_{p_i}(t, \cdot, \cdot, u)$ are twice continuously differentiable functions and every

$$L_{p_i}(t, q, \dot{q}, \cdot)$$

is a continuously differentiable function. For a fixed admissible control $u(\cdot)$ we obtain for all $p_i \in \mathcal{P}$ the above hybrid mechanical system with

$$\tilde{L}_{p_i}(t, q, \dot{q}) \equiv L_{p_i}(t, q, \dot{q}, u(t)).$$

It is also assumed that $L_{p_i}(t, q, \cdot, u)$ are strongly convex functions, i.e., for any

$$(t, q, \dot{q}, u) \in \mathbb{R} \times \mathbb{R}^n \times \mathbb{R}^n \times \mathbb{R}^m$$

and $\xi \in \mathbb{R}^n$ the inequality

$$\sum_{\lambda,\theta=1}^{n} \frac{\partial^2 L_{p_i}(t, q, \dot{q}, u)}{\partial \dot{q}_\lambda \partial \dot{q}_\theta} \xi_\lambda \xi_\theta \geq \alpha \sum_{\lambda=1}^{n} \xi^2_\lambda, \ \alpha > 0,$$

holds for all $p_i \in \mathcal{P}$. This natural convexity condition is a direct consequence of the classical representation for the kinetic energy of a conventional mechanical system. Under the abovementioned assumptions, the two-point boundary value problem (7.47) has a solution for every admissible control $u(\cdot) \in \mathcal{U}$ [18]. We assume that every equation of the type (7.47) has a unique absolutely continuous solution for every $u(\cdot) \in \mathcal{U}$. For an admissible control $u(\cdot)$ the full solution to the boundary value problem (7.47) is next denoted by $q^u(\cdot)$. We call (7.47) the hybrid *Euler–Lagrange control system.*

Example 7.2. *We consider a variable linear mass-spring system attached to a moving frame. The considered control*

$$u(\cdot) \in \mathcal{U} \bigcap \mathbb{C}^1_1(0, 1)$$

is the velocity of the frame. By ω_{p_i} we denote the variable masses of the system. The kinetic energy

$$K = 0.5\omega_{p_i}(\dot{q} + u)^2$$

depends directly on $u(\cdot)$. Therefore,

$$L_{p_i}(q, \dot{q}, u) = 0.5(\omega_{p_i}(\dot{q} + u)^2 - \kappa q^2),$$

where $\kappa \in \mathbb{R}_+$ and

$$\frac{d}{dt}\frac{\partial L_{p_i}(t,q,\dot{q},u)}{\partial \dot{q}} - \frac{\partial L_{p_i}(t,q,\dot{q},u)}{\partial q} = \omega_{p_i}(\ddot{q}+\dot{u}) + \kappa q = 0.$$

By κ we denote here the elasticity coefficient of the spring system.

Note that some important hybrid controlled mechanical systems have Lagrangian functions of the following type:

$$L_{p_i}(t,q,\dot{q},u) = L^0_{p_i}(t,q,\dot{q}) + \sum_{\nu=1}^{m} q_\nu u_\nu,\ m < n.$$

In this special case we easily obtain

$$\frac{d}{dt}\frac{\partial L^0_{p_i}(t,q,\dot{q})}{\partial \dot{q}_\lambda} - \frac{\partial L^0_{p_i}(t,q,\dot{q})}{\partial q_\lambda} = \begin{cases} u_\lambda & \lambda = 1,...,m, \\ 0 & \lambda = m+1,...,n. \end{cases}$$

The control input $u(\cdot)$ is interpreted here as an external force.

We now deal with the Hamiltonian reformulation for the Euler–Lagrange control system (7.47). For every location p_i from $\mathcal{P}$ we introduce the generalized momenta

$$s_\lambda := L_{p_i}(t,q,\dot{q},u)/\partial \dot{q}_\lambda$$

and define the Hamiltonian function $H_{p_i}(t,q,s,u)$ as a Legendre transform applied to every $L_{p_i}(t,q,\dot{q},u)$, i.e.,

$$H_{p_i}(t,q,s,u) := \sup_{\dot{q}}[\sum_{\lambda=1}^{n} s_\lambda \dot{q}_\lambda - L_{p_i}(t,q,\dot{q},u)\ ].$$

In the case of hyperregular Lagrangians $L_{p_i}(t,q,\dot{q},u)$ (see, e.g., [1]) the Legendre transform, namely,

$$Leg_{p_i} : (t,q,\dot{q},u) \to (t,q,s,u),$$

is a diffeomorphism for every $p_i \in \mathcal{P}$. Using the introduced Hamiltonian $H(t,q,s,u)$ and the vector of generalized momenta

$$s := (s_1, ...s_n)^T,$$

we can rewrite system (7.47) in the following Hamilton-type form:

$$\dot{q}_\lambda(t) = \frac{\partial H_{p_i}(t, q, s, u)}{\partial s},$$
$$\dot{s}_\lambda(t) = -\frac{H_{p_i}(t, q, s, u)}{\partial q}, \tag{7.48}$$
$$q(0) = c_0, \; q(1) = c_1.$$

Under the abovementioned assumptions, the boundary value problem (7.48) has a solution for every $u(\cdot) \in \mathcal{U}$. We will call (7.48) a *Hamilton control system*. The main advantage of (7.48) in comparison with (7.47) is that (7.48) immediately constitutes a control system in standard state space form with state variables (q, s). In physics it is usually called the "phase variable."

Consider the system from Example 7.2 and compute

$$H_{p_i}(q, s, u) = \sup_{\dot{q}}[\, s\dot{q} - 0.5(\omega_{p_i}(\dot{q} + u)^2 - \kappa q^2)\,].$$

The maximization procedure applied to

$$(s\dot{q} - 0.5\omega_{p_i}\dot{q}^2 - \omega_{p_i}\dot{q}u)$$

implies the following relation:

$$\dot{q} = s/\omega_{p_i} - u, \;\; H_{p_i}(q, s, u) = s^2/\omega_{p_i} - su - 0.5s^2/\omega_{p_i} + 0.5\kappa q^2.$$

The Hamilton equations can now be written in an explicit form, i.e.,

$$\dot{q} = \frac{\partial H_{p_i}(q, s, u)}{\partial s} = \frac{1}{\omega_{p_i}} s - u,$$
$$\dot{s} = -\frac{\partial H_{p_i}(q, s, u)}{\partial q} = -\kappa q.$$

Note that for

$$L_{p_i}(t, q, \dot{q}, u) = L^0_{p_i}(t, q, \dot{q}) + \sum_{\nu=1}^{m} q_\nu, u_\nu$$

we obtain the associated Hamilton functions in the form

$$H_{p_i}(t, q, s, u) = H^0_{p_i}(t, q, s) - \sum_{\nu=1}^{m} q_\nu u_\nu,$$

where $H^0_{p_i}(t, q, s)$ is the Legendre transform of $L^0_{p_i}(t, q, \dot{q})$.

Let us now introduce the class of HOCPs associated with the mechanical systems determined above. We write

$$\begin{aligned} &\text{minimize } J := \int_0^1 \sum_{i=1}^r \beta_{[t_{i-1},t_i)}(t) f^0_{p_i}(q^u(t), u(t))dt \\ &\text{subject to } u(t) \in U \; t \in [0,1], \; t_i \in \tau, \; i = 1, ..., r. \end{aligned} \tag{7.49}$$

Here

$$f^0_{p_i} : [0,1] \times \mathbb{R}^n \times \mathbb{R}^m \to \mathbb{R}$$

is a continuous and convex function. We have assumed that the boundary value problems (7.47) have a unique solution $q^u(\cdot)$ and that the optimization problem (7.49) also possesses a solution. Let $(q^{opt}(\cdot), u^{opt}(\cdot))$ be an optimal solution of (7.49). Note that we can also use the associated Hamiltonian-type representation of the initial OCP (7.49).

We mainly focus our attention on the application of direct numerical algorithms to an adequate relaxation of the hybrid optimization problem (7.49). Evidently, a general OCP involving ordinary differential equations can be formulated in various ways as an optimization problem in a suitable abstract space and solved by some standard numerical algorithms.

Example 7.3. *Using the* Euler–Lagrange *control system from* Example 7.2, *we now examine the following HOCP:*

$$\begin{aligned} &\text{minimize } J := -\int_0^1 \sum_{i=1}^r \beta_{[t_{i-1},t_i)}(t) k_{p_i}(u(t) + q(t))dt \\ &\text{subject to } \ddot{q}(t) = -\frac{\kappa}{\omega_{p_i}} q(t) = -\dot{u}(t), \; i = 1, ..., r, \\ &q(0) = 0, \; q(1) = 1, \;\; u(\cdot) \in \mathbb{C}^1_1(0,1), \; 0 \le u(t) \le 1 \; \forall t \in [0,1], \end{aligned}$$

where k_{p_i} are given (variable) coefficients associated with every location. The solution $q^u(\cdot)$ of the above boundary value problem can be written as follows:

$$q^u(t) = C^u_i \sin(t\sqrt{\kappa/\omega_{p_i}}) - \int_0^t \sqrt{\kappa/\omega_{p_i}} \sin(\sqrt{\kappa/\omega_{p_i}}(t-l))\dot{u}(l)dl,$$

where $t \in [t_{i-1}, t_i)$, $i = 1, ..., r$, and

$$C^u_i = \frac{1}{\sin\sqrt{\kappa/\omega_{p_i}}}[\, 1 + \int_0^1 \sqrt{\kappa/\omega_{p_i}} \sin(\sqrt{\kappa/\omega}(t-l))\dot{u}(l)dl \,]$$

is a constant in every location. Consequently, we have

$$J = -\int_0^1 \sum_{i=1}^{r} \beta_{[t_{i-1},t_i)}(t) k_{p_i} [u(t) + q^u(t)\,] dt =$$

$$-\int_0^1 \sum_{i=1}^{r} \beta_{[t_{i-1},t_i)}(t) k_{p_i} [\, u(t) + C_i^u \sin(t\sqrt{\kappa/\omega_{p_i}}) -$$

$$\int_0^t \sqrt{\kappa/\omega_{p_i}} \sin(\sqrt{\kappa/\omega_{p_i}}(t-l)) \dot{u}(l) dl\,] dt.$$

Let now $k_{p_i} = 1$ *for all* $p_i \in \mathcal{P}$. *Using the* hybrid version of the Pontryagin maximum principle (*see* [4]), *we conclude that the admissible control*

$$u^{opt}(t) \equiv 0.5$$

is an optimal solution of the given OCP. Note that this result is also consistent with the Bauer maximum principle *[6]. For* $u^{opt}(\cdot)$ *we can compute the corresponding optimal trajectory given as follows:*

$$q^{opt}(t) = \frac{\sin(t\sqrt{\kappa/\omega_{p_i}})}{\sin\sqrt{\kappa/\omega_{p_i}}}, \quad t \in [t_{i-1}, t_i), \quad i = 1, ..., r.$$

Note that the optimal trajectory obtained above is not an absolutely continuous function. Evidently, we have

$$q^{opt}(t_i^-) \neq q^{opt}(t_i^+)$$

and the optimal dynamics is of impulsive nature. Otherwise, all restrictions of function $q^{opt}(\cdot)$ *on every time interval* $[t_{i-1}, t_i)$ *are absolutely continuous functions.*

As we can see from the above example, an optimal trajectory $q(\cdot)$ is not obligatorily an absolutely continuous function on the full time interval $[0, 1]$. The hybrid Pontryagin maximum principle mentioned above guarantees the absolute continuity of trajectories only on the time intervals associated with the corresponding locations. In general, an optimal hybrid system of a mechanical nature can have jumps in the state (impulsive dynamics).

An effective numerical procedure, as a rule, uses the specific structure of the problem under consideration. Our aim is to consider the variational structure of the HOCP (7.49). Let

$$\Gamma_i := \{\gamma(\cdot) \in \mathbb{C}_n^1([t_{i-1}, t_i]) \;:\; \gamma(t_{i-1}) = c_{i-1},\; \gamma(t_i) = c_i\},$$

where $i = 1, ..., r$. The vectors c_i, where $i = 1, ..., r$, are defined by the corresponding switching mechanism of a concrete hybrid system. We refer to [3,4,26] for some possible switching rules determined for various classes of hybrid control systems. We now present an immediate consequence of the classical Hamilton principle from analytical mechanics.

Theorem 7.18. *Let all* Lagrangians $L_{p_i}(t, q, \dot{q}, u)$ *be a strongly convex function with respect to* $\dot{q}_i$, $i = 1, ..., n$. *Assume that every boundary value problem from* (7.47) *has a unique solution for every*

$$u(\cdot) \in \mathcal{U} \bigcap \mathbb{C}^1_m(0, 1).$$

A piecewise absolutely continuous function $q^u(\cdot)$, *where* $u(\cdot) \in \mathcal{U} \bigcap \mathbb{C}^1_m(0, 1)$, *is a solution of the sequence of boundary value problems* (7.47) *if and only if a restriction of this function on* $[t_{i-1} t_i)$, $i = 1, ..., r$, *can be found as follows:*

$$q^u_i(\cdot) = \text{argmin}_{q(\cdot) \in \Gamma_i} \int_{t_{i-1}}^{t_i} L_{p_i}(t, q(t), \dot{q}(t), u(t)) dt.$$

For an admissible control function $u(\cdot)$ from $\mathcal{U}$ we now introduce the following two functionals:

$$T_{p_i}(q(\cdot), z(\cdot)) := \int_{t_{i-1}}^{t_i} [L_{p_i}(t, q(t), \dot{q}(t), u(t)) - L_{p_i}(t, z(t), \dot{z}(t), u(t))] dt,$$

$$V_{p_i}(q(\cdot)) := \max_{z(\cdot) \in \Gamma_i} \int_{t_{i-1}}^{t_i} [L_{p_i}(t, q(t), \dot{q}(t), u(t)) - L_{p_i}(t, z(t), \dot{z}(t), u(t))] dt,$$

for all indices $p_i \in \mathcal{P}$. Generally, we define every restriction of $q^u(\cdot)$ on intervals $[t_{i-1}, t_i)$ as an element of the corresponding Sobolev spaces

$$\mathbb{W}^{1,\infty}_n(t_{i-1}, t_i),$$

i.e., the space of absolutely continuous functions with essentially bounded derivatives. Let us give a variational interpretation of the admissible solutions $q^u(\cdot)$ to a sequence of problems (7.47).

Theorem 7.19. *Let all* Lagrangians $L_{p_i}(t, q, \dot{q}, u)$ *be strongly convex functions with respect to* $\dot{q}_i$, $i = 1, ..., n$. *Assume that every boundary value problem from* (7.47) *has a unique solution for every*

$$u(\cdot) \in \mathcal{U} \bigcap \mathbb{C}^1_m(0, 1).$$

A piecewise absolutely continuous function $q^u(\cdot)$, *where*

$$u(\cdot) \in \mathcal{U} \bigcap \mathbb{C}^1_m(0, 1),$$

is a solution of the sequence of problems (7.47) *if and only if every restriction of this function on* $[t_{i-1} t_i)$, $i = 1, ..., r$, *can be found as follows:*

$$q^u_i(\cdot) = \text{argmin}_{q(\cdot) \in \mathbb{W}^{1,\infty}_n(t_{i-1}, t_i)} V_{p_i}(q(\cdot)). \tag{7.50}$$

Proof. Let

$$q^u(\cdot) \in \mathbb{W}_n^{1,\infty}(t_{i-1}, t_i)$$

be a unique solution of a partial problem (7.47) on the corresponding time interval, where $u(\cdot) \in \mathcal{U} \bigcap \mathbb{C}_m^1(0,1)$. Using the Hamilton principle in every location $p_i \in \mathcal{P}$, we obtain the following relations:

$$\min_{q(\cdot)\in \mathbb{W}_n^{1,\infty}(t_{i-1},t_i)} V_{p_i}(q(\cdot)) =$$

$$\min_{q(\cdot)\in \mathbb{W}_n^{1,\infty}(t_{i-1},t_i)} \max_{z(\cdot)\in\Gamma_i} \int_{t_{i-1}}^{t_i} [L_{p_i}(t, q(t), \dot{q}(t), u(t)) -$$

$$\int_{t_{i-1}}^{t_i} L_{p_i}(t, z(t), \dot{z}(t), u(t))]dt =$$

$$\min_{q(\cdot)\in \mathbb{W}_n^{1,\infty}(t_{i-1},t_i)} \int_{t_{i-1}}^{t_i} L_{p_i}(t, q(t), \dot{q}(t), u(t))dt -$$

$$\min_{z(\cdot)\in\Gamma_i} \int_{t_{i-1}}^{t_i} L_{p_i}(t, z(t), \dot{z}(t), u(t))dt = \int_{t_{i-1}}^{t_i} L_{p_i}(t, q^u(t), \dot{q}^u(t), u(t))dt -$$

$$\int_{t_{i-1}}^{t_i} L_{p_i}(t, q^u(t), \dot{q}^u(t), u(t))dt = V_{p_i}(q^u(\cdot)) = 0.$$

If the condition (7.50) is satisfied, then $q^u(\cdot)$ is a solution of the sequence of the boundary value problem (7.47). This completes the proof. □

Theorem 7.18 and Theorem 7.19 make it possible to express the initial HOCP (7.49) as a multiobjective optimization problem over the set of admissible controls and generalized coordinates. Then we have

$$\begin{aligned} &\text{minimize } J(q(\cdot), u(\cdot)) \text{ and } P(q(\cdot)) \\ &\text{subject to} \\ &(q(\cdot), u(\cdot)) \in (\bigcup_{i=1,\dots,r} \Gamma_i) \times (\mathcal{U} \bigcap \mathbb{C}_m^1(0,1)) \end{aligned} \tag{7.51}$$

or

$$\begin{aligned} &\text{minimize } J(q(\cdot), u(\cdot)) \text{ and } V(q(\cdot)) \\ &\text{subject to} \\ &(q(\cdot), u(\cdot)) \in (\bigcup_{i=1,\dots,r} \Gamma_i) \times (\mathcal{U} \bigcap \mathbb{C}_m^1(0,1)), \end{aligned} \tag{7.52}$$

where

$$P(q(\cdot)) := \int_0^1 \sum_{i=1}^{r} \beta_{[t_{i-1},t_i)}(t) L_{p_i}(t, q(t), \dot{q}(t), u^{opt}(t)) dt$$

and

$$V(q(\cdot)) := \beta_{[t_{i-1},t_i)}(t) V_{p_i}(q(\cdot)).$$

The auxiliary minimization problems (7.51) and (7.52) are multiobjective optimization problems (see, e.g., [16,29]). The set of restrictions

$$\Gamma \times \{\mathcal{U} \bigcap \mathbb{C}^1_m(0,1)\}$$

is a convex set. Since

$$f_0(t, \cdot, \cdot),\ t \in [0,1],$$

is a convex function, $J(q(\cdot), u(\cdot))$ is also convex. If $P(\cdot)$ (or $V(\cdot)$) is a convex functional, then we deal with the convex multiobjective minimization problems.

The variational representation of the solution of the two-point boundary value problem (7.47) eliminates the differential equations from the consideration. The minimization problems (7.51) and (7.52) constitute in fact a convex relaxation of the initially given HOCP (7.49) and also provide a theoretical basis for numerical algorithms. The auxiliary optimization problem (7.51) has two objective functionals. Consider (7.51) and introduce the Lagrange function [29]

$$\begin{aligned}\Lambda(t, q(\cdot), u(\cdot), \mu, \mu_3) &:= \mu_1 J(q(\cdot), u(\cdot)) + \mu_2 P(q(\cdot)) + \\ &\mu_3 |\mu| \text{dist}_{(\bigcup_{i=1,\dots,r} \Gamma_i) \times (\mathcal{U} \bigcap \mathbb{C}^1_m(0,1))}\{(q(\cdot), u(\cdot))\},\end{aligned}$$

where

$$\text{dist}_{(\bigcup_{i=1,\dots,r} \Gamma_i) \times (\mathcal{U} \bigcap \mathbb{C}^1_m(0,1))}\{\cdot\}$$

denotes the distance function

$$\text{dist}_{(\Gamma_i) \times (\mathcal{U} \bigcap \mathbb{C}^1_m(0,1))}\{(q(\cdot), u(\cdot))\} := \inf\{||(q(\cdot), u(\cdot)) - \varrho||_{\mathbb{C}^1_n(0,1) \times \mathbb{C}^1_m(0,1)},$$

and

$$\varrho \in (\bigcup_{i=1,\dots,r} \Gamma_i) \times (\mathcal{U} \bigcap \mathbb{C}^1_m(0,1))\},\quad \mu := (\mu_1, \mu_2)^T \in \mathbb{R}^2_+.$$

Note that the above distance function is associated with the following Cartesian product:

$$(\bigcup_{i=1,\dots,r} \Gamma_i) \times (\mathcal{U} \bigcap \mathbb{C}^1_m(0,1)).$$

Recall that a feasible point $(q^*(\cdot), u^*(\cdot))$ is called *weak Pareto optimal* for the multiobjective problem (7.52) if there is no feasible point $(q(\cdot), u(\cdot))$ for which

$$J(q(\cdot), u(\cdot)) < J(q^*(\cdot), u^*(\cdot)) \text{ and } P(q(\cdot)) < P(q^*(\cdot)).$$

A necessary condition for $(q^*(\cdot), u^*(\cdot))$ to be a weak Pareto optimal solution to (7.52) in the sense of the Karush–Kuhn–Tucker condition is that for every $\mu_3 \in \mathbb{R}$ there exist $\mu^* \in \mathbb{R}^2_+$ such that

$$0 \in \partial_{(q(\cdot),u(\cdot))} \Lambda(t, q^*(\cdot), u^*(\cdot), \mu^*, \mu_3). \tag{7.53}$$

By $\partial_{(q(\cdot),u(\cdot))}$ we denote here the generalized gradient of the Lagrange function Λ introduced above.

We refer to [29] for further theoretical details. If $P(\cdot)$ is a convex functional, then the necessary condition (7.53) is also sufficient for $(q^*(\cdot), u^*(\cdot))$ to be a weak Pareto optimal solution for (7.52).

7.7 Weak Relaxation of the Singular HOCPs

Let us observe that various OCPs of the type (6.8) and (7.24), namely, problems with the affine-type control systems usually possess the singularity property (see, e.g., [192,193]). Let us now give an exact definition of a singular HOCP of the type (6.8) or (7.24).

Definition 7.5. *The mapping*

$$\Xi_{x_0,t_f} : u(\cdot) \in \mathcal{B}_I \bigotimes (\mathcal{Q}_K)^I \to x^u(t_f)$$

is called an end point mapping of a switched mode system (6.6). *The admissible control* $u(\cdot)$ *and the corresponding trajectory* $x^u(\cdot)$ *defined both on* $[0, t_f]$ *are said to be singular if the* Fréchet *derivative*

$$D_v \Xi_{x_0,t_f}$$

is a nonsurjective mapping.

Here $x^u(\cdot)$ denotes a trajectory of system (6.6) "closed" by an admissible control $u(\cdot)$. Note that under the main smoothness conditions for system (6.6) the end point mapping is continuously Fréchet differentiable (see, e.g., [192]). Let us also note that Ξ_{x_0,t_f} is defined on a neighborhood of the admissible control $u(\cdot)$ in the sense of the corresponding norm.

Motivated by the abovementioned problem complexities, namely, the specific control restrictions and possible singularity, we are interested in relatively simply implementable computational procedures even in the case of a singular OCP associated with the switched mode dynamics. Naturally these numerical procedures need to be free from the use of the "strong" variant of the Pontryagin maximum principle. Recall that a "singular case" is characterized by an impossibility to apply a first-order optimality condition in the classic form of a Pontryagin maximum principle (see [96,132,192]). Otherwise, it is well known that the convexity assumptions (global and/or local) play a key role in computational optimization. In fact, the generic convexity property implies the numerical tractability and applicability of the first-order optimization schemes for various minimization problems in abstract spaces.

We now assume that the initially given HOCP (6.8) associated with the switched mode dynamics is singular in the sense of Definition 7.5. The following result constitutes a justification of the "weak" version of the Pontryagin maximum principle in the case of a weakly relaxed OCP (7.24).

Theorem 7.20. *Assume that all hypotheses for the switched mode system (6.6) and for the HOCP (6.8) are satisfied and that the initial* HOCP (6.8) *is singular. Then the objective functional $J_\lambda(\nu(\cdot))$ of the weakly relaxed* OCP (7.24) *is locally convex in a neighborhood of an optimal solution $u_0^{opt}(\cdot)$.*

Proof. Consider an ϵ-ball $\mathcal{W}_\epsilon(u_0^{opt}(\cdot))$ around the optimal control function $u_0^{opt}(\cdot)$. Similar to the proof of Theorem 7.10 the ϵ-neighborhood $\mathcal{W}_\epsilon(u_0^{opt}(\cdot))$ is determined here by the norm

$$||\cdot||_{\mathbb{L}^2\{[0,t_f];\mathbb{R}^{I+K\times I}\}}$$

of the Hilbert space $\mathbb{L}^2\{[0,t_f];\mathbb{R}^{I+K\times I}\}$. Let us consider the difference

$$J(u_2(\cdot)) - J(u_1(\cdot))$$

for some

$$u_1(\cdot),\ u_2(\cdot) \in \mathcal{W}_\epsilon(u_0^{opt}(\cdot)).$$

The gradient

$$\nabla J_\lambda(u(\cdot))(\cdot)$$

in problem (7.24) with a $\lambda \geq 0$ can be calculated as follows:

$$\nabla J_\lambda(u(\cdot))(t) = -\frac{\partial H_\lambda(t, x(t), u(t), p(t), p_{n+1})}{\partial u},$$

where function $p(\cdot)$ and constant p_{n+1} can formally be determined from the following boundary value problem:

$$\frac{dp(t)}{dt} = -\frac{\partial H_\lambda(t, x(t), u(t), p(t), p_{n+1})}{\partial x},$$
$$\tilde{p}(t_f) = -\frac{\partial \phi(\tilde{x}(t_f))}{\partial \tilde{x}}.$$

Since

$$\mathrm{conv}\{\mathcal{B}_I \bigotimes (\mathcal{Q}_K)^I\}$$

contains the admissible control set of the initial HOCP, the weakly relaxed problem is also singular. From the weak version of the conventional maximum principle we next obtain

$$\partial H_\lambda(x_0^{opt}(t), u_0^{opt}(t), p(t), p_{n+1}(t))/\partial u = 0.$$

Therefore,

$$\langle \nabla J(u_0^{opt}(\cdot)), u_1(\cdot) - u_2(\cdot)\rangle_{\mathbb{L}^2\{[0,t_f];\mathbb{R}^{I+K\times I}\}} = 0$$

for all admissible $u_1(\cdot)$, $u_2(\cdot)$. From the Lipschitz continuity of the control state mapping $x^u(\cdot)$ and taking into consideration the boundedness of the obtained convexified admissible control set, we deduce the inequality

$$J(u_2(\cdot)) - J(u_1(\cdot)) < \frac{r}{2}||u_1(\cdot) - u_2(\cdot)||^2_{\mathbb{L}^2\{[0,t_f];\mathbb{R}^{I+K\times I}\}}$$

for a suitable constant $r > 0$. The combination of the two last relations implies the prox-regularity property of the functional $J(u(\cdot))$ at $u_0^{opt}(\cdot)$ (see Definition 7.1 and Definition 7.2).

Since $J(u(\cdot))$ is continuous, the prox-regularity property implies the paraconvexity of $J(u(\cdot))$ in a neighborhood of an optimal solution $u_0^{opt}(\cdot)$. The proof is completed. □

Let us note that Theorem 7.20 and the established convexity property make it possible to apply various first-order numerical optimization methods (see, e.g., [261]) to the singular “switched mode”–type HOCP (7.24). This computational remark is specifically important in the absence of the possibility to use the corresponding maximum principle in that case, namely, in the case of a singular OCP.

7.8 Notes

Let us note that the number of publications related to the relaxation techniques for HSs and SSs is extremely small. This fact is a direct consequence of the sophisticated mathematical character of this topic. Exactly this situation was one of the main motivations for this book. The existing literature devoted to the relaxation theory of HOCPs and SOCPs is summarized here [63,68,81,319,320,335].

CHAPTER 8

Applications of the Relaxation-Based Approach

8.1 On the Existence of Optimal Solutions to OCPs Involving Hybrid and Switched Systems

The main application of the general relaxation theory (RT) in mathematical optimization theory consists in establishing the existence results for optimization problems. The same is also true with respect to the dynamic optimization/optimized control and the corresponding relaxed optimal control problems (OCPs). That implies some existence results associated with the well-determined relaxation schemes. This section contains some of the existence results for hybrid OCPs (HOCPs) and switched OCPs (SOCPs) considered in the previous parts of this book.

Let us now complete some auxiliary results that are immediately used in the existence proofs in the case of the generic Gamkrelidze relaxation (proposed in Chapter 7).

Theorem 8.1. *Assume that all the technical hypotheses for the HOCP (7.1) from Chapter 7 are satisfied. Then the set of relaxed controls*

$$\aleph(n+1) \times \mathcal{U}^{n+1}$$

is convex and weakly compact. In other words, for any sequence

$$\{\nu_l(\cdot)\} \subset \aleph(n+1) \times \mathcal{U}^{n+1}, \; l \in \mathbb{N},$$

there exists an element

$$\hat{\nu}(\cdot) \in \aleph(n+1) \times \mathcal{U}^{n+1}$$

such that the following is satisfied:

$$\lim_{s\to\infty} \int_0^{t_f} \int_U g(t, \eta(t), u) \mu_{l_s}(t)(du)dt = \int_0^{t_f} \int_U g(t, \eta(t), u) \hat{\mu}(t)(du)dt$$

for all continuous functions $g(\cdot,\cdot,\cdot)$. Here

$$\{\nu_{l_s}\} \subset \aleph(n+1) \times \mathcal{U}^{n+1}$$

A Relaxation-Based Approach to Optimal Control of Hybrid and Switched Systems
https://doi.org/10.1016/B978-0-12-814788-7.00014-X

is a subsequence of $\{\nu_l(\cdot)\}$ *and*

$$\mu(\cdot) \in R_c([0, t_f], U)$$

is a relaxed control that corresponds to the Gamkrelidze control

$$\nu(\cdot) \in \aleph(n+1) \times \mathcal{U}^{n+1}.$$

The proof of this theorem can be found in [173]. Note that convexity of the set $\aleph(n+1) \times \mathcal{U}^{n+1}$ follows directly from the assumptions associated with the HOCP (7.1) (convexity of U) and the constructive definition of the set $\aleph(n+1)$. Weak compactness of the set $\aleph(n+1)$ can also be easily shown.

Recall that the concrete relaxed dynamics with respect to the generalized controls analyzed in Theorem 8.1 was comprehensively discussed in Chapter 7. This relaxed dynamics expressed by the Young relaxation (7.5) or equivalently (in the sense of the celebrated equivalence theorem, Chapter 7) by the relaxed differential inclusion (RDI) (7.6) or by system (7.3) possesses some fundamental continuity properties that are summarized in the following result.

Theorem 8.2. *Assume that all the technical hypotheses for the HOCP (7.1) from Chapter 7 are satisfied. Let*

$$\{\nu_l(\cdot)\} \subset \aleph(n+1) \times \mathcal{U}^{n+1}$$

be a sequence that converges in the sense of Theorem 8.1 to

$$\hat{\nu}(\cdot) \in \aleph(n+1) \times \mathcal{U}^{n+1}.$$

Then, there exists a subsequence

$$\{\nu_{l_s}\} \subset \aleph(n+1) \times \mathcal{U}^{n+1}$$

of $\{\nu_l(\cdot)\}$ *such that the trajectories* $\{y_{l_s}(\cdot)\}$ *of the Gamkrelidze system (7.3) generated by* $\{\nu_l(\cdot)\}$ *converge strongly (in the* sup-norm*) to a solution* $\hat{y}(\cdot)$ *of the same system (7.3) generated by the limit control* $\hat{\nu}(\cdot)$.

Note that the proof of this result is based on the fundamental approximability result from [50].

The fundamental Theorem 8.1 and Theorem 8.2 give rise to the existence result for the Gamkrelidze relaxed HOCP (7.7).

Theorem 8.3. *Assume that the conditions of Theorem 8.1 are satisfied. Then there exists an optimal solution*

$$\nu^{opt}(\cdot) \in \aleph(n+1) \times \mathcal{U}^{n+1}$$

of the Gamkrelidze relaxed HOCP (7.7) such that

$$\bar{\mathcal{J}}(\nu^{opt}(\cdot)) \equiv \inf_{\nu(\cdot)\in\aleph(n+1)\times\mathcal{U}^{n+1}} \bar{\mathcal{J}}(\nu(\cdot)).$$

The proof of this result is based (as mentioned above) on the auxiliary Theorem 8.1 and Theorem 8.2 and, moreover, uses the celebrated dominated convergence theorem (see, e.g., [6,20]).

We now discuss the existence of an optimal solution to the β-relaxed HOCP (or SOCP) (7.22). First let us note that the relaxed β-system (7.16) has a linear structure with respect to the "new controls," namely, with respect to the β-controls

$$\gamma_M(\cdot) \in \aleph(M).$$

Moreover, the set of admissible controls $\aleph(M)$ is convex. Let us first examine a specific case, namely, the relaxed HOCP of the type (7.22) with a terminal functional $\mathcal{J}$ and with the specific control-affine structure (compare with a simple affine switched system (ASS) from Definition 1.4, Chapter 1, with a unique function $a_{q_i}(t, x_i(t))$). We have

$$\begin{aligned} \dot{x}(t) &= A(t)x(t) + \int_U B_{q_i}(t,u)\mu(t)(du) \text{ a.e. on } [0,t_f], \\ x(0) &= x_0. \end{aligned} \tag{8.1}$$

Here $A(\cdot)$ and $\{B_q(\cdot,\cdot)\}$ for $q \in Q$ are smooth system control matrices of suitable dimensions. We also assume that the control set U is convex and compact. In that case the above Gamkrelidze relaxed HOCP constitutes a convex minimization problem in a very sophisticated Banach space of all relaxed controls, namely, in the space

$$(\mathbb{L}^1([0,t_f],\mathbb{C}(U)))^*.$$

On the other hand, the associated β-relaxed system

$$\begin{aligned} \dot{z}(t) &= A(t)z(t) + \sum_{k=1}^{M} \gamma^k(t)B_{q_i}(t,v^k) \text{ a.e. on } [0,t_f], \\ z(0) &= x_0 \end{aligned}$$

is linear with respect to the vector of new controls $\gamma_M(\cdot) \in \aleph(M)$. Using the main results of this chapter, we can prove the following result.

Theorem 8.4. *Consider a simple ASS with a unique function* $a_{q_i}(t, x_i(t))$. *Let the initially given ASS involving* HOCP *be stable with respect to the* β*-relaxation* (8.1). *If the functionals* $\phi(\cdot)$, $h(\cdot)$ *are convex and the functional* $q(t, \cdot)$ *is convex for all* $t \in [0, t_f]$, *then the corresponding* β*-relaxed problem* (8.1) *is a convex OCP.*

Note that the proof of Theorem 8.4 is trivial. It follows from the convex structure of the relaxed system (8.1) and the results of Chapter 5 and Chapter 7 for convex control systems and convex OCPs. In a similar manner, one can obtain the next (more general) result.

Theorem 8.5. *Consider the hybrid (or switched) control system from* (7.1) *with the following "separated" structure:*

$$f_{q_i}(t, x, u) = f_{q_i}^1(t, x) + f_{q_i}^2(t, u),$$

where f_q^1, f_q^2, $q \in Q$, *are continuous functions and for all* $t \in [0, t_f]$ *and all* $q \in Q$ *and we have*

$$||f_q^1(t, x_1) - f_q^1(t, x_2)|| \leq L||x_1 - x_2||, \quad \forall x_1, x_2 \in \mathbb{R}^n, \; q \in Q.$$

Let

$$f_{q,k}^1(t, \cdot), \; f_{q,k}^2(t, \cdot), \; k = 1, ..., n,$$

be convex and monotonically nondecreasing functionals for every $t \in [0, t_f]$. *Assume that the functionals* $\phi(\cdot)$, $h(\cdot)$ *are convex, monotonically nondecreasing and the function* $q(t, \cdot)$ *is convex, monotonically nondecreasing for every* $t \in [0, t_f]$. *Then the corresponding* β*-relaxed problem (7.22) is a convex OCP.*

The proof of the above result is similar to the proof of Theorem 8.4. Finally note that the theory of the β-relaxation presented in Chapter 7 makes it clear that the practical solvability of the general β-relaxed problem depends on the existence of an optimal solution to the associated Gamkrelidze relaxation (see theorems from Chapter 7). Once the existence of an optimal solution to the Gamkrelidze relaxed HOCP is established (for example, by Theorem 8.3) the practical solvability of HOCP (7.22), namely, the existence of a minimizing sequence (see Chapter 3), is guaranteed. A similar observation is also true with respect to the weakly relaxed HOCPs involving switched mode systems (see Chapter 6 and Chapter 7).

8.2 Necessary Optimality Conditions and Relaxed Controls

We now consider the concrete formulations of the first-order necessary optimality conditions for some of the relaxed HOCPs and SOCPs studied in Chapter 7. The first part of this section contains applications of the advanced Pontryagin-type maximum principle to relaxed problems. The second part discusses a general form of constraint qualifications for the relaxed OCPs of hybrid and switched nature.

8.2.1 Application of the Pontryagin Maximum Principle to Some Classes of Relaxed Hybrid and Switched OCPs

Once the existence problem has been studied (see Section 8.1) we can take into consideration the necessary optimality conditions for the relaxed HOCPs and SOCPs from Chapter 7. Let us firstly examine the Gamkrelidze relaxation (7.7) of the basic HOCP (7.1). Let us rewrite here the relaxed version (7.7) in the complete form, i.e.,

$$\begin{aligned} &\text{minimize } \tilde{\mathcal{J}}(\nu(\cdot)) = \int_0^{t_f} \int_U f_0(t, x(t), u)\mu(t)(du)dt \\ &\text{subject to} \\ &\dot{y}(t) = \sum_{j=1}^{n+1} \alpha^j(t)\tilde{f}(t, y(t), u^j(t)) \text{ a.e. on } [0, t_f], \\ &y(0) = x_0, \\ &\nu(\cdot) \in \aleph(n+1) \times \mathcal{U}^{n+1}, \\ &q_i \in Q, \; i = 1, ..., r+1. \end{aligned} \tag{8.2}$$

Recall that the complete control input for the relaxed problem (8.2) is determined by

$$\nu(t) = (\alpha(\cdot), u(\cdot)) \in \aleph(n+1) \times \mathcal{U}^{n+1}.$$

Let us also recall the definition of the complete admissible control set (the Cartesian product)

$$\begin{aligned} &\aleph(n+1) := \Big\{(\alpha^1(\cdot), ..., \alpha^{n+1}(\cdot))^T \; : \; \alpha^j(\cdot) \in \mathbb{L}^1_1([0, t_f]), \; \alpha^j(t) \geq 0, \\ &\sum_{j=1}^{n+1} \alpha^j(t) = 1 \; \forall t \in [0, t_f]\Big\}, \; \alpha(\cdot) := (\alpha^1(\cdot), ..., \alpha^{n+1}(\cdot))^T, \end{aligned}$$

and

$$\mathcal{U} := \{v(\cdot) \in \mathbb{L}^2_m([0, t_f]) \; : \; v(t) \in U \text{ a.e. on } [0, t_f]\},$$

where $U \subseteq \mathbb{R}^m$. An optimal solution to the relaxed HOCP (8.2) is next denoted by

$$(y^{opt}(\cdot), \nu^{opt}(\cdot)).$$

We now reformulate the known Pontryagin maximum principle for hybrid systems (HSs) (see, e.g., [39,255,296]) and apply it to the relaxed HOCP (8.2). We consider here the generic Gamkrelidze relaxation from Chapter 7.

Theorem 8.6. *Let the functions*

$$f_0, \; f_{q_i}, \; q_i \in Q,$$

be continuously differentiable and the relaxed HOCP (8.2) be Lagrange regular. Then there exist a function $\psi_i(\cdot)$ from $\mathbb{W}_n^{1,\infty}(0, t_f)$ (an adjoint vector) and a vector

$$a = (a_1 \dots a_{r-1})^T \in \mathbb{R}^{r-1}$$

such that

$$\begin{aligned} \dot{\psi}_i(t) &= -\frac{\partial H_{q_i}(y_i^{opt}(t), \nu^{opt}(t), \psi(t))}{\partial y} \text{ a.e. on } (t_{i-1}^{opt}, t_i^{opt}), \\ \psi_r(t_f) &= 0, \end{aligned} \tag{8.3}$$

and

$$\psi_i(t_i^{opt}) = \psi_{i+1}(t_i^{opt}) + \left(a_i \frac{\partial m_{q_i, q_{i+1}} y^{opt}(t_i^{opt})}{\partial y} \right), \; i = 1, ..., r-1. \tag{8.4}$$

Moreover, for every admissible control $\nu(\cdot)$ the following inequality is satisfied:

$$\begin{aligned} &\left(\frac{\partial H_{q_i}(y^{opt}(t), \nu^{opt}(t), \psi(t))}{\partial \nu}, (\nu(t) - \nu^{opt}(t)) \right) \leq 0 \\ &\text{a.e. on } [t_{i-1}^{opt}, t_i^{opt}], \; i = 1, ..., r, \end{aligned} \tag{8.5}$$

where

$$H_{q_i}(y, \nu, \psi) := (\psi_i, \sum_{j=1}^{n+1} \alpha^j(t) \tilde{f}(t, y(t), u^j(t))) - f_0(t, y, u)$$

is a "partial" Hamiltonian for the location $q_i \in Q$ and $(\cdot, \cdot)$ denotes the corresponding scalar product.

In the case of some additional "effects" in the hybrid dynamic models of the type (8.2) one needs to adapt these effects correctly and in conformity of the general Pontryagin maximum principle from Chapter 6. For example, in the case of impulsive HSs (see Chapter 5 and Chapter 6) we have to consider the extended generalized control vector (in comparison with (8.2))

$$(\nu, \theta),$$

where the additional parameter θ describes jumps (see Chapter 6).

Let us also note that in the case of a relaxed HOCP or SOCP that is equivalent to a convex programming in a real Hilbert space (see, for example, Chapter 7) the necessary optimality conditions are also sufficient [190,275]. For example, the convexified linear quadratic (LQ)-type OCP (7.36)

$$\begin{aligned} &\text{minimize } J(u(\cdot)) \\ &\text{subject to } u(\cdot) \in \text{conv}(\mathcal{S}) \end{aligned}$$

constitutes a convex optimization problem in the generic Hilbert space $\mathbb{L}^2[t_0, t_f; \mathbb{R}^m]$ (a result from Chapter 7) and the optimality conditions in that case can be formulated in the form of the following theorem.

Theorem 8.7. *Assume that the switched-type OCP (7.36) is Lagrange regular. Then there are Lagrange multipliers*

$$p^{opt} \in \mathbb{L}^2[t_0, t_f; \mathbb{R}^m]$$

such that

$$\min_x L(x, p^{opt}) = L(x^{opt}, p^{opt}).$$

Here $L(x, p)$ is the Lagrange function associated with the OCP (7.36). Evidently, Theorem 8.6 expresses the generalized Karush–Kuhn–Tucker optimality conditions. We refer to Chapter 2 and Chapter 3 for the formal proof and further mathematical details.

Let us now consider the β-relaxations of HOCPs studied in Chapter 7. The β-relaxation scheme (7.22) for the originally given HOCP (or SOCP) (7.15) induces an extended control vector

$$\gamma_M(\cdot) \in \aleph(M).$$

We can apply the hybrid version of the Pontryagin-like maximum principle from Chapter 6 to the unconstrained version of the β-relaxed HOCP (7.22), i.e.,

$$\begin{aligned} &\text{minimize } \tilde{J}_M(\gamma_M(\cdot)) \\ &\text{subject to } \gamma_M(\cdot) \in \aleph(M), \\ &q_i \in Q,\ i = 1, ..., r+1,\ x(0) = x_0 \in M_{q_1}. \end{aligned} \tag{8.6}$$

The following theorem summarizes the Pontryagin maximum principle for the β-relaxed Mayer-type HOCPs (8.6).

Theorem 8.8. *Let the functions*

$$f_0, \ f_{q_i}, \ q_i \in Q,$$

be continuously differentiable and the β-relaxed Mayer-type HOCPs (8.6) be Lagrange regular. Then there exist a function $\psi_i(\cdot)$ from $\mathbb{W}_n^{1,\infty}(0, t_f)$ (an adjoint vector) and a vector

$$a = (a_1 \ldots a_{r-1})^T \in \mathbb{R}^{r-1}$$

such that

$$\begin{aligned} \dot{\psi}_i(t) &= -\frac{\partial H_{q_i}(z_i^{opt}(t), \gamma_M^{opt}(t), \psi(t))}{\partial z} \text{ a.e. on } (t_{i-1}^{opt}, t_i^{opt}), \\ \psi_r(t_f) &= 0, \end{aligned} \tag{8.7}$$

and

$$\psi_i(t_i^{opt}) = \psi_{i+1}(t_i^{opt}) + \left(a_i \frac{\partial m_{q_i,q_{i+1}} z^{opt}(t_i^{opt})}{\partial z} \right), \ i = 1, ..., r-1. \tag{8.8}$$

Moreover, for every admissible control $\gamma_M(\cdot)$ the following inequality is satisfied:

$$\begin{aligned} &\left(\frac{\partial H_{q_i}(z^{opt}(t), \gamma_M^{opt}(t), \psi(t))}{\partial \gamma_M}, (\gamma_M(t) - \gamma_M^{opt}(t)) \right) \leq 0 \\ &\text{a.e. on } [t_{i-1}^{opt}, t_i^{opt}], \ i = 1, ..., r, \end{aligned} \tag{8.9}$$

where

$$H_{q_i}(z, \gamma_M, \psi) := \big(\psi_i, \sum_{k=1}^{M} \gamma^k(t) \tilde{f}(t, z(t), v^k)\big) - f_0(t, y, u)$$

is a "partial" Hamiltonian for the location $q_i \in Q$, $(\cdot, \cdot)$ denotes the corresponding scalar product, and $z(\cdot)$ is a solution of the β-system (7.16).

Finally note that the necessary (and sufficient) optimality conditions for the weakly relaxed switched mode HOCP (6.8) (associated with the switched mode system (6.6) from Chapter 6) are in fact given by establishing the minimizing sequence for the obtained weak relaxation (7.24). As mentioned in Chapter 3 the existence of a minimizing sequence for an optimization problem constitutes in fact a "numerical solvability" of this problem. For the theory of switched-type LQ OCPs and the corresponding optimality conditions see Chapter 7.

8.2.2 On the Constraint Qualifications

Consider now an HS from Chapter 7 (see system (7.2))

$$\dot{x}(t) = \tilde{f}(t, x(t), u(t)) \text{ a.e. on } [0, t_f],$$
$$\tilde{f}(t, x, u) := \sum_{i=1}^{r+1} \beta_{[t_{i-1}, t_i)}(t) f_{q_i}(t, x, u),$$
$$x(0) = x_0$$

and the Mayer version of the HOCP (7.1). As we can see from Section 8.2.1, the necessary optimality conditions make sense only under the Lagrange regularity assumptions. These regularity assumptions (constraint qualifications = CQs) do not constitute a part of the optimality conditions and need to be verified before these conditions can be applied. This theoretical paradigm implies a question of general interest. How to verify the applicability conditions for the obtained Pontryagin-like first-order necessary optimality conditions? Clearly, for the problem of dynamic optimization of an HS this important question is a very sophisticated analytic task. We refer to [23–25] for some additional discussions in that direction.

In this section we develop a specific CQ for a class of HOCPs. Let us firstly consider a smooth optimization problem with equality constraints given in an abstract operator form

$$\begin{aligned} &\text{minimize } \psi_0(\xi) \\ &\text{subject to } \xi \in Q := \{\xi \in X \,:\, A(\xi) = a\}, \end{aligned} \tag{8.10}$$

where $\psi_0 : X \to \mathbb{R}$ is a sufficiently smooth function in a real Banach space X and

$$A : X \to Y$$

is an operator between real Banach spaces X and Y. We assume A to be Fréchet differentiable on X (see Chapter 2). For (8.10) we introduce the associated abstract Lagrange function

$$\mathcal{L}(\lambda_0, \lambda, \xi) = \lambda_0 \psi_0(\xi) + \langle \lambda, A(\xi) - a \rangle_Y,$$

where

$$\lambda_0 \in \mathbb{R}$$

and λ is a linear functional from the dual space Y^*. Here we have

$$\langle \lambda, \cdot \rangle_Y : Y \to \mathbb{R}.$$

Let ξ^{opt} be a *local solution* of (8.10). We are firstly interested to consider the case when the equality constraints are regular.

Definition 8.1. *Let a mapping $A : X \to Y$ be* Fréchet *differentiable at some point $\xi \in X$. We say the mapping A is regular at this point ξ if*

$$R(A'(\xi)) = Y.$$

The mapping A is called nonregular (sometimes also called irregular, abnormal, or degenerate) if the regularity condition is not satisfied. Various regularity conditions are comprehensively discussed by Mangasarian and Fromowitz [236], Robinson [274], Zowe and Kurcyusz [337], Auslender [23], and Malanowski [234,235] and in the books of Ioffe and Tikhomirov [192], Bertsekas [87], Fletcher [162], and many other authors.

We now formulate necessary optimality conditions for the abstract problem (8.10) in the form of the generalized Lagrange multiplier rule (for the analytical background see [192,87] and [198]).

Theorem 8.9. *Let x^{opt} be a local solution to* (8.10). *Assume that the objective functional ψ_0 and the constraint mapping A are* Fréchet *differentiable at ξ^{opt}. Additionally assume that the range*

$$R(A'(\xi^{opt}))$$

of the mapping

$$\xi \to A'(\xi^{opt})\xi$$

is closed. Then there exist Lagrange *multipliers*

$$\lambda_0^{opt} \geq 0, \; \lambda^{opt} \in Y^*$$

not all equal to zero such that

$$\mathcal{L}'_\xi(\lambda_0^{opt}, \lambda^{opt}, \xi^{opt}) = \lambda_0^{opt} \psi'_0(\xi) + \langle \lambda^{opt}, A'(\xi^{opt}) \rangle_Y = 0.$$

If, moreover, A is regular at x^{opt} and the mapping

$$A'(\xi^{opt})$$

is continuous, then

$$\lambda_0^{opt} \neq 0.$$

Clearly, in the case $\lambda_0 \neq 0$ one can put

$$\lambda_0 = 1.$$

Various supplementary conditions have been proposed under which it is possible to assert that the celebrated Lagrange multiplier rule holds with

$$\lambda_0 = 1.$$

These conditions are called "constraint qualifications" (or regularity conditions) (see, e.g., [130,198]). We mention just a few: Mangasarian–Fromovitz constraint qualifications [102], Kurcyusz–Robinson–Zowe regularity conditions [337], and Ioffe regularity conditions [192].

The abstract OCP (8.10) constitutes in fact a theoretical framework for many particular optimization problems. For example, some of the HOCPs and SOCPs from Chapter 7 are in fact concrete "realizations" of the abstract framework (8.10). Note that in the case of a convex optimization problem (8.10) (see Chapter 6 and Chapter 7) we have the well-known Slater regularity conditions [198]. This specific QC for the convex case can be directly applied to convex OCPs studied in the previous chapters.

Observe that in the case the Fréchet derivative

$$A'(\xi^{opt})$$

is not onto, then the Lagrange optimality conditions are nonconstructive and trivially satisfied with

$$\lambda_0 = 0.$$

The Lagrange approach in that case provides no additive information about solutions of the optimization problem (8.10). Therefore the development of optimality conditions for a priori given nonregular problems (8.10) has also become an active research topic (see, e.g., [107] and references therein).

Let us now apply the known abstract surjectivity results related to the expanding operators (discussed in Chapter 2) to the general regularity problem. These results in combination with Theorem 8.9 characterize the expected CQ.

Theorem 8.10. *Let $A : X \to Y$ be expanding and continuously* Fréchet *differentiable on a real* Banach *space X. If*

$$A'(\tilde{\xi})$$

is regular at an arbitrary point $\tilde{\xi} \in X$, then the mapping $A'(\xi)$ is regular at every point $\xi \in X$.

Since the range

$$R(A'(\xi))$$

of the expanding continuous mapping $A'(\xi)$ is a closed set, the Lagrange optimality conditions are satisfied. Moreover, A is regular at ξ^{opt} and $A'(\xi^{opt})$ is continuous. It is important to keep in mind that Theorem 8.10 makes it possible to verify the "total" surjectivity of the operator $A'(\xi^{opt})$ by using a "pointwise" surjectivity at an arbitrary admissible point (!)

$$\xi \in Q.$$

Moreover, under the assumptions of Theorem 8.10, the necessary regularity of the operator $A'(\xi^{opt})$ at the optimal solution ξ^{opt} follows from the regularity of $A'(\xi)$ at an arbitrary point $\xi \in Q$.

Let us also briefly consider a nonregular case. The main idea of a constructive approach to the Lagrange-type multiplier rule (and to the necessary optimality conditions in optimal control) in that case is based on a formal construction proposed in [107]. One replaces the operator $A'(\xi^{opt})$ (assumed to be nononto) by a linear operator

$$\Psi_p(\xi^{opt})$$

associated with the pth Taylor polynomial of A at ξ^{opt}. The last one is an onto operator. Let now

$$A : X \to Y$$

be a p-times continuously Fréchet differentiable operator. Moreover, we assume that the space Y is decomposed into a direct sum

$$Y = Y_1 \oplus ... \oplus Y_p,$$

where

$$\begin{aligned}
&Y_1 = \mathrm{cl}\{R(A'(\xi))\},\\
&Y_i = \mathrm{cl}\{\mathrm{Sp}(R(P_{Z_i} A^{(i)}(\xi))(\cdot)^i)\},\ i = 2, ..., p-1,\\
&Y_p = Z_p,
\end{aligned}$$

and Z_i is a closed complementary subspace for

$$(Y_1 \oplus ... \oplus Y_{i-1})$$

with respect to Y, $i = 2, ..., p$. Here

$$P_{Z_i} : Y \to Z_i$$

is the projection operator onto Z_i along

$$(Y_1 \oplus ... \oplus Y_{i-1})$$

with respect to Y, $i = 2, ..., p$. Define the following mappings (see [107])

$$\psi_i : X \to Y_i, \ \psi_i(\xi) = P_{Y_i} A(\xi),$$

where $P_{Y_i} : Y \to Y_i$ is the projection operator onto Y_i along

$$(Y_1 \oplus ... \oplus Y_{i-1} \oplus Y_{i+1} \oplus ... \oplus Y_p)$$

with respect to Y, $i = 1, ..., p$.

Definition 8.2. *The linear operator*

$$\Psi_p(\xi_0)(\xi) \in L(X, Y), \ \xi, \xi_0 \in X,$$

with

$$\Psi_p(\xi_0)(x) = \psi_1'(\xi_0) + \frac{1}{2}\psi_2''(\xi_0)(\xi) + ... + \frac{1}{p!}\psi_p^{(p)}(\xi_0)(\xi)^{p-1},$$

is called a p-factor operator. We say the mapping A is p-regular at $\xi_0 \in X$ *along an element* $x \in X$ *if*

$$R(\Psi_p(\xi_0)(\xi)) = Y.$$

Consider now the optimization problem (8.10) with a nonregular operator A. We define the *p-factor Lagrange function*

$$\mathcal{L}^p(\lambda_0(\xi), \lambda(\xi), \xi_0, \xi) = \lambda_0(\xi)\psi_0(\xi_0) + \sum_{i=1}^{p} \langle \lambda_i(\xi), \psi_i^{(i-1)}(\xi_0)(\xi)^{i-1} \rangle_Y,$$

where

$$\xi, \xi_0 \in X, \ \lambda_0(\xi) \in \mathbb{R},$$

and

$$\lambda_i(\xi) \in Y_i^*, \ i = 1, ..., p.$$

The function $\mathcal{L}^p$ is a generalization of the conventional Lagrange function $\mathcal{L}$ and it reduces to the Lagrange function for the regular case. Using $\mathcal{L}^p$, one can prove the following necessary optimality conditions (see [107]).

Theorem 8.11. *Let X and*

$$Y = Y_1 \oplus ... \oplus Y_p$$

be real Banach *spaces,* $\psi_0 : X \to \mathbb{R}$ *a twice continuously* Fréchet *differentiable function, and* $A : X \to Y$ *be a p-times continuously* Fréchet *differentiable mapping. Assume that for an element*

$$\xi \in \bigcap_{i=1}^{p} \operatorname{Ker}\{\psi_i^{(i)}(\xi^{opt})\}$$

the set

$$R(\Psi_p(\xi^{opt})(\xi))$$

is closed in Y. If ξ^{opt} *is a local solution to problem* (8.10), *then there exist multipliers*

$$\lambda_0^{opt}(\xi) \in \mathbb{R}$$

and

$$\lambda_i^{opt}(\xi) \in Y_i^*, \ i = 1, ..., p,$$

such that they do not all vanish, and

$$\mathcal{L}_\xi^{\prime\, p}(\lambda_0^{opt}(\xi), \lambda^{opt}(\xi), \xi^{opt}, \xi) = \lambda_0^{opt}(\xi)\psi_0(\xi^{opt}) + \\ + \sum_{i=1}^{p} \langle (\psi_i^{(i)}(\xi^{opt})(\xi)^{i-1})^* \lambda_i^{opt}(\xi) \rangle_{Y^*} = 0.$$

If, moreover,

$$R(\Psi_p(\xi^{opt})(\xi)) = Y,$$

then $\lambda_0^{opt}(\xi) \neq 0$.

Using the results from Chapter 2, we are now ready to formulate our p-regularity condition for the case of differentiable expanding operators.

Theorem 8.12. *Let* $A : X \to Y$ *be expanding and continuously* Fréchet *differentiable on a real* Banach *space X. If A is p-regular at an arbitrary point*

$$\xi_0 \in X$$

along an element $\xi \in X$, *then the mapping A is p-regular at every point of X along the element* $x \in X$. *In particular,*

$$R(\Psi_p(\xi^{opt})(\xi)) = Y.$$

This analytic result is a direct consequence of Theorem 8.11 and of the basic properties of expanding operators studied in Chapter 2.

We now consider the generalization of the basic optimization problem (8.10). This generalization includes more specific details similar to the OCPs and involves the control variable u and the state variable x, i.e.,

$$\begin{aligned} &\text{minimize } J(x,u) \\ &\text{subject to} \\ &F(x,u) = 0_Z, \\ &(x,u) \in S. \end{aligned} \tag{8.11}$$

Here X, Y are real Banach spaces, $J : X \times Y \to R$ is an objective functional, and

$$F : X \times Y \to Z$$

is a given state-control mapping. One can observe that the specific state-control constraint in (8.11) also expresses (among others) the HS of the type (7.2). Here Z is a real Banach space. By S we denote in (8.11) a nonempty subset of $X \times Y$. Let us now consider the abstract OCP (8.11) with an expanding state-control mapping F. Recall the corresponding definition from Chapter 2 in the case of the concrete mapping F.

Definition 8.3. *The mapping $F(\cdot,\cdot)$ is called expanding if there is a number $d > 0$ such that*

$$||F(x_1,u_1) - F(x_2,u_2)||_Z \geq d(||x_1 - x_2||_X + ||u_1 - u_2||_Y)$$

for all

$$(x_1,u_1), (x_2,u_2) \in X \times Y.$$

Note that we use the standard norm

$$||\cdot||_X + ||\cdot||_Y$$

in the product space $X \times Y$. Let us introduce the Lagrangian for the general OCP (8.11), i.e.,

$$\mathcal{L}(\lambda_0, \lambda, x, u) := \lambda_0 J(x,u) + \langle \lambda, F(x,u) \rangle_Z,$$

where $\lambda_0 \in \mathbb{R}$ and $\lambda \in Z^*$. Assume that F and J are Fréchet differentiable. By F' and J' we denote here the corresponding Fréchet derivatives with respect to the variable (pair) (x,u). The next theorem generalizes the known Lagrange multiplier rule.

Theorem 8.13. *Let* (x^{opt}, u^{opt}) *be a local solution of problem* (8.11). *Let the functional* $J(\cdot,\cdot)$ *be* Fréchet *differentiable at* (x^{opt}, u^{opt}) *and the mapping*

$$F(\cdot,\cdot)$$

be Fréchet *differentiable in a neighborhood of* (x^{opt}, u^{opt}). *Suppose that*

$$F'(\cdot,\cdot)$$

is continuous at (x^{opt}, u^{opt}) *and the range*

$$R(F'(x^{opt}, u^{opt}))$$

is a closed set. Then there are a real number

$$\lambda_0^{opt} \geq 0$$

and a continuous linear functional $\lambda^{opt} \in Z^*$ *with*

$$(\lambda_0^{opt}, \lambda^{opt}) \neq (0, 0_{Z^*})$$

and

$$\begin{aligned} \mathcal{L}'(\lambda_0^{opt}, \lambda^{opt}, x^{opt}, u^{opt}) = (\lambda_0^{opt} J'(x^{opt}, u^{opt}) + \\ + \langle \lambda^{opt}, F'(x^{opt}, u^{opt}) \rangle_Z)(x - x^{opt}, u - u^{opt}) \geq 0, \end{aligned}$$

for all $(x, u) \in S$. *If, in addition to the assumptions given above, some regularity condition is fulfilled, then*

$$\lambda_0^{opt} > 0.$$

Note that if the mapping

$$F'(x^{opt}, u^{opt})$$

is surjective, then we have the regular case, namely,

$$\lambda_0^{opt} > 0.$$

Evidently, the regular case is of prime interest for applications. Note that the regularity conditions are also frequently used in proofs of convergence of some numerical schemes for OCPs [239,269]. Using the obtained theoretical result, namely, Theorem 8.13, we can prove the following useful result.

Theorem 8.14. *Let X, Y, Z be real* Banach *spaces. Let*

$$F : X \times Y \to Z$$

be expanding and continuously Fréchet *differentiable on $X \times Y$. If*

$$F'(\tilde{x}, \tilde{u})$$

is regular at an arbitrary point

$$(\tilde{x}, \tilde{u}) \in X \times Y,$$

then the mapping $F'(x, u)$ is regular at every admissible point $(x, u) \in X \times Y$.

Recall that some optimality criteria for the abstract OCP (8.11) contain the regularity condition of the mapping

$$x \to F(x, u^{opt})$$

at the point x^{opt} [192]. Now we assume that the mapping

$$F(\cdot, u^{opt}) : X \to Z$$

is expanding and continuously Fréchet differentiable on X with

$$R(F'_x(\tilde{x}, u^{opt})) = Z, \ \tilde{x} \in X,$$

where F'_x is the partial derivative of F with respect to x. Using the abstract theorems presented above, we can now deduce the regularity condition of the mapping $F(\cdot, u^{opt})$ at the optimal point x^{opt}.

We finally want to apply the abstract regularity theory developed for the abstract OCP (8.11) to the Mayer version of the main HOCP (7.1). We realize this "comeback" to the practically applicable theory using the following important example.

Example 8.1. *Consider the following constrained HOCP of the type (7.1):*

$$\text{minimize } J(x(\cdot), u(\cdot)) = \int_0^1 f_0(t, x(t), u(t))dt$$

subject to

$$\dot{x}(t) = \tilde{f}(t, x(t), u(t)) \text{ a.e. on } [0, 1],$$

$$q_i \in Q, \ i = 1, ..., r + 1, \ x(0) = x_0 \in M_{q_1},$$

$$u(t) \in U \text{ a.e. on } [0, 1],$$

$$g(x(t)) = 0.$$

In addition to the generic assumptions we suppose that $\tilde{f}(t,\cdot,\cdot)$ is an expanding function for all $t \in [0,1]$ and that $g(\cdot)$ is a differentiable expanding function.

Recall that the usual regularity conditions for OCPs of the above type are based on the controllability or local controllability assumptions for the linearized equality constraint [198]. *In practice, the controllability or local controllability conditions constitute a difficult task. Moreover, the conditions of the standard criterion use the unknown information about the optimal pair (x^{opt}, u^{opt}). The linearization techniques can be applied to switched systems (SSs). In the case of general HSs the linearization procedure cannot be implemented because of the dependence of the switching times on the trajectory (discussed in the previous chapters)*

$$t_i(x(\cdot)).$$

Using Theorem 8.14, *we can deduce the regularity condition for some special cases of* HOCPs. *We first express the original problem given above as an infinite-dimensional minimization problem, i.e.,*

$$\begin{aligned}&\text{minimize } \tilde{J}(u(\cdot))\\ &\text{subject to } u(\cdot) \in \mathcal{U},\ \tilde{g}(u(\cdot))(t) = 0,\ \forall t \in [0,1],\end{aligned}$$

with the aid of functions

$$\tilde{J} : \mathbb{L}^2_m([0,1]) \to \mathbb{R},\ \tilde{g} : \mathbb{L}^2_m([0,1]) \to \mathbb{C}^1([0,1],\mathbb{R}),$$

$$\tilde{J}(u(\cdot)) := \int_0^1 f_0(t, x^u(t), u(t))dt,\ \tilde{g}(u(\cdot))(t) := g(x^u(t))\ \forall t \in [0,1].$$

The function $g(\cdot)$ is continuously differentiable. Evidently, we have

$$\begin{aligned}&||g(u_1(\cdot))(\cdot) - g(u_2(\cdot))(\cdot)||_{\mathbb{C}^1([0,1],\mathbb{R})} \geq\\ &\max_{t\in[0,1]} ||g(x^{u_1}(t)) - g(x^{u_2}(t))|| \geq d_1 \max_{t\in[0,1]} ||x^{u_1}(t) - x^{u_2}(t)|| =\\ &= d_1 \max_{t\in[0,1]} || \int_0^t \tilde{f}(\tau, x^{u_1}(\tau), u_1(\tau)) - \tilde{f}(\tau, x^{u_2}(\tau), u_2(\tau))d\tau ||,\end{aligned}$$

where $d_1 > 0$. The linear operator

$$\rho(\cdot) \to \int_0^t \rho(\tau)d\tau,$$

where

$$\rho(\cdot) \in \mathbb{C}([0,1],\mathbb{R}^n),$$

is a linear homeomorphism (see [6]). *Therefore*

$$d_1 \max_{t\in[0,1]} \left|\left| \int_0^t \tilde{f}(\tau, x^{u_1}(\tau), u_1(\tau)) - \tilde{f}(\tau, x^{u_2}(\tau), u_2(\tau))d\tau \right|\right| \geq$$
$$d_2 \max_{t\in[0,1]} ||\tilde{f}(t, x^{u_1}(t), u_1(t)) - \tilde{f}(t, x^{u_2}(t), u_2(t))||$$

for a positive number d_2. *Since* $\tilde{f}(t, \cdot, \cdot)$ *is an expanding function for all* $t \in [0, 1]$, *we obtain*

$$d_2 \max_{t\in[0,1]} ||\tilde{f}(t, x^{u_1}(t), u_1(t)) - \tilde{f}(t, x^{u_2}(t), u_2(t))|| \geq$$
$$d_3 \max_{t\in[0,1]} (||x^{u_1}(t) - x^{u_2}(t)|| + ||u_1(t) - u_2(t)||) \geq$$
$$d_3 \max_{t\in[0,1]} ||u_1(t) - u_2(t)|| \geq d_3||u_1(\cdot) - u(\cdot)||_{\mathbb{L}^2_m([0,1])},$$

where $d_3 > 0$. *Hence*

$$||g(u_1(\cdot))(\cdot) - g(u_2(\cdot))(\cdot)||_{\mathbb{C}^1([0,1],\mathbb{R})} \geq d||u_1(\cdot) - u(\cdot)||_{\mathbb{L}^2_m([0,1])}$$
$$\forall u_1(\cdot), u_2(\cdot) \in \mathbb{L}^2_m([0, 1]),$$

where $d > 0$. *Under the assumptions given above, the function* $\tilde{g}(\cdot)$ *is* Fréchet *differentiable* [200,201]. *In some specific cases the* Fréchet *derivative*

$$\tilde{g}'(u(\cdot))$$

of $\tilde{g}(\cdot)$ *can be computed explicitly* [107,269]. *If*

$$R(\tilde{g}'(u(\cdot))) = \mathbb{C}^1([0, 1], \mathbb{R})$$

for a control function $u(\cdot) \in \mathbb{L}^2_m([0, 1])$, *then the operator* $\tilde{g}$ *is regular at every*

$$u(\cdot) \in \mathbb{L}^2_m([0, 1])$$

and satisfies all conditions of the basic Theorem 8.14.

Finally note that, in connection with the classic controllability results from the hybrid control theory, the question arises which conditions are necessary or sufficient for

$$R(\mathcal{A}) = Y,$$

where $\mathcal{A} : X \to Y$ is a linear operator between two real Banach spaces. We refer to [226] for the basic theoretical facts in that connection. It seems to be possible to use the methodology proposed above and obtain some new controllability conditions for HSs and SSs by applying the above analytical results from the theory of differentiable stable operators. Moreover, the main Example 8.1 can be easily extended to a concrete relaxation of the basic HOCP (7.1).

8.3 Well-Posedness and Regularization of the Relaxed HOCPs

This section is devoted to the regularization techniques for the relaxed realizations of the HOCPs and SOCPs. Let us start by considering an abstract situation that explains the ill-posedness in the general framework. Examine a proper extended real-valued functional (see Chapter 2)

$$\Upsilon : \mathcal{W} \rightarrow (-\infty, \infty]$$

on an abstract convergence space $\mathcal{W}$. The (global) minimization problem

$$\min_{w \in \mathcal{W}} \Upsilon(w) \tag{8.12}$$

is called Tikhonov well-posed if and only if there exists exactly one global minimizer w^{opt} and every minimizing sequence for (8.12) converges to w^{opt}. Otherwise, the abstract optimization problem (8.12) is specified as Tikhonov ill-posed. Note that in general uniqueness of a global minimizer to (8.12) does not imply well-posedness. Recall that the historically first specific ill-posedness concept for optimization problems was proposed by Tikhonov in [303]. In [304] Tikhonov pointed out that many OCPs (involving ordinary differential equations [ODEs]) are ill-posed with respect to the norm convergence of the minimizing sequence. Essential progress in the development of a solution concept and solution procedures for ill-posed optimization problems was initiated by the work of Mosco [250]. A regularization method using the stabilizing properties of the classic proximal mapping (see [275,153]) was introduced by Martinet in [237].

Since the generic relaxation schemes for HOCPs and SOCPs discussed in Chapter 7 finally involve the "modified" (relaxed) ODEs, the above observation of Tikhonov is also true in the case of relaxed HOCPs and SOCPs. Let us now give some easy illustrative examples of well- and ill-posed HOCPs and SOCPs involving ODEs. We also refer to [144,308,336] and [26,28] for the examples of ill-posed conventional OCPs.

Example 8.2. *Consider the following ill-posed OCP:*

$$\begin{aligned}
&\text{minimize } J(u(\cdot)) := \int_0^1 x^2(t)dt \\
&\text{subject to } \dot{x}_1(t) = u(t), \ \text{ a.e. on } [0, t_{sw}], \\
&x(0) = 0, \\
&\dot{x}_2(t) = 2u(t), \ \text{ a.e. on } [t_{sw}, 1], \\
&u(\cdot) \in \mathbb{L}^2([0,1]), \ |u(t)| \leq 1.
\end{aligned} \tag{8.13}$$

By $\mathbb{L}^2([0,1])$ *we denote the standard* Lebesgue *space of all square integrable functions* $u : [0,1] \rightarrow \mathbb{R}$. *The switching time* t_{sw} *in (8.13) can be defined by a switching manifold (an HS)* *or by any time-driven rule (an SS). Note that the indexed state variable x has two modes in Example 8.13 for $q_1 = 1$ and $q_2 = 2$.*

In both cases the unique optimal control in problem (8.13) is

$$u^{opt}(t) = 0$$

(considered a.e. on the full time interval $[0, 1]$). However the "combined" minimizing sequence $\{u_r(\cdot)\}$, $r \in \mathbb{N}$,

$$u_r(t) = \mathbf{1}_{[0,t_{sw}]} \sin(2\pi rt) + \mathbf{1}_{[t_{sw},1]} 1/2 \sin(2\pi rt),$$

where $\mathbf{1}_{[a,b]}$ is a characteristic function of a time interval $[a, b]$ for the same OCP (8.13), does not converge in the sense of the norm of the space $\mathbb{L}^2([0, 1])$. We evidently have

$$x_r(t) = \frac{1}{2\pi r}(1 - \cos(2\pi rt)), \; x_r(1) = 0$$

and

$$\lim_{r\to\infty} J(u_r(\cdot)) = \lim_{r\to\infty} \frac{3}{8\pi^2 r^2} = J(u^{opt}(\cdot)) = 0.$$

Note that one can extend the ill-posed OCP from Example 8.2 or construct similar examples. In particular one can include the additional target condition $x(1) \leq 0$ or similar. The resulting OCPs are also ill-posed in the sense of the abstract concept (8.12) or in the sense of the general concept studied in Chapter 3. It is evident that well- or ill-posedness properties of an optimization problem depend strongly on the functional spaces and the corresponding norms under consideration.

Example 8.3. *Consider the minimization of the following functional:*

$$\int_0^1 (x^2(t) + u^2(t))dt$$

subject to the switched-type dynamic system

$$\begin{aligned}
&\dot{x}_1(t) = u(t) \text{ a.e. on } [0, t_{sw}],\\
&x(0) = 0,\\
&\dot{x}_2(t) = 2u(t) \text{ a.e. on } [t_{sw}, 1],\\
&|u(t)| \leq 1 \text{ a.e.}
\end{aligned}$$

This optimization problem has a unique optimal solution

$$u^{opt}(t) = 0$$

(considered a.e. on $[0,1]$*). The OCP under consideration is well-posed with respect to the norm* $||\cdot||_{\mathbb{L}^2([0,1])}$. *If one strengthens the convergence requirement and considers the convergence with respect to the norm of the space* $\mathbb{L}^\infty(0,1)$*, then the same problem becomes ill-posed. For example, the minimizing sequence* $\{u_r(\cdot)\}$, $r \in \mathbb{N}$, *with*

$$u_r(t) = \begin{cases} 0 & if\, t > 1/r, \\ s & 0 \le t \le 1/r \le t_{sw}, \end{cases}$$

where $s =$ const, *does not converge in the sense of the* ∞*-norm*

$$||u_r(\cdot) - u^{opt}||_{\mathbb{L}^\infty([0,1])} = s \quad \forall r \in \mathbb{N}.$$

Consider now the β-relaxed system (system (7.16) from Chapter 7) for the switched-type dynamics from Example 8.2 and Example 8.3, i.e.,

$$\dot{z}(t) = \sum_{k=1}^{M} \gamma^k(t) \tilde{f}(t, z(t), v^k) \text{ a.e. on } [0, t_f],$$
$$z(0) = x_0, \; q \in \{1, 2\},$$

where (see Chapter 7)

$$\tilde{f}(t, x(t), U) := \{\tilde{f}(t, x(t), u) \; : \; u \in U\}.$$

For the above two examples we evidently have

$$\dot{z}(t) = \sum_{k=1}^{M} \gamma^k(t)(\beta_{[0,t_{sw})}(t) + 2\beta_{[t_{sw},1]}(t))u(t),$$

where $v_k \in U_M$ (as described in Chapter 7). Since the set U of admissible controls in the above examples is convex, the argumentation from Example 8.2 and Example 8.3 can also be applied to the corresponding β-relaxed OCPs that correspond to the initially given OCPs from the above two examples. This means that the β-relaxed OCPs in Example 8.2 and Example 8.3 are also ill-posed.

One of the historically most prominent OCPs, namely, the celebrated Fuller problem, is also an ill-posed OCP (see, e.g., [166]). Using the elements of the HS and SS definitions, one can extend the conventional Fuller construction to the hybrid and the switched case. Let us also note that the ill-posed OCPs governed by partial differential equations are of a frequent occurrence. For further examples of well- and ill-posed OCPs we recommend the work of T. Zolezzi [336] and the references therein. In our book we deal with OCPs of hybrid and switched nature in the presence of some additional constraints. It is necessary to stress that usually a formal proof of the well-posedness for a constrained HOCP or SOCP constitutes a very sophisticated theoretical task. Therefore, we follow in this book the idea of effective and "numerically stable" computational methods for the above classes of OCPs. This is the main motivation to a priori include a regularization method into the resulting numerical approach. The reader can observe that we use the celebrated proximal point method for this purpose.

Note that a complex OCP with different additional constraints does not need to have only one optimal solution. On the other hand, our prime interest is to study minimizing sequences generated by a concrete numerical method. In practice, one has to focus the attention on the convergence properties of the applied numerical procedure. These expected convergence properties mathematically express the numerical consistence of the solution method. A general theoretic investigation (in the sense of Tikhonov) of the convergence properties for every minimizing sequence is usually not necessary. Therefore, in some of our works (see [26–29]) we have modified the strict Tikhonov concept and used the "numerical stability" paradigm. The concept of numerical stability mentioned above is weaker in comparison to the formal Tikhonov well-posedness definition and is in fact specified by a concrete computational method in action. In [27,29] we proposed some modifications of the classic proximal point approach and constructed numerically stable approximations for constrained OCPs.

The same approach, namely, the necessary generalization of the "numerical stability" approach mentioned above can be examined not only in the context of conventional OCPs but also in the framework of HSs and SSs discussed in this book. Note that the classic useful numerical methods and optimization techniques (see, e.g., [291]) can usually be combined with a regularization approach for the ill-posed OCPs. It should be mentioned that the ill-posedness theory in the context of the OCPs involving HSs and SSs constitutes a new topic. The same is also true with respect to ill-posedness and regularization of the relaxed HOCPs and SOCPs. These theoretic and numeric tasks constitute in fact a future work.

Finally let us note that in the literature alternative definitions of well-posedness in optimization can be found. These concepts are naturally based on the different viewpoints. For example, the well-posedness and ill-posedness concepts for abstract variational and optimization problems are comprehensively discussed in the works of Dontchev and Zolezzi [144], Kaplan and Tichatschke [202], and Vasil'ev [308]. For the theory of locally ill-posed problems see [191]. The abstract optimization problem of the type (8.12) is called Hadamard well-posed if there exists exactly one global minimizer w^{opt} and, roughly speaking, w^{opt} depends continuously upon the parameters of the problem under consideration. The notation of Hadamard well-posedness in optimization reminds us of the analogous concept for the boundary value problems from mathematical physics [182]. More importantly than the mere similarity, there are significant results, showing that many linear operator equations, or variational inequalities, are well-posed in the classical sense of Hadamard if and only if an associated minimization problem has a unique solution, which depends continuously on the parameters of the problem. There are many links between Tikhonov and Hadamard definitions of well-posedness in the optimization theory [145]. In some works of the author we also considered the linkages between both well-posedness concepts in the framework of stable operators in Banach spaces (see Chapter 2).

The notations of well-posedness in the optimization of hybrid and switched dynamic systems are significant as far as the numerical solution procedures are involved. Ill-posed problems in the sense of Tikhonov or Hadamard in the context of the more sophisticated HOCPs and SOCPs (in comparison with the conventional OCPs) should be handled with special care, since numerical methods will fail in general, and regularization techniques will be required. The same argument is evidently true with respect to all the relaxed HOCPs and SOCPs studied in this book.

8.4 Numerical Treatment of the HOCPs

In this section we describe, very briefly, some basic effective numerical methods developed for general OCPs. We can consider, for example, the basic (Bolza or Mayer) HOCP from Chapter 7 as well as a relaxed version of this problem and also include some additional practically motivated target and state constraints (see, e.g., [179] for the useful paradigm of constrained systems).

The short survey presented here is written taking into consideration the consistency of the resulting numerical schemes. We try to make the main conceptual difficulties in connection with this concept clear. Consider the basic problem (7.1) in the presence of some additional state and target constraints. We have

$$\begin{aligned}
&\text{minimize } J(x(\cdot), u(\cdot)) = \int_0^{t_f} f_0(t, x(t), u(t))dt \\
&\text{subject to } \dot{x}(t) = f_{q_i}(t, x(t), u(t)) \text{ a.e. on } [t_{i-1}, t_i], \\
&q_i \in Q,\ i = 1, ..., r+1, \\
&x(0) = x_0 \in M_{q_1}, \\
&u(t) \in U \text{ a.e. on } [0, t_f], \\
&h_j(x(t_f)) \leq 0\ \forall j \in I, \\
&g(t, x(t)) \leq 0\ \forall t \in [0, t_f].
\end{aligned} \tag{8.14}$$

Here $f_0 : [0, t_f] \times \mathbb{R}^n \times \mathbb{R}^m \to \mathbb{R}$ is a continuously differentiable function,

$$\begin{aligned}
&h_j : \mathbb{R}^n \to \mathbb{R} \text{ for } j \in I, \\
&g : [0, t_f] \times \mathbb{R}^n \to \mathbb{R},
\end{aligned}$$

and $x_0 \in \mathbb{R}^n$ is a fixed initial state. By I we denote a finite set of index values. We assume that the functions $f, h_j(\cdot), \; j \in I$, and

$$g(t, \cdot), \; t \in [0, t_f],$$

are continuously differentiable and the function f_0 is integrable. The control set U is a compact and convex subset of $\mathbb{R}^m$. For example, one can restrict the consideration to the following control sets:

$$U := \{u \in \mathbb{R}^m \; : \; b_-^i \leq u_i \leq b_+^i, \; i = 1, ..., m\},$$

where $b_-^i, b_+^i, i = 1, ..., m$, are constants. The admissible control functions $u : [0, t_f] \to \mathbb{R}^m$ are square integrable functions in time. Let

$$\mathcal{U} := \{v(\cdot) \in \mathbb{L}_m^2([0, t_f]) \; : \; v(t) \in U \text{ a.e. on } [0, t_f]\}$$

be the set of admissible control functions. Without loss of generality we consider in (8.14) only one state constraint. In addition, we assume that for each $u(\cdot) \in \mathcal{U}$ the initial value problem in (8.14) has a unique absolutely continuous solution $x^u(\cdot)$.

Using the results of Chapter 7, we can conclude that the Gamkrelidze relaxation of the initially given HOCP (7.1) can now be determined as follows:

$$\begin{aligned}
&\text{minimize } \bar{\mathcal{J}}(\nu(\cdot)) \\
&\text{subject to} \\
&\dot{y}(t) = \sum_{l=1}^{n+1} \alpha^l(t) \tilde{f}(t, y(t), u^j(t)) \text{ a.e. on } [0, t_f], \\
&y(0) = x_0, \\
&\nu(\cdot) \in \aleph(n+1) \times \mathcal{U}^{n+1}, \\
&q_i \in Q, \; i = 1, ..., r+1, \\
&\tilde{h}_j(y(t_f)) \leq 0 \; \forall j \in I, \\
&\tilde{g}(t, y(t)) \leq 0 \; \forall t \in [0, t_f],
\end{aligned} \tag{8.15}$$

where

$$\bar{\mathcal{J}}(\nu(\cdot)) := \int_0^{t_f} \int_U f_0(t, x(t), u) \mu(t)(du) dt$$

and

$$\alpha(\cdot) \in \aleph(n+1), \; u^j(\cdot) \in \mathcal{U},$$

for $j = 1, ..., n+1$. The constraint functions $\tilde{h}_j(\cdot)$ and $\tilde{g}$ are suitable convexifications of the originally given constraints in (8.14). Let us also assume that the initially given problem (8.14) as well as the Gamkrelidze relaxed problem (8.15) have an optimal solution.

It is well known that the class of OCPs of the type (8.14) is broadly representative in modern hybrid/switched control engineering. Recall that an OCP in the hybrid and switched setting can also be considered as a problem with a terminal objective functional

$$\mathcal{J}(u(\cdot)) := \phi(x^u(t_f)),$$

where $\phi(\cdot)$ is a differentiable function. Note that the initially given HOCP (8.14) with an integral functional J as well as its suitable relaxation (for example, the Gamkrelidze relaxation (8.15)) can be reformulated as an equivalent HOCP with a modified terminal functional $\mathcal{J}$.

As mentioned in Chapter 7 and also in this chapter the relaxed problem can be used not only for the existence theory but also in numerical solution procedures for the initially given OCP. Roughly speaking one needs to propose some adequate solution procedures for the above Gamkrelidze relaxation (8.15) and then interpret the obtained (relaxed) optimal solutions in the framework of the originally given HOCP (8.14).

Recall that the computational methods based on the Bellman optimality principle were among the first proposed for OCPs [78,110,111]. These methods are especially attractive when an OCP is discretized a priori and the discrete version of the Bellman equation is used to solve it. Dynamic programming algorithms are widely used, specifically in the engineering and operations research (see [88,229]). Different approximation schemes have been proposed to cope with the "curse of dimensionality" (Bellman's own phrase) with rather limited success. Moreover, the Bellman-type methods are usually inefficient for OCPs with constraints. The same inefficiency and the increasing problems with the dimensionality naturally occur in the case of more sophisticated relaxed HOCPs and SOCPs.

The constructive use of the first-order necessary conditions of optimal control theory, specifically of the suitable version of a Pontryagin maximum principle, yields a boundary value problem with ODEs. This is also true with respect to the relaxed HOCPs and SOCPs. The necessary optimality conditions and the corresponding boundary value problems play an important role in numerics of optimal control. There are a lot of pioneering works in Russian focusing on the numerical treatment of the boundary value problem for the Hamilton–Pontryagin system of differential equations (see, e.g., [155,244] and the references therein). Some of the first numerical methods for the conventional OCPs were based on the shooting method [106]. This technique was extended to a class of problems with state constraints and is now known as a multiple shooting method (see [112,113,253]). It usually guarantees very accurate solutions provided that good initial conditions for the adjoint variable $p(0)$ are available. In practice, one needs to integrate the Hamilton–Pontryagin system and then adjust the chosen initial value of the adjoint variable such that the computed $p(\cdot)$ satisfies all terminal conditions determined for the final point $p(t_f)$. In the view of the necessary convergence properties, the multiple shooting scheme has all advantages and disadvantages of the family of the Newton methods. This numerical technique can evidently be applied to the relaxed OCP of the type (8.15).

The general disadvantage of the so-called "indirect numerical methods" (based on the necessary optimality conditions for OCPs) is the possible local nature of the obtained quasioptimal solution. On the other hand there are a lot of historical and recent professional publications focusing on the so-called Sakawa-type algorithms (see, e.g., [284,101]). This group of numerical methods uses, in fact, the idea of augmented Hamiltonian for some specific necessary regularizations of the auxiliary Pontryagin maximization problem. For the closely related family of Chernousko methods see also [128,286] and references therein.

It is common knowledge that a general OCP involving ODEs can be formulated in various ways as an optimization problem in a suitable function space (see, e.g., [148,149,192, 193]). The original problem (8.14) as well as the relaxed version (8.15) can be expressed as infinite-dimensional optimization problems. Clearly, the infinite-dimensional reformulations of the above HOCPs also provide a theoretic fundament for direct applications of the existing optimization methods. For example, a method of sequential linearizations (proposed by Fedorenko in [155]), the feasible direction algorithm (see [269]), and the function space algorithm proposed by Pytlak in [269] substantially use the abovementioned infinite-dimensional reformulation of a given OCP. The convergence analysis for a general class of methods for the conventional OCPs with state constraints is presented in [269]. In [27,29] we also apply the proximal-type techniques to the OCPs written in the abstract form mentioned above. Note that an effective computational algorithm based on the infinite-dimensional form of a given OCP includes an integrating procedure for the corresponding initial value problem. Note that there is a variety of concrete numerical algorithms for classic OCPs based on a suitable reformulation of the initially given OCPs. One possibility is that an OCP is represented as a minimization problem with respect to the variable (pair)

$$(x(\cdot), u(\cdot)).$$

The variables $x(\cdot)$ and $u(\cdot)$ may be treated as independent variables. We refer to the celebrated Balakrishnan ϵ-method from [72]. The reformulations of the initially given dynamic optimization problem (mentioned above), the Pytlak method, and the ϵ-method can evidently be applied to the relaxed problem (8.15). These applications constitute a future work that belongs to the area of numerical analysis and to the area of computational algorithms for HOCPs and SOCPs.

An OCP (conventional, hybrid, and switched) with state constraints can also be solved by applying the modern numerical algorithms of nonlinear programming. For example, the implementation of the interior point method is presented in [322]. In [82] the proximal point methods are used for solving stochastic OCPs. For a general survey of applications of nonlinear programming to optimal control see [90]. Although there is extensive literature on computational optimal control, relatively few results have been published on the application of the sequential quadratic programming (SQP) optimization algorithms to the conventional OCPs (see [252,118] and references therein). Note that necessary calculation of second-order derivatives of the objective functional and constraints in HOCPs (8.14) and (8.15) can be avoided by applying an SQP-based scheme in which these derivatives are approximated by quasi-Newton formulae. It is evident that in this case one deals with all usual disadvantages of the Newton-based methods.

The family of gradient-type algorithms (see [259]) are the methods based on the evaluation of the gradient for the objective functional in an OCP. The first applications of the gradient methods in optimal control are discussed in [72]. See also the work of Bryson and Denham [111] and the book of Teo, Goh, and Wong [300]. The gradient of the objective functional can be computed by solving the state equations such that the obtained trajectory is then used to integrate the adjoint equations backward in time. The state and the adjoint variables are used to calculate the gradient of the functional with respect to control variable only. This is possible due to the fact that the trajectories of a conventional and hybrid/switched control system are uniquely determined by controls. The gradient methods are widely used and are regarded as the most reliable, if not very accurate, methods for the classic OCPs [149,204]. The same is also true with respect to the HOCPs and SOCPs.

There are two major sources of the inaccuracies in the gradient-type methods: errors caused by integration procedures applied to the state equations and the errors introduced by the optimization algorithm. In some of the author's works we apply the evaluation of the gradient for the objective functional and the gradient method to the various types of HOCPs and SOCPs. Note that the computational gradient-type methods are also studied in connection with a possible numerical approach to the relaxed versions of the classic OCPs [301].

When solving an OCP with ODEs we deal with functions and systems which, except in very special cases, are to be replaced by some concrete numerically tractable approximations. Resulting implementations of the numerical schemes for original and relaxed OCPs are usually based on the suitable discrete approximations of the control set and on the finite-difference approximations of the given dynamics. Finite-difference methods turn out to be a powerful tool for theoretical and practical analysis of general control problems (see, e.g., [247,260, 148]). Discrete approximations can be applied directly to the problem at hand or to auxiliary problems used in the solution procedure [246,145,269]. The numerical methods for OCPs with constraints (with the exception of some works) are either the methods based on the full discretizations (full parametrization of the state and control variables), or they are function space algorithms. The first group of methods assumes a priori discretization of the conventional or HS/SS equations. The second group of methods is, in fact, a theoretical work on the convergence of algorithms.

The major drawback of some numerical schemes from the first group is the lack of the corresponding convergence analysis. This is especially true in connection to the multiple shooting method [112,113] and to some "collocation methods" (see, e.g., the work of Stryk [294]). For the alternative model of the collocation method see, e.g., [89]. Note that collocation methods are similar to the gradient algorithms applied to a priori discretized problems with the exception that gradients are not calculated using the adjoint equations. In that case the both variables, namely, $x(\cdot)$ and $u(\cdot)$, are considered as optimization parameters. We refer to [146, 183,184] for high-order numerical schemes applied to unconstrained problems and for the corresponding error estimates associated with the discrete approximations of OCPs. Note that the interesting discrete approximations to the dual OCPs are studied in [184]. There are a number of results scattered in the literature on discrete approximations that are very often closely related, although apparently independent. For a survey of some recent works on computational optimal control, including discrete approximations, see [145] and [32]. Finally let us note that the convergence analysis of discrete approximations of OCPs is of primary importance in computational optimal control (see, e.g., [149,204,239]). This remark is specifically true for the more sophisticated HOCPs and SOCPs.

8.5 Examples

This section is devoted to some numerical illustrations of the relaxations methods for hybrid and switched-type OCPs discussed in Chapter 7.

Let us firstly give an introductive example that contains the method of β-relaxation (see Chapter 7) and consider the celebrated Goddard-type problem of a rocket vertical ascent (see [177]). Note that this problem constitutes a historically first application of the optimal control to the rocket dynamics.

Example 8.4. *Consider the following OCP:*

$$\begin{aligned}
&\text{minimize } -x_2(100)\\
&\text{subject to}\\
&\dot{x}_1(t) = -u(t),\ \dot{x}_2(t) = x_3(t),\\
&\dot{x}_3(t) = -0.01 + \frac{2u(t) - 0.05x_3^2(t)\exp(-0.1x_2(t))}{x_1(t)},
\end{aligned}$$

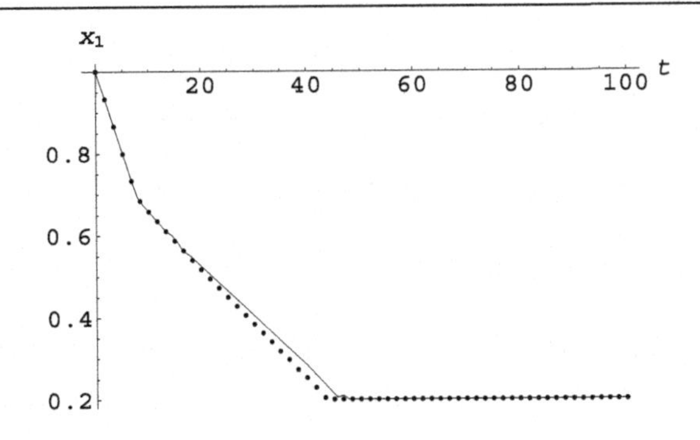

Figure 8.1: Optimal mass of the rocket.

$$\begin{aligned} & x_1(0) = 1, \\ & x_2(0) = x_3(0) = 0, \\ & x_1(100) = 0.2, \\ & 0 \leq u(t) \leq 0.04, \; t \in [0, 100], \\ & u(\cdot) \in \mathbb{L}^1([0, 100]), \end{aligned} \tag{8.16}$$

where $x_1(t)$, $x_2(t)$, *and* $x_3(t)$ *denote the mass, the altitude, and the vertical velocity of the rocket, respectively. The original Goddard problem has been studied theoretically and numerically by many authors* (see, e.g., [238] and the references therein). *This famous problem differs slightly from our test problem* (8.16). *In the original setting the final time* t_f *is free and, in addition, the value*

$$x_3(t_f) = 0$$

is prescribed.

From the qualitative point of view, the aim is to maximize the attained altitude $x_2(100)$ *of the rocket. Moreover, the terminal mass should be equal to 0.2.*

For

$$M = 100$$

we now apply the β*-relaxations and the gradient-based method from Chapter 7.*

For the aim of comparison, the originally given Goddard-type OCP was also solved using the direct collocation method and the DIRCOL package [294]. *The calculated mass, altitude, and vertical velocity of the rocket are shown in* Fig. 8.1, Fig. 8.2, *and* Fig. 8.3.

The dotted line on the figures denotes the computational results obtained in the original problem. The β*-relaxation techniques in combination with the gradient-based algorithm generate the graphical results presented by continuous lines. The constraints are fulfilled with the tolerance* 10^{-5}.

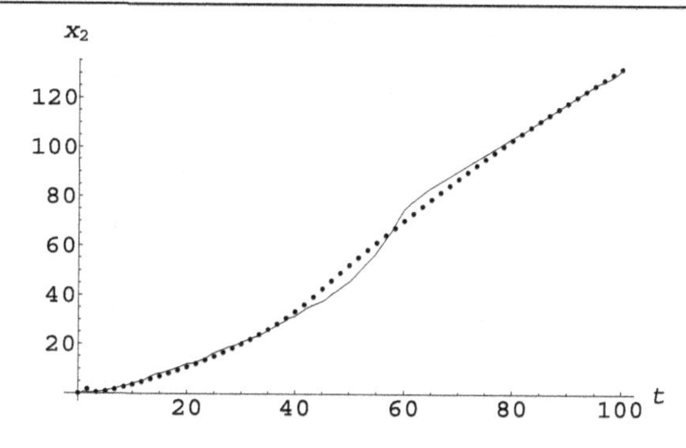

Figure 8.2: Optimal altitude of the rocket.

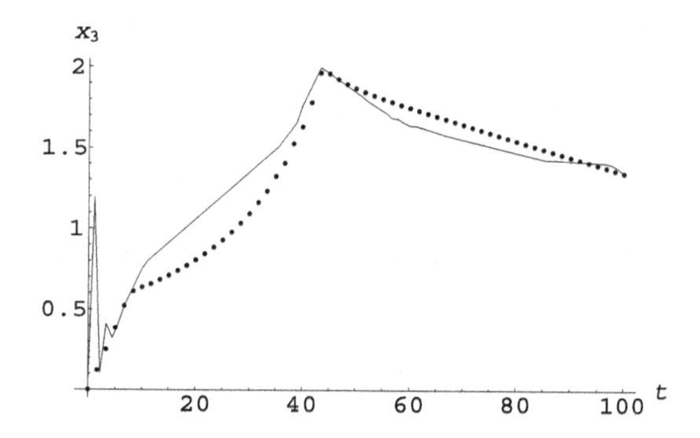

Figure 8.3: Optimal vertical velocity of the rocket.

Using the β-relaxation, we are also able to calculate the quasioptimal (numerically optimal) value of the objective functional: We get the following value: -131.901. *The optimal objective value calculated by the DIRCOL package is equal to* -132.1961. *Note that the numerical consistency is guaranteed by the corresponding theoretical results presented in Chapter 7.*

We next consider an example of the specific SOCP (5.17). The numerical solution procedure for this optimization problem is based on the newly elaborated infimal convolution-based OCP (5.23) (see Chapter 5).

Example 8.5. *Consider the following applied dynamic model of a unicycle robot studied in* [212]*:*

$$\begin{aligned} \dot{x}_1(t) &= u_1(t)\cos(x_3(t)), \\ \dot{x}_2(t) &= u_1(t)\sin(x_3(t)), \\ \dot{x}_3(t) &= u_2(t), \\ x(0) &= \begin{bmatrix} 15 & 15 & 180\,\text{deg} \end{bmatrix}^T, \; t_0 = 0, \; t_f = 1. \end{aligned} \tag{8.17}$$

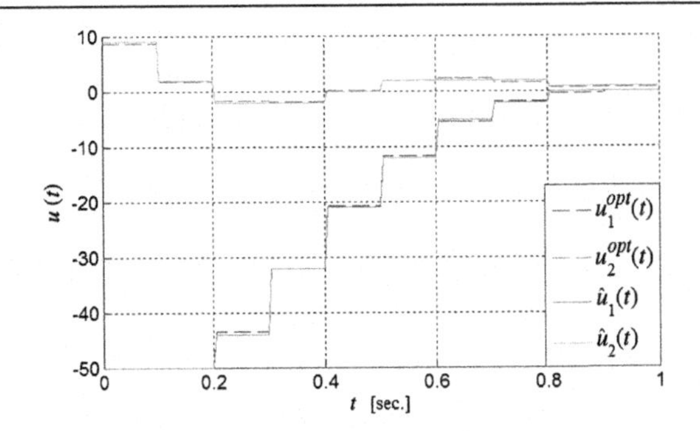

Figure 8.4: Quasioptimal control inputs $\hat{u}_1(t)$ and $\hat{u}_2(t)$.

The control variable $u_1(t)$ reflects the linear velocity of the vehicle while the control $u_2(t)$ determines its orientation. The state variables

$$x_1(t),\ x_2(t)$$

in the above model (8.17) denote the coordinates of the robot on the plane and $x_3(t)$ denotes the corresponding orientation. We next consider the minimization of the following objective functional:

$$J(u(\cdot)) = \frac{1}{2}\int_0^1 \left(x_1^2(t) + x_2^2(t) + x_2^3(t)\right) dt.$$

Moreover, we consider the given finite set of admissible constant control values

$$\mathcal{Q} = \{-50, -49, -48, \ldots, 48, 49, 50\}.$$

The computational results of application of the basic Armijo gradient algorithm (5.24) from Chapter 5 are presented in Fig. 8.4.

The controls $u_1^{opt}(\cdot)$ and $u_2^{opt}(\cdot)$ for the corresponding weakly relaxed version (5.23) of the initially given OCP involving (8.17) are indicated by the dashed lines. The solid lines represent

$$\hat{u}_1(\cdot),\ \hat{u}_2(\cdot),$$

namely, the approximating quasioptimal controls for the originally given OCP driven by system (8.17). The resulting quasioptimal trajectories are shown in Fig. 8.5.

The dashed lines correspond to the trajectory generated by $u^{opt}(\cdot)$ and the solid lines indicate the trajectory that corresponds to the control $\hat{u}(\cdot)$. The initial controls for the basic Armijo gradient algorithm are selected here as proposed in [212]. *The evolution of the values of the objective functional $J(\hat{u}(\cdot))$ in the originally given OCP is presented in* Fig. 8.6. *The computed controls are obtained in* 68 *iterations and the calculated optimal cost is equal to* 85.17.

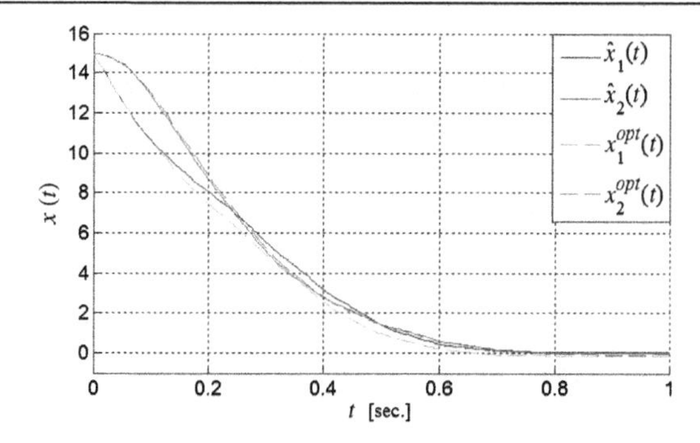

Figure 8.5: Trajectories components $\hat{x}_1(t)$ and $\hat{x}_2(t)$.

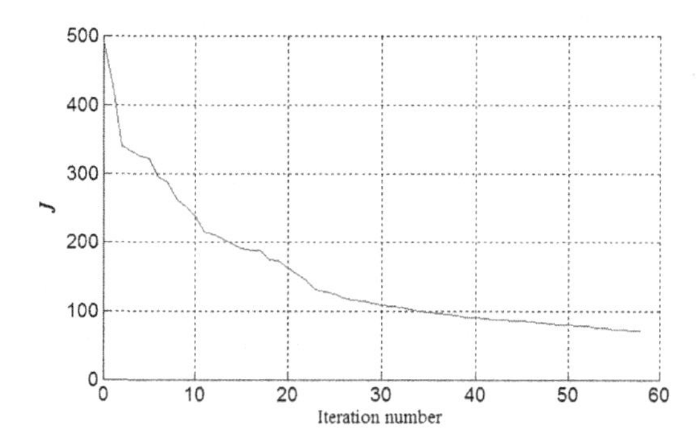

Figure 8.6: Performance measure evaluation.

Let us now consider the switched mode control system problem (6.6) from Chapter 6 and the corresponding HOCP (6.8). Recall that the full relaxation of (6.8), namely, the convexified HOCP (7.9), was studied in Chapter 7. We now consider the novel infimal convolution-based relaxation (7.24) proposed in Chapter 7 for the initially given HOCP (6.8). This approach was called a "weak relaxation." Let us also recall that the main motivation for the weak relaxation approach was the minimal-gap argument (see Chapter 7). Our next example illustrates this weak relaxation technique in a concrete calculation of an optimal solution.

Example 8.6. *Consider the celebrated double-tank system*

$$\begin{aligned} \dot{x}_1(t) &= v(t) - \sqrt{x_1(t)}, \\ \dot{x}_2(t) &= \sqrt{x_1(t)} - \sqrt{x_2(t)} \end{aligned} \tag{8.18}$$

with the initial conditions

$$x_1(0) = x_2(0) = 2.0.$$

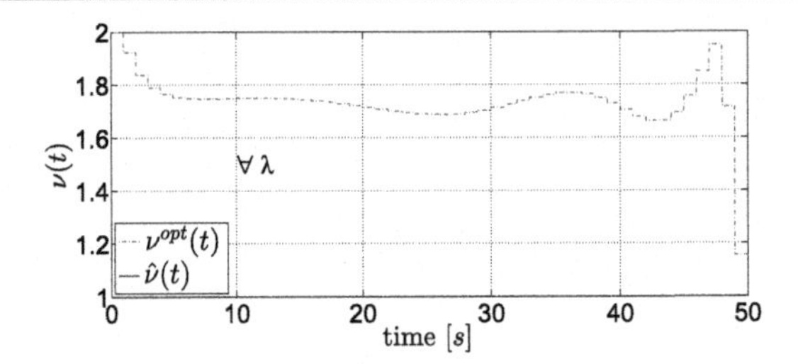

Figure 8.7: Calculated optimal control strategy $v^{opt}(t)$.

The system input $v(\cdot)$ describes the inflow rate to the upper tank, controlled by the value of $v(t)$ and having discrete possible values between

$$v_{\min} = 1, \; v_{\max} = 2.$$

The corresponding control discretization is assumed to be given with the step

$$\Delta v = 0.01.$$

The objective of the optimization problem is to have the fluid level in the lower tank close to the given value 3.0. Let $t_f = 50$. The corresponding cost functional has the following form:

$$J(v(\cdot)) = 2 \int_0^{t_f} (x_2(t) - 3)^2 dt.$$

We use the Armijo-based projected gradient method (see Chapter 6) for the weakly relaxed HOCP involving system (8.18) and put

$$\gamma = 1, \; \alpha = 0.1.$$

The prox parameter λ for the weakly relaxed HOCP that corresponds to the initially given problem was taken as a varying parameter (for various small positive numbers). The necessary integration procedure was realized by the standard Runge–Kutta methods. Results of a typical run, consisting of 29 iterations of the main gradient algorithm, are shown in Figs. 8.7–8.9. The calculated optimal control strategy $v^{opt}(t)$ is presented in Fig. 8.7.

Let us note that the dynamic behavior of the calculated optimal state component

$$x_1^{opt}(t)$$

is presented in Fig. 8.8.

The computed solution is poorly behaved at the end of the time horizon. Note that the computational results obtained in [319] also contain the same effect. This behavior is an evident

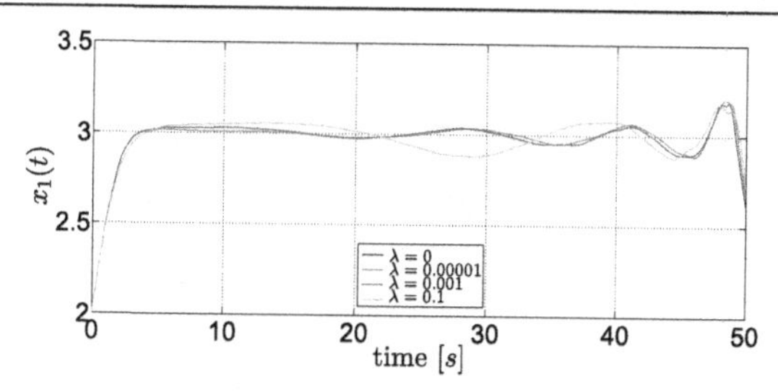

Figure 8.8: Calculated optimal control strategy $x_1^{opt}(t)$ for various values of parameter λ.

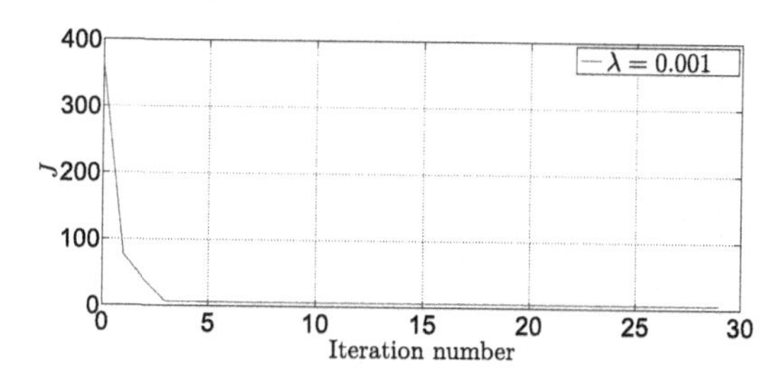

Figure 8.9: Calculated values of the optimal cost functional.

consequence of the given system nonlinearities in combination with the nontrivial switching dynamics. The numerical evaluation of the optimal cost functional is given in Fig. 8.9. Let us note that the value of the calculated quasioptimal cost we obtained,

$$J(u^{quasiopt}(\cdot)) = 4.71,$$

is near to the optimal value

$$J(u^{opt}(\cdot)) = 4.74$$

and the corresponding relaxation gap is equal to 0.03. *Note that here we have* $u^{quasiopt}(t) = u_{(29)}(t)$.

A similar comparative calculation can also be realized for the fully relaxed version of the HOCP associated with the given double-tank system (8.18). *The numerical evaluation of the optimal (convexified) cost functional*

$$\bar{co}\{J(u^{relaxed}(\cdot))\}$$

leads to the following result:

$$\bar{co}\{J(u^{relaxed}(\cdot))\} = 4.32.$$

As we can observe the relaxation gap associated with the fully relaxed OCP *is equal to* 0.42. *Moreover, the relaxation gap of the weakly relaxed HOCP is less than the gap associated with the fully relaxed HOCP. Finally note that the fully relaxed scheme can be obtained here using the simple structure of the dynamic system under consideration and the corresponding quadratic cost functional.*

Finally note that in our book we have mostly discussed the theoretic aspects of the newly elaborated relaxation-based approaches. The simple numerical example we present in this section illustrates a conceptual applicability of the proposed analytical techniques. The combined solution methodology we propose needs a further comprehensively numerical examination that includes simulations of several HOCPs and SOCPs involving various hybrid, switched, and switched mode dynamic systems.

8.6 A Remark About the Practical Stabilization of a Class of Control-Affine Dynamic Systems

The material of our book mainly contains the OCPs of hybrid and switched nature and the related (or auxiliary) topics. The relaxed hybrid and switched dynamic systems under consideration constitute a formal "extension" of the originally given dynamic. This theoretic fact can also be used for studying the qualitative behavior of the resulting systems.

In this section we consider the practical stability (stabilization) of a class of control-affine systems that are similar to the β-relaxed systems from Chapter 7. Note that the general stability is of prime interest for the control engineering and systems theory [205]. We use here the practical stability approach proposed in [266,267] and apply it to a specific dynamic system.

Consider the following initial value problem for ODEs with a general control-affine structure:

$$\begin{aligned} \dot{x}(t) &= f(t, x(t)) + B(t)u(t) + \xi(t) \text{ a.e. on } \mathbb{R}_+, \\ x(0) &= x_0, \end{aligned} \tag{8.19}$$

where $x_0 \in \mathbb{R}^n$ is a fixed initial state. The given function $f : \mathbb{R}_+ \times \mathbb{R}^n \to \mathbb{R}^n$ is assumed to be continuous on $\mathbb{R}_+$ and uniformly Lipschitz continuous on an open bounded set $\mathcal{R} \subseteq \mathbb{R}^n$. By

$$B(t) \in \mathbb{R}^{n \times m}, \ t \in \mathbb{R}_+,$$

we denote here a control matrix. We next assume that $B(\cdot)$ is a continuous matrix function. The uncertainties $\xi(\cdot)$ are assumed to be uniformly bounded

$$\sup_{t \in \mathbb{R}_+} ||\xi(t)|| \leq M \in \mathbb{R}_+.$$

By $x(t) \in \mathbb{R}^n$ and $u(t) \in \mathbb{R}^m$ we denote here the state and the control vector, respectively. Let us firstly consider the basic system (8.19) over a control set $\mathcal{U}$ of essentially bounded measurable control inputs. In parallel with (8.19) we examine the corresponding linearized control system

$$\begin{aligned} \dot{y}(t) &= f_x(t, x^u(t))y(t) + B(t)v(t) + \xi(t) \text{ a.e. on } \mathbb{R}_+, \\ y(0) &= 0, \end{aligned} \tag{8.20}$$

where $u(\cdot) \in \mathcal{U}$ and $x^u(\cdot)$ is the absolutely continuous solution to the initial system (2.1) generated by an admissible $u(\cdot)$. Let us also note that the celebrated Rademacher theorem guarantees the (almost everywhere) differentiability property of the function $f(t, \cdot)$ (see, e.g., [130]).

We next suppose that the pair $(A(t), B(t))$, where

$$A(t) := f_x(t, x^u(t)),$$

is controllable for every $t \in \mathbb{R}_+$. We also assume that

$$||f_x(t, x)|| \leq C \tag{8.21}$$

for all

$$(t, x) \in \mathbb{R}_+ \times \mathbb{R}^n.$$

We also assume that a class of locally Lipschitz (feedback) control functions $w(\cdot, \cdot)$ such that

$$w(t, x^u(t)) = v(t)$$

in (8.20) constitutes an admissible set of inputs. This family of functions $w(\cdot, \cdot)$ is next denoted by $\mathcal{L}$. For each

$$u(\cdot) \in \mathcal{U}, \ w(\cdot, \cdot) \in \mathcal{L}$$

the initial value problem (8.20) has a unique solution denoted by $y^v(\cdot)$. We refer to [185] for the necessary existence and uniqueness results. Consider the practically important "proportional" control design of the following type:

$$w(t, y(t)) = K(t)y(t), \ \ K(t) \in \mathbb{R}^{m \times n}, \ t \in \mathbb{R}_+.$$

Here $K(\cdot)$ is a gain matrix function. This unknown gain matrix constitutes a free parameter of the control design under consideration and the linear closed-loop system can be written as

$$\begin{aligned} \dot{y}(t) &= (f_x(t, x^u(t)) + B(t)K(t))y(t) + \xi(t), \text{ a.e. on } \mathbb{R}_+, \\ y(0) &= 0. \end{aligned} \tag{8.22}$$

Let us firstly describe the desired control design for the linearized system (8.22) in a qualitative manner: the trajectory $y^v(\cdot)$ of the closed-loop linearized system (8.22) with a concrete matrix function $K(\cdot)$ needs to stay for $t \in \mathbb{R}_+$ in an ellipsoidal region (with the center at the origin), i.e.,

$$\mathcal{E} := \{y \in \mathbb{R}^n \mid y^T P y \leq 1\}.$$

Here P is a positive defined symmetrical $n \times n$-dimensional matrix. Our main idea is to study the auxiliary linearized system (8.22) and propose a constructive geometrical characterization of the corresponding minimal-size attractive ellipsoid (AE) $\mathcal{E}$. Various linearization methods associated with the dynamic models given by ODEs have been long recognized as a powerful tool for stabilization of the conventional control systems. The geometric stability criterion for the linearized system (8.20) mentioned above is next used in the robust feedback control design procedure for the originally given nonlinear system (8.19).

We now explain (qualitatively) some advances in the celebrated AE method (AEM; see [266, 267]) we develop in this section. The nonstationary character of the dynamic systems (8.19) and (8.22) makes a direct application of the conventional LMI-based robust control design that involves the AEM techniques impossible.

Recall that the classic AEM was developed under the strict assumptions of stationarity for the given dynamic system. Moreover, it also involves the restrictive "quasi-Lipschitz" condition for the right-hand side of the initial differential equation (see [266]). However, many modern engineering control systems involve a sophisticated nonstationary systems modeling framework. These nonstationary dynamic models with time-dependent parameters are adequately modeled by time-dependent systems of the type (8.19) closed by a nonstationary feedback (for example, by the law $w(t,x)$ given above).

We next need an exact analytic result that establishes the quality of the linear approximation (8.22) for system (8.19). We use here the notation $\mathbb{L}_r^\infty$ for a standard Lebesgue space of measurable essentially bounded r-dimensional vector functions defined on a "sufficiently big" time interval $I \subset \mathbb{R}_+$.

Theorem 8.15. *Assume that the initial system (8.19) given on a time interval I satisfies all the above technical assumptions. Then there exists a function*

$$o : \mathbb{R}_+ \to \mathbb{R}_+$$

such that $s^{-1}o(s) \to 0$ *as* $s \downarrow 0$ *and*

$$||x^{u+v}(\cdot) - (x^u(\cdot) + y^v(\cdot))||_{\mathbb{L}_n^\infty} \leq o(||v(\cdot)||_{\mathbb{L}_m^\infty})$$

for all

$$u(\cdot) \in \mathcal{U}, \ v(\cdot) \in \mathbb{L}_m^\infty.$$

Proof. Assume $u(\cdot) \in \mathbb{L}^{\infty}_{m}$. For a function $w(\cdot,\cdot) \in \mathcal{L}$ we have $v(\cdot) \in \mathbb{L}^{\infty}_{m}$. From the well-known comparison theorem (see [205]) with the selected comparison functions

$$z(t) := x^{u}(t) + y^{v}(t),$$
$$\psi(t,x) = f(t,x) + B(t)u + v(t),$$

where $t \in \mathbb{R}_{+}$, we obtain

$$\begin{aligned}
&||x^{u+v}(\cdot) - (x^{u}(\cdot) + y^{v}(\cdot))||_{\mathbb{L}^{\infty}_{n}} \leq \\
&e^{C} \int_{I} ||\dot{x}^{u}(t) + \dot{y}^{v}(t) - f(t, x^{u}(t) + y^{v}(t)) - \\
&B(t)(u(t) + v(t))||dt = \\
&e^{C} \int_{I} ||\langle (f_{x}(t, x^{u}(t)), B(t)), (y^{v}(t), v(t))\rangle - \\
&[f(t, x^{u}(t) + y^{v}(t)) + B(t)v(t) - f(t, x^{u}(t)]||dt.
\end{aligned} \tag{8.23}$$

Here C is a constant in (8.21). From the componentwise variant of the mean value theorem (see [6]) we next deduce

$$\begin{aligned}
&(f_{i}(t, x^{u}(t)) + y^{v}(t)) + B(t)(u(t) + v(t)) - (f_{i}(t, x^{u}(t)) + B(t)u(t)) = \\
&\langle (f_{i})_{x}(t, x^{u}(t) + \nu_{i}(t)), B(t))(y^{v}(t), v(t))\rangle
\end{aligned}$$

for $i = 1, ..., n$ and a suitable bounded function $\nu(\cdot)$. Assumption (8.21) and the Lipschitz continuity of $f(t,\cdot)$ on a bounded set $\mathcal{R}$ imply the existence of a (continuous) function

$$o_{1} : \mathbb{R}_{+} \to \mathbb{R}_{+}$$

such that

$$s^{-1} o_{1}(s) \to 0$$

as $s \downarrow 0$ and

$$\begin{aligned}
&||\langle (f_{x}(t, x^{u}(t)), B(t)), (y^{v}(t), v(t))\rangle - \\
&[f(t, x^{u}(t) + y^{v}(t)) + B(t)v(t) - f(x^{u}(t))]|| \leq o_{1}(||v(\cdot)||_{\mathbb{L}^{\infty}_{m}})
\end{aligned}$$

for all $t \in I$. From (8.23) we finally deduce the expected estimation

$$||x^{u+v}(\cdot) - (x^{u}(\cdot) + y^{v}(\cdot))||_{\mathbb{L}^{\infty}_{n}} \leq o(||v(\cdot)||_{\mathbb{L}^{\infty}_{m}})$$

with

$$o(s) := \exp(C) \times o_{1}(s).$$

The proof is finished. □

Theorem 8.15 will next be used in a concrete robust control design procedure for the original nonlinear system (8.19).

Let us now discuss the celebrated Clarke flow invariance concept from [132].

Definition 8.4. *A smooth manifold S in a Euclidean space is called flow invariant in the sense of a given dynamic system*

$$\begin{aligned} \dot{z}(t) &= \phi(z(t)), \ t \in \mathbb{R}_+, \\ z(0) &= 0 \end{aligned} \tag{8.24}$$

if $z(t) \in S$ for all $t \geq T \in \mathbb{R}_+$.

The next abstract result gives a general criterion of a flow invariant manifold.

Theorem 8.16. *A smooth manifold S is flow invariant for system (8.24) if and only if $\phi(x)$ belongs to the tangent space T_S of S for all x from the given Euclidean space.*

The formal proof of this basic theorem is based on an extended Lyapunov-type technique. We refer to [132] for the additional mathematical details. Note that Theorem 8.16 has a natural geometrical interpretation.

Let us also recall the general invariance concept: a set $\mathcal{D}$ in the state space of a dynamic system is called (positively) invariant if an admissible trajectory initiated in this set remains inside the set at all future time instants. Let us denote by $\Omega(z(0))$ a (positive) limit set of system (8.22) (the set of all positive limit points; see, e.g., [205]). We now give a related Lyapunov set stability concept (see [267]).

Definition 8.5. *A compact invariant set $\mathcal{D} \subset \mathbb{R}^n$ of the closed-loop dynamic system* (8.22) *is called asymptotically* Lyapunov *stable if*

$$\Omega(z(0)) \subset \mathcal{D}$$

and

- *for all $\epsilon > 0$ there exists $\delta_1 > 0$ such that the initial condition*

  $$\text{dist}[z(0), \mathcal{D}] \leq \delta_1$$

  implies

  $$\text{dist}[z(t), \mathcal{D}] \leq \epsilon$$

  for all $t \in \mathbb{R}_+$ (the Lyapunov *stability of the set);*

- *there exists* $\delta_2 > 0$ *such that*

$$\text{dist}[z(0), \mathcal{D}] \leq \delta_1$$

implies

$$\lim_{t\to\infty} \text{dist}[z(t), \mathcal{D}]$$

(the attraction property of the set).

Here

$$\text{dist}[z; \mathcal{D}] := \min_{\tilde{z}\in\mathcal{D}} ||z - \tilde{z}||_{R^n}$$

is the usual Euclidean distance between a point $z \in \mathbb{R}^n$ and $\mathcal{D}$. As mentioned above, the existence and a constructive characterization of an invariant set for a general dynamical system (8.24) constitute very sophisticated theoretic questions. Our aim is to specify an invariant set for the closed-loop linear system (2.3) in the form of an ellipsoid $\mathcal{E}$. Applying the classic concept of an asymptotically stable set, namely, Definition 8.5, we now introduce the main concept of an AE.

Definition 8.6. *An ellipsoid* $\mathcal{E}$ *is called an* AE *for system* (8.22) *if it is an asymptotically stable invariant set of this system.*

It is evident that an AE $\mathcal{E}$ for system (8.22) is determined by the matrices P and $K(\cdot)$. The chosen gain matrix function $K(\cdot)$ determines in fact the resulting dynamic behavior of the trajectory $y^v(t)$ such that the basic inequality

$$(y^v(t))^T Py(t) \leq 1, \; t \in R_+,$$

is satisfied. From the point of view of a practical engineering robust control design, the last condition can of course be considered in a (suitable) approximative sense. The resulting $\mathcal{E}$-restricted dynamic behavior of the linear system (8.22) closed by the nonstationary linear feedback $w(\cdot, \cdot)$ can finally be interpreted as a practical stability of this dynamic system.

The previously discussed general facts and concepts are used in this section for a constructive geometric interpretation of the basic AEM in the context of the linearized system (8.22). We first introduce an auxiliary dynamic variable $\theta(\cdot)$, $\theta(0) = 0$ and determine the smooth manifold in the (extended) Euclidean state space $\mathbb{R}^{n+1}$. We have

$$S_P := \{z \in \mathbb{R}^{n+1} \mid y^T(t)Py(t) - 1 + \theta(t) = 0\}.$$

Here

$$z := (y, \theta)^T,$$

and $y(\cdot)$ corresponds to system (8.22). Introduce the following additional notation:

$$h(z) := z^T P z - 1 + \theta.$$

The necessary and sufficient condition for the flow invariance of S_P can now be determined using Theorem 8.16, i.e.,

$$\begin{aligned} &\langle \nabla h(z), \phi(z) \rangle = 0, \\ &\theta(t) \geq 0, \ \forall t \in \mathbb{R}_+. \end{aligned} \tag{8.25}$$

By $\nabla h(\cdot)$ we denote the gradient of the function $h(\cdot)$ introduced above. The vector field $\phi(\cdot)$ corresponds to the right-hand side of the following system of equations:

$$\begin{aligned} &\dot{y}(t) = (f_x(t, x^u(t)) + B(t)K(t))y(t) + \xi(t), \\ &\dot{\theta}(t) = -2y^T(t)P\dot{y}(t), \\ &y(0) = 0, \quad \theta(0) = 0. \end{aligned} \tag{8.26}$$

We now use the dynamics of system (8.22) and rewrite the second differential equation from (8.26) in the following equivalent form:

$$\dot{\theta}(t) = -2y^T(t)P\big[(f_x(t, x^u(t)) + B(t)K(t))y(t) + \xi(t)\big]. \tag{8.27}$$

The above observations can be summarized as the following result.

Theorem 8.17. *The ellipsoidal set $\mathcal{E}$ is an invariant set for the closed-loop system* (8.22) *if and only if the variable $\theta(t)$ determined by the initial value problem* (8.26)–(8.27) *is nonnegative for all $t \in \mathbb{R}_+$.*

Proof. The scalar product in (8.25)

$$\langle \nabla h(z), \phi(z) \rangle \big|_{z=z(t)}$$

can be easily calculated, i.e.,

$$\begin{aligned} &\langle (2y^T(t)P, 1), (\dot{y}(t), -2y^T(t)P\dot{y}(t)) \rangle = \\ &2y^T(t)P\dot{y}(t) - 2y^T(t)P\dot{y}(t) = 0. \end{aligned}$$

Evidently, the first condition from (8.25) is true. Therefore, (8.25) is reduced to the second condition, namely, to the nonnegativity of the variable $\theta(t)$ determined by system (3.2). The proof is completed. □

The formal solution of the initial value problem on a time interval I for the artificial variable $\theta(\cdot)$ can be written as

$$\theta(t) = -\int_I [2y^T(t)P[f_x(t, x^u(t)) + B(t)K(t)]y(t) + y^T(t)P\xi(t)]\,dt.$$

Therefore the nonnegativity condition for $\theta(t)$ required by Theorem 8.16 implies the inequality

$$\int_I [y^T(t)P[f_x(t, x^u(t)) + B(t)K(t)]y(t) + y^T(t)P\xi(t)]\,dt \leq 0. \tag{8.28}$$

Taking into account the robust control applications we are naturally interested to construct an AE $\mathcal{E}$ of a minimal "size" (volume). This natural requirement and the corresponding selection of the matrix P need to be adequately formalized. Let us recall the well-established formalization procedure in the form of a specific linear matrix inequality (LMI)-constrained minimization problem [266,267]. In the case of a stationary dynamic system one can use a specific optimization problem for a minimal-size AE design and for the calculation of the corresponding gain matrix function $K(\cdot)$, i.e.,

$$\begin{aligned} &\text{minimize } \operatorname{tr}(P^{-1}) \\ &\text{subject to } P^T = P,\ P > 0,\ \{K(t)\} \in \mathcal{K}_t, \end{aligned} \tag{8.29}$$

where $t \in \mathbb{R}_+$ and $\mathcal{K}_t \subset \mathbb{R}^{m\times n}$ is the set of admissible matrices that ensure invariance of the AE $\mathcal{E}$ for system (8.22). A constructive description of the set $\mathcal{K}_t$ constitutes a sophisticated mathematical problem. This problem can be constructively solved in the case of stationary dynamic systems and, moreover, under some additional hard conditions.

The same observation is also true for the originally given nonlinear system (8.19). Control systems discussed above, namely, dynamic models (8.19) and (8.22), have a nonstationary nature. Therefore, in that case the nature of the resulting systems makes a direct application of the celebrated LMI-based approach impossible. Roughly speaking one can not give a constructive description of the set $\mathcal{K}$ of admissible gain matrices in the form of a unique LMI. The restrictions set $\mathcal{K}_t$ in (3.5) evidently has a dynamic structure and contains time-dependent admissible matrices. A possible solution of the optimization problem (8.29) with the time-dependent restriction $\mathcal{K}_t$ evidently involves a very massive calculation associated with every time instant.

The critical analysis of the conventional AEM implies the necessity to generalize the usual AEM and to develop a conceptually new approach to the parameter selection of an AE (the ellipsoid matrix P). The same observation is also true with respect to the resulting feedback control design, namely, to the gain matrix $K(t)$. In this section we follow a "guaranteed" approach and select (a priori) a suitable matrix $\hat{P}$ that involves a relative "small" ellipsoidal invariant set $\mathcal{E}$ for the closed-loop linear system (8.22).

Theorem 8.18. *For a given (admissible) ellipsoid matrix*

$$\hat{P}$$

with

$$\hat{P} = \hat{P}^T > 0$$

associated with the linearized system (8.22) *the suitable gain matrix*

$$K(t) \in \mathcal{K}_t, \ t \in \mathbb{R}_+,$$

can be found as solutions of the following LMI:

$$\hat{P}[f_x(t, x^u(t)) + B(t)K(t)] + \sqrt{M}||\hat{P}|| \times E \leq 0, \tag{8.30}$$

where E is an $n \times n$-dimensional unit matrix.

Proof. The first summand in (8.30) evidently coincides with the matrix of the first summand in the integrand in (8.28). Let us now estimate the expression

$$y^T(t)P\xi(t)$$

from (8.28). We get

$$(y^T(t)P\xi(t))^2 = y^T(t)P\xi(t)\xi^T(t)Py \leq$$
$$||\hat{P}||^2 y^T(t)\xi(t)\xi^T(t)y(t).$$

For the Frobenius norm

$$||\xi(t)\xi^T(t)||_{Fr}$$

of the matrix $\xi(t)\xi^T(t)$ we obtain

$$||\xi(t)\xi^T(t)||_{Fr} = ||\xi(t)||^2 \leq \sup_{t \in \mathbb{R}_+} ||\xi(t)||^2 \leq M.$$

Therefore,

$$(y^T(t)P\xi(t))^2 \leq y^T(t)(M||\hat{P}||^2 E)y(t)$$

and the matrix condition (8.30) implies the integral inequality (8.28). Using (8.28) and the above theorems, we deduce that the obtained LMI (8.30) determines the admissible gain matrices

$$K(t) \in \mathcal{K}_t.$$

By definition of the set $\mathcal{K}_t$ the above inclusion guarantees invariance of the AE $\mathcal{E}$ for system (8.22). The proof is completed. □

Results for linear systems obtained above can now be applied to the robust control design of the originally given nonlinearly affine control system (8.19). Using the feedback input $w(\cdot, \cdot)$ implemented in the linearized system (8.22), we consider the following simple control strategy associated with the initial dynamics (8.19):

$$\begin{aligned} u(t) &:= u^{ref}(t) + w(t, y^v(t)), \\ w(t, y^v(t)) &= K(t) y^v(t), \end{aligned} \tag{8.31}$$

where $y^v(\cdot)$ is a solution to the linearized system (8.22) and $u^{ref}(\cdot)$ is a selected reference (tracking) control that corresponds to a reference trajectory $x^{ref}(\cdot)$ of (8.19).

We assume that $u^{ref}(\cdot)$ is an essentially bounded measurable function from $\mathcal{U}$. This assumption evidently guarantees the admissibility

$$u(\cdot) \in \mathcal{U}.$$

An absolutely continuous solution of the initial system (8.19) generated by $u^{ref}(\cdot)$ is denoted by $x^{ref}(\cdot)$. We next choose a "combined" control input $u(\cdot)$ in (8.31) such that the closed-loop variant of system (8.19),

$$\begin{aligned} \dot{x}(t) &= f(t, x(t)) + B(t)u^{ref}(t) + B(t)K(t)y^v(t) + \xi(t), \\ x(0) &= x_0, \end{aligned} \tag{8.32}$$

possesses the required robustness property. This basically means that (8.32) admits an AE, i.e.,

$$\mathcal{E}_0, \; 0 \in \mathcal{E}_0.$$

One can also use here the meaningful concept of "practical stability" as an alternative terminology. Note that the feedback-type control design (8.31) depends on the state vector of system (8.22). Our aim is to establish the practical stability (in the AE framework) of the original nonlinear system (8.19) using the robust control design developed for the linearized model (8.22). For a given admissible matrix $\hat{P}$ we next give a constructive estimation of the AE $\mathcal{E}_0$ associated with the nonlinearly ASS (8.19).

Theorem 8.19. *Consider a system* (2.1) *that satisfies all the basic assumptions from* Section 8.2 *and the corresponding closed-loop realization* (8.32). *Assume that matrices* $\hat{P}$ *and* $K(t)$ *are determined as mentioned above. Let the reference trajectory* $x^{ref}(\cdot)$ *be uniformly bounded, i.e.,*

$$||x^{ref}(t)|| \leq \chi, \; t \in \mathbb{R}_{+}.$$

Then the AE $\mathcal{E}_0$ *associated with the closed-loop version* (8.32) *of system* (8.19) *admits the following estimation:*

$$\mathcal{E}_0 := (||K(\cdot)|| + 1)\, \mathcal{E} + \chi. \tag{8.33}$$

Proof. The fundamental theorems above imply the following simple estimation:

$$||x^u(\cdot) - (x^{ref}(\cdot) + y(\cdot))||_{\mathbb{L}_n^{\infty}} \leq o(||K(\cdot)y(\cdot)||_{\mathbb{L}_m^{\infty}}).$$

The last inequality leads to the following result:

$$\begin{aligned} &||x^u(\cdot) - x^{ref}(\cdot)||_{\mathbb{L}_m^{\infty}} \leq o(||K(\cdot)y^v(\cdot)||_{\mathbb{L}_m^{\infty}}) + \\ &||y(\cdot)||_{\mathbb{L}_n^{\infty}} \leq (||K(\cdot)|| + 1)||y(\cdot)||_{\mathbb{L}_n^{\infty}}. \end{aligned}$$

Finally we deduce

$$\begin{aligned} &||x^u(\cdot)|| \leq (||K(\cdot)|| + 1)||y(\cdot)||_{\mathbb{L}_n^{\infty}} + \\ &||x^{ref}(\cdot)||_{\mathbb{L}_m^{\infty}} \leq (||K(\cdot)|| + 1)||y(\cdot)||_{\mathbb{L}_n^{\infty}} + \chi, \end{aligned} \tag{8.34}$$

where $t \in \mathbb{R}_{+}$.

Since $x^{ref}(\cdot)$ is assumed to be bounded and $\mathcal{E}$ is an invariant ellipsoid for (8.22), the inequality (8.34) implies the required estimation (8.33). From (8.34) we also deduce the invariance property of the ellipsoidal set

$$\mathcal{E}_0$$

for the closed-loop system (8.32). The proof is completed. □

Theorem 8.19 makes it possible to characterize an invariant ellipsoidal set

$$\mathcal{E}_0$$

associated with the sophisticated nonlinearly affine system (8.19). We use the minimal-size invariant ellipsoid $\mathcal{E}$ for this purpose. Evidently, the robust control design $w(t, y)$ for the linearized system (8.22) has a simple feedback-type structure. Theorem 8.19 opens up the possibility to use this simplified linear control design for robust control of the sophisticated nonlinear system (8.19). Note that the size of the ellipsoid $\mathcal{E}_0$ depends on the size of $\mathcal{E}$ and in fact on an adequate choice of the matrix $\hat{P}$.

In our opinion the AEM-based practical stabilization theory presented above constitutes an adequate analytical tool for studying the stability and stabilization problems associated with the relaxed HSs and SSs with affine structures. This class of challenging problems can give a new impulse to the classical Lyapunov-based stability theory for modern control systems. Let us finally note that this family of problems (stability of relaxed HSs and SSs) constitutes nowadays an absolutely new research area.

8.7 Notes

The RT and related aspects (for example, computational aspects) for HOCPs and SOCPs constitute a relative new theoretical methodology. That is the reason we cannot present here a wide list of the existing works. In this section we have included some of the papers and books that are relevant for applications of the RT-based analytical and numerical techniques to optimization of hybrid and switched dynamic systems [61,64,68,81,266,267,301,317–320,335].

CHAPTER 9

Conclusion and Perspectives

As we can see through this book (see the main Definitions 1.1, 1.2, 1.3, 1.4, 4.7, 4.8, 6.4, 6.6, 6.7, 6.8, 6.9 and systems (6.6), (6.7)) the HSs and SSs under consideration are conceptually different dynamic systems. The same is also true with respect to the OCPs generated by these systems (for example, HOCP (1.1) and SOCP (1.2) from Chapter 1, Chapter 6, and Chapter 7). As stated in Chapter 1 the switching times in an HS are determined by some switching manifolds in the state space $\mathbb{R}^n$. Roughly speaking the switching times of an HS are functions of states (for example, $t_i = t_i(x(\cdot))$ for a system as in Definition 1.2). The above observation implies that the switching times for HSs depend on the a posteriori information about the given system and cannot be determined before the definition of the causal (controlled) dynamics. As mentioned this situation is similar to the methodology of the feedback control, namely, to the state-dependent control inputs, in a conventional control system. In stark contrast to the HS formalism discussed above, the SS concept involves a time-dependent switching mechanism. From the above "comparison" we can deduce a methodological similarity between the "time-triggered" SSs and open-loop systems abstractions (time-dependent control inputs) in the classic control theory.

The state dependence of the switching mechanism in HSs implies many mathematical consequences for this class of dynamical models. For example, the formal proof of the hybrid Pontryagin maximum principle (see Chapter 8) is technically more complex compared with the classic case. Moreover, the proposed generalizations of the conventional RT techniques for HSs (see Chapter 7) constitute a mathematically difficult task.

Let us recall that the switching times τ in an HS are a part of the state output ("hybrid trajectory") $\mathcal{X}^u$ introduced in Definition 1.3. Consider a generic HS from Definition 1.1 and an admissible control function $u(\cdot) \in \mathcal{U}$ that generates the associated hybrid trajectory $\mathcal{X}^u$. For the intervals of the type $[t_{i-1}, t_i]$, where $t_i \in \tau$, we have introduced the following characteristic function (see Chapter 1):

$$\beta_{[t_{i-1},t_i)}(t) = \begin{cases} 1 & \text{if } t \in [t_{i-1}, t_i), \\ 0 & \text{otherwise.} \end{cases}$$

Recall that the formal (mathematical) expression of the above conceptual observations we have mentioned is the following dependency (on the state $x^u(\cdot)$) of the characteristic function:

$$\beta_{[t_{i-1}(x(\cdot)),t_i(x^u(\cdot)))}(t).$$

A Relaxation-Based Approach to Optimal Control of Hybrid and Switched Systems
https://doi.org/10.1016/B978-0-12-814788-7.00015-1

This sophisticated nonlinear (implicit) dependency evidently affects all the usual first-order techniques from the classic control and systems theory. However, many authors exclude this fact from the consideration of concrete classes of HSs. This exclusion has grave consequences. First of all it implies a formal change of the abstract modeling approach. Secondly this "replacement" can provoke an inclusion of the general HSs into the tighter class of dynamic systems, namely, into the class of the conventional ODEs involved in control systems. In this chapter we give an example of such an "inclusion" of HSs into the conventional control systems and analyze the obtained formal consequences.

We illustrate the above contradictory situation with the example of results discussed in [142, 143]. Let us present the main definitions and the system replacement proposed in this work.

The authors of [143] consider an ordered sequence of positive numbers

$$t_0 < t_1 < \cdots < t_v$$

and introduce the corresponding intervals

$$\Delta_k := [t_{k-1}, t_k], \ k = 1, ..., v \in \mathbb{N}.$$

This sequence is not determined a priori and is similar to the switching time sequence τ introduced in our book. Moreover, a collection $\{x_k\}$ of continuous functions

$$x_k : \Delta_k \to \mathbb{R}^{n_k}, \ k = 1, ..., v,$$

is determined. This collection generates the associated set of vectors

$$p := (t_0, (t_1, x_1(t_0), x_1(t_1)), (t_2, x_2(t_1), x_2(t_2)), ..., (t_v, x_v(t_{v-1}), x_v(t_v))).$$

The hybrid-type OCP in [143] is next stated as follows (we preserve here the original notation of [142,143]):

$$\begin{aligned}
&\text{minimize } J := \phi(p) \\
&\text{subject to} \\
&\dot{x}_k = f_k(t, x_k, u_k), u_k \in U_k, \\
&t \in \Delta_k, \ k = 1, ..., v, \\
&\eta_j(p) = 0, \ j = 1, ..., q, \\
&\phi_i(p) \leq 0, \ i = 1, ..., m,
\end{aligned} \tag{9.1}$$

where $x_k \in R^{n_k}$, $u_k \in R^{r_k}$. Additionally it is assumed that the control inputs $u_k(t)$ are measurable and essentially bounded on the intervals Δ_k. Additionally, the authors of [142] introduce some usual technical "smoothness" assumptions for problem (9.1). Finally the "hybrid"-type

OCP (9.1) is similar to the corresponding HOCPs and SOCPs from Chapters 1, 7, and 8. Note that in our book we have restricted our consideration to the case

$$p = x_v(t_v)$$

and do not consider the possible dependence of the objective functional on the switching times.

Let us now analyze the main dynamic systems concept proposed by the authors.

Definition 9.1. *The tuple* $W = (t_0; t_k, x_k(t), u_k(t), \ k = 1, ..., v)$ *is called an admissible process in (9.1) if it satisfies all the constraints and for every* $k = 1, ..., v$ *there exists a compact set* Ω_k *such that*

$$(t, x_k(t), u_k(t)) \in int\{\Omega_k\}$$

a.e. on Δ_k.

Note that the existence of a specific compact set Ω_k for each index k and the technical inclusion condition are required (in Definition 9.1) taking into consideration the future use of variations. These variations are finally introduced for the formal proof of the specific "hybrid maximum principle" the authors proposed. Let us note that the authors in fact consider some functions $x(\cdot)$, $u(\cdot)$, that take values x_k, u_k, respectively, for $t \in (t_{k-1}, t_k)$, where $k = 1, ..., v$. For the internal time instants

$$t_k, \ k = 1, ..., v - 1,$$

they admit that two possible values:

$$x(t_k - 0) = x_k(tk)$$

and

$$x(t_k + 0) = x_{k+1}(t_k)$$

for the trajectory $x(\cdot)$. For the measurable control function $u(\cdot)$ this uncertainty is inessential. An admissible process (in the sense of Definition 9.1) W can be now written as

$$W = (\theta, x(\cdot), u(\cdot)),$$

for

$$\theta := \{t_0, t_1, ..., t_v\}.$$

As one can see from the last formal observation the concept of an admissible process from [143] is equivalent to the concept of a "hybrid trajectory"

$$\mathcal{X}^u := (\tau, x(\cdot), \mathcal{R})$$

proposed in this book (see Chapter 1, Definition 1.3).

Let us now continue our analysis and present the concrete optimality concepts used in [143]. Consider an admissible process W from the basic Definition 9.1. The generic optimality definitions from Chapter 6 can also be specified in the case of systems from Definition 9.1. Let us present these concepts.

Definition 9.2. *An admissible process*

$$W^{opt} := (\theta^{opt}, x^{opt}(\cdot), u^{opt}(\cdot))$$

is called globally optimal in (9.1) if

$$J(W^{opt}) \leq J(W)$$

for any admissible process W.

Definition 9.3. *We say that a process $W^{opt} := (\theta^{opt}, x^{opt}(\cdot), u^{opt}(\cdot))$ determined on a time interval*

$$\Delta^0 := [t_0^0, t_\upsilon^0]$$

gives a strong minimum in (9.1) if there exists an $\epsilon > 0$ such that for any admissible process W defined on a time interval $[0, t_\upsilon]$ and satisfying the conditions

$$\begin{aligned} &||x_k^{opt} - x_k||_{\mathbb{C}} < \epsilon, \ \forall k = 1, ..., \upsilon, \\ &|t_0 - t_k| < \epsilon \ \forall k = 0, ..., \upsilon, \end{aligned}$$

we have $J(W^{opt}) \leq J(W)$.

Definition 9.4. *We say that a process W^{opt} gives a* Pontryagin *minimum in problem (9.1) if for any constant N there exists an*

$$\epsilon = \epsilon(N) > 0$$

such that for any admissible process $W = (\theta, x(\cdot), u(\cdot))$ that satisfies

$$\begin{aligned} &||x_k^{opt} - x_k||_{\mathbb{C}} < \epsilon, \ \forall k = 1, ..., \upsilon, \\ &|t_0 - t_k| < \epsilon \forall k = 0, ..., \upsilon, \\ &||u_k^{opt} - u_k||_{\mathbb{L}_1} < \epsilon, \ ||u_k^{opt} - u_k||_{\mathbb{L}_\infty} \leq N \ \forall k = 1, ..., \upsilon, \end{aligned}$$

we have $J(W^{opt}) \leq J(W)$.

Let us repeat that the above definitions are similar to the concepts introduced in Chapter 6 for HSs and SSs. If a process gives the global minimum, then it gives a strong minimum, and if W^{opt} is a strong minimum, then it is also a Pontryagin minimum. The authors in [142,143] claim that the main "hybrid-type" problem (9.1) can be reduced to a conventional (canonical) OCP for a Pontryagin type on a fixed time interval $[0, T]$. The Pontryagin OCP mentioned above has the following simple form:

$$\begin{aligned}
&\text{minimize } J := \phi(p) \\
&\text{subject to} \\
&\dot{x} = f(x, u), \\
&u \in U, \ (x, u) \in \mathcal{Q} \supset \Omega, \\
&\eta_j(p) = 0, \ j = 1, ..., q, \\
&\phi_i(p) \leq 0, \ i = 1, ..., m.
\end{aligned} \tag{9.2}$$

Following the original notation of the paper we are analyzing let us define $p := (x(0), x(T)) \in R^{2n}$ as a vector of terminal values of the trajectory $x(\cdot)$. Moreover, $\mathcal{Q}$ is an open set of the Euclidean space of a suitable dimension. The authors emphasize that the above condition $(x, u) \in \mathcal{Q}$ should not be regarded as a constraint, but as a definition of an open domain where the problem is defined. Evidently, problem (9.2) is a special case of (9.1) for $v = 1$, i.e., when the intermediate points are absent. It is natural to suppose that elements of problem (9.2) satisfy assumptions similar to the technical (smoothness) assumptions assumed for problem (9.1).

The main aim of [143] is to pass from the "hybrid" problem (9.1) to an OCP problem of the classic type (9.2) and, moreover, to establish a correspondence between the admissible and optimal processes in these two problems. They try to reduce all the state and control variables to a "common fixed time interval" (for example, to [0, 1]). For this aim a specific "time replacement" is proposed. Let us analyze this "equivalence" approach in more details.

Assume $(\theta, x(\cdot), u(\cdot))$ is an arbitrary admissible process for problem (9.1). The authors of [143] introduce a "new time" $s \in [0, 1]$ and determine functions

$$\rho_k : [0, 1] \to \Delta_k, \ k = 1, ..., v,$$

from the following differential equations:

$$\frac{d\rho_k}{ds} = z_k(s), \ \rho_k(0) = t_{k-1}. \tag{9.3}$$

Here $z_k(s) > 0$ are arbitrary measurable essentially bounded functions on the time interval $[0, 1]$ characterized by the following condition:

$$\rho_k(1) = t_k, \ \int_0^1 z_k(s)ds = |\Delta_k|.$$

The newly introduced functions ρ_k play the role of the "conventional" time t on the corresponding intervals Δ_k. Define also the following combined functions:

$$y_k(s) = x_k(\rho_k(s)), \; v_k(s) = u_k(\rho_k(s)), \; k = 1, ..., v,$$

for $s \in [0, 1]$. The new "state" and "control" functions naturally satisfy the differential equations, i.e.,

$$\begin{aligned} &\frac{dy_k}{ds} = z_k f(\rho_k, y_k, v_k), \; v_k \in U_k, \\ &d\rho_k ds = zk, k = 1, ..., v, \\ &(\rho_k, y_k, v_k) \in Q_k, \; z_k > 0. \end{aligned} \tag{9.4}$$

Moreover, the "new time" ρ_k satisfies the natural "boundary conditions"

$$\begin{aligned} &\rho_{k+1}(0) - \rho_k(1) = 0, \; k = 1, ..., v - 1, \\ &\eta_j(\hat{p}) = 0, \; j = 1, ..., q, \\ &\phi_i(\hat{p}) \leq 0, \; i = 1, ..., m, \end{aligned} \tag{9.5}$$

where vector $\hat{p}$ is defined similar to p with replacement x by y and is adopted to the normalized time interval [0, 1] under consideration. Using the time replacement introduced above and the corresponding redefined state and control variables, the rewritten version of the originally given OCP (9.1) has the following form:

$$\begin{aligned} &\text{minimize } J := \phi(p) \\ &\text{subject to} \\ &(9.4)\text{–}(9.5). \end{aligned} \tag{9.6}$$

Note that the specific OCP (9.6) is formally determined on the set of admissible processes

$$\tilde{W} := (\rho(\cdot), y(\cdot), v(\cdot), z(\cdot)).$$

Following the formal definitions of the new time, state, and control the authors establish some expected correspondences between the originally given OCP (9.1) and the transformed problem (9.6). That means any admissible process $W = (\theta, x(\cdot), u(\cdot))$ of problem (9.1) can be transformed to an admissible process

$$\tilde{W} = (\rho(\cdot), y(\cdot), v(\cdot), z(\cdot))$$

of (9.6). We fully agree with this statement of the authors. Moreover, an important property of both transformation mappings (see [143]) is that they preserve the value of the cost functional. The authors construct a mapping F that transforms process W into an admissible process $\tilde{W}$

and also a mapping G that transforms $\tilde{W}$ into W. Using the properties that there exist two mappings which transform any admissible process of one problem into an admissible process of another problem with the same value of the cost functional, the authors deduce one of the main results of the paper presented below.

Theorem 9.1. *If a process W^{opt} is optimal (i.e., globally minimal) in problem (9.1), then the process $\tilde{W}^{opt} = F(W^{opt})$ is optimal in (9.6), and vice versa, if a process $\tilde{W}^{opt}$ is optimal in problem (9.6), then the process*

$$W = G(W^{opt})$$

is optimal in problem (9.1).

The second (stronger) result of the paper [143] is summarized in the next theorem.

Theorem 9.2. *If a process W^{opt} gives a strong (Pontryagin) minimum in problem (9.1), then the process*

$$\tilde{W}^{opt} = F(W^{opt})$$

gives a strong (respectively, Pontryagin) minimum in (9.6), and vice versa, if a process $\tilde{W}^{opt}$ gives a strong (Pontryagin) minimum in (9.6), then the process

$$W^{opt} = G(w\tilde{W}^{opt})$$

gives a strong (respectively, Pontryagin) minimum in problem (9.1).

Finally, the authors of [142,143] claim that the study of optimality (in the sense of Definition 9.2, 9.3, or 9.4) of a process W^{opt} in the initially given "hybrid-type" problem (9.1) reduces to the study of optimality of the corresponding process $\tilde{W}^{opt}$ in the transformed problem (9.6).

Our aim now is to show that the above statement is incorrect for the wide classes of HSs studied in this book. First let us consider the proposed time transformation (9.3) in detail and try to apply it to a concrete HS from Definitions 1.1–1.3 (see Chapter 1).

Consider an HS from Definitions 1.1–1.3 and the "hybrid system" from (9.1). In that case every time instant t_k, $k = 1, ..., v$, is a functional $t_k(x(\cdot))$ of the trajectory $x(\cdot)$ (see Definition 1.2, Chapter 1). The same is also true for the time interval

$$\Delta_k = \Delta_k(x(\cdot))$$

determined above. Note that the natural number $v \in \mathbb{N}$ in the above formalism applied to the HS from Definitions 1.1–1.3 also depends on the switching mechanism. Taking into consideration the initial condition in (9.3) we now can write

$$\rho_k(0)[x(\cdot)] = t_{k-1}(x(\cdot)),\ k = 1, ..., v.$$

Moreover,

$$\rho_k(1)[x(\cdot)] = t_k(x(\cdot)),\ \int_0^1 z_k(s)ds = |\Delta_k(x(\cdot))|$$

and we obtain the (implicit) dependence of the introduced times $\rho_k(s),\ k = 1, ..., v$, on the hybrid trajectory $x(\cdot)$ of the system from (9.1). This dependence is a simple consequence of (9.3). We have

$$\rho_k(s)[x(\cdot)] = \rho_k(0) + \int_0^s z_k(\xi)d\xi.$$

The above fact, namely, the implicit dependence of the newly introduced time in (9.3) on the initial hybrid trajectory $x(\cdot)$ makes it impossible to follow correctly with the formalism (9.4) and adequately describe (by an ODE) the "new state" $y(\cdot)$. Summarizing we can decide that the time replacement proposed in [143] cannot be applied to the HSs discussed in this book. The class of HSs we studied is strictly characterized by the state triggered location transitions.

However, the system transformation proposed in [143] can smoothly be applied to the wide classes of SSs. We refer to Chapters 1, 6, 7, and 8 for the necessary concepts. Roughly speaking the auxiliary system representation developed in [143] is an adequate mathematical tool for the dynamic systems with switches triggered by time. This statement can be easily illustrated with the help of an example from [143]. The above example falls into the formal concept of a SS proposed in our book (see, e.g., Chapter 6).

The analysis of the particular highly professional paper presented above makes the possible conceptual errors and misunderstandings related to the various theoretical aspects of hybrid and switched control systems clear. Let us complete this book by completing some perspectives of the RT we presented. From our point of view the RT for hybrid and switched dynamic systems has the following useful features:

- RT constitutes mathematically rigorous foundations of the self-closed existence theory in hybrid and switched OCPs;
- it is involved in the numerical analysis associated with the dynamic optimization of HSs and SSs;
- a suitable relaxation step can be an important part of the concrete implementable algorithms for a constructive numerical treatment of HOCPs and SOCPs.

Except for the existing hybrid versions of the PMM, the further development of the necessary and sufficient optimality conditions for the relaxed hybrid and switched dynamic systems is in fact an open question. The same is also true with respect to the corresponding computational extensions of these (expected) optimality conditions.

Let us finally note that also the usual questions related to the qualitative theory for dynamic systems (involving ODEs), for example, the celebrated Lyapunov stability aspects in the context of the relaxed hybrid and switched control systems, constitute a very interesting research area. One can easily see that a Lyapunov stable control design applied to a relaxed HS also implies the stable dynamic behavior of the initially given (nonrelaxed) hybrid model. This fact opens the door for an effective use of the relaxed HSs and SSs in the stable and "practically stable" (for example, in the sense of [267]) control design procedures for complex engineering systems.

Bibliography

A

[1] R.A. Adams, Sobolev Spaces, Academic Press, New York, 1973.
[2] A. Ahmed, E.I. Verriest, Nonlinear systems evolving with state suprema as multi-mode multi-dimensional systems: analysis and observation, in: Proceedings of the 5th IFAC Conference on Analysis and Design of Hybrid Systems, Atlanta, USA, 2015, pp. 242–247.
[3] A. Ahmed, E.I. Verriest, Estimator design for a subsonic rocket car (soft landing) based on state-dependent delay measurement, in: Proceedings of the 52nd IEEE Conference on Decision and Control, Florence, Italy, 2013, pp. 5698–5703.
[4] Y.I. Alber, R.S. Burachik, A.N. Iusem, A proximal point method for nonsmooth convex optimization problems in Banach spaces, Abstract and Applied Analysis 2 (1997) 97–120.
[5] V.M. Alekseev, V.M. Tikhomirov, S.V. Fomin, Optimal Control, Plenum Publishing Co., New York, 1987.
[6] C.D. Aliprantis, K.C. Border, Infinite Dimensional Analysis, Springer, New York, 1999.
[7] R. Alur, D.L. Dill, Automata for modeling real-time systems, in: Lecture Notes in Computer Science, vol. 443, Springer, Berlin, 1990, pp. 32–335.
[8] F. Alvarez, On the minimizing property of a second order dissipative system in Hilbert spaces, SIAM Journal on Control and Optimization 38 (2000) 1102–1119.
[9] F. Amato, Robust Control of Linear Systems Subject to Time-Varying Parameters, Springer, Berlin, 2005.
[10] A.D. Ames, Y. Or, Stability and completion of Zeno equilibria in Lagrangian hybrid systems, IEEE Transactions on Automatic Control 56 (2011) 1322–1336.
[11] E.J. Anderson, P. Nash, Linear Programming in Infinite-Dimensional Spaces, Wiley, Chichester, 1987.
[12] D. Angeli, E.D. Sontag, Monotone control systems, IEEE Transactions on Automatic Control 48 (2003) 1684–1698.
[13] G. Anger, Inverse and Improperly Posed Problems in Differential Equations, Akademie-Verlag, Berlin, 1979.
[14] L. Angermann, A posteriori error estimates for approximate solutions of nonlinear equations with weakly stable operators, Numerical Functional Analysis and Optimization 5 (1997) 447–459.
[15] N. Aoki, K. Hiraide, Topological Theory of Dynamical Systems, North-Holland, Amsterdam, 1994.
[16] L. Armijo, Minimization of functions having Lipschitz continuous first partial derivatives, Pacific Journal of Mathematics 16 (1966) 1–3.
[17] A.V. Arutyunov, S.M. Aseev, Investigation of the degeneracy phenomenon in the Maximum Principle for optimal control with state constraints, SIAM Journal on Control and Optimization 35 (1997) 930–952.
[18] E. Asplund, R. Rockafellar, Gradients of convex functions, Transactions of the American Mathematical Society 139 (1969) 443–467.
[19] S.A. Attia, V. Azhmyakov, J. Raisch, On an optimization problem for a class of impulsive hybrid systems, Discrete Event Dynamic Systems 20 (2010) 215–231.
[20] K. Atkinson, W. Han, Theoretical Numerical Analysis, Springer, New York, 2009.
[21] H. Attouch, G. Buttazzo, G. Michaille, Variational Analysis in Sobolev and BV Spaces: Applications to PDEs and Optimization, SIAM and MPS, Philadelphia, 2006.

[22] J.P. Aubin, A. Cellina, Differential Inclusions. Set-Valued Maps and Viability Theory, Springer, Berlin, 1984.

[23] A. Auslender, Numerical methods for nondifferentiable convex optimization, Mathematical Programming Study 30 (1987) 102–126.

[24] A. Auslender, J.P. Crouzeix, P. Fedit, Penalty proximal methods in convex programming, Journal of Optimization Theory and Applications 55 (1998) 1–21.

[25] A. Auslender, Regularity theorems in sensitivity theory with nonsmooth data, in: J. Guddat, H.Th. Jongen, B. Kummer, F. Nozicka (Eds.), Parametric Optimization and Related Topics, Akademie-Verlag, Berlin, 1987, pp. 9–15.

[26] V. Azhmyakov, A constructive method for solving stabilization problems, Discussiones Mathematicae Differential Inclusions, Control and Optimization 20 (2000) 51–62.

[27] V. Azhmyakov, Minimax robust estimation and control in stochastic systems, Stability and Control: Theory and Applications 3 (2000) 253–262.

[28] V. Azhmyakov, W.H. Schmidt, Strong convergence of a proximal-based method for convex optimization, Mathematical Methods of Operations Research 57 (2003) 393–407.

[29] V. Azhmyakov, W.H. Schmidt, Explicit approximations of relaxed optimal control processes, in: W.H. Schmidt, G. Sachs (Eds.), Optimal Control, Hieronymus Bücherproduktion GmbH, München, 2003, pp. 179–192.

[30] V. Azhmyakov, A numerical method for optimal control problems using proximal point approach, Aportaciones Matematicas 18 (2004) 13–31.

[31] V. Azhmyakov, A numerically stable method for convex optimal control problems, Journal of Nonlinear and Convex Analysis 5 (2004) 1–18.

[32] V. Azhmyakov, W.H. Schmidt, Approximations of relaxed optimal control problems, Journal of Optimization Theory and Applications 130 (2006) 61–78.

[33] V. Azhmyakov, J. Raisch, A gradient-based approach to a class of hybrid optimal control problems, in: Proceedings of the 2nd IFAC Conference on Analysis and Design of Hybrid Systems, Alghero, Italy, 2006, pp. 89–94.

[34] V. Azhmyakov, Optimal control of hybrid and switched systems, in: Proceedings of the IX International Chetaev Conference, Irkutsk, Russia, 2007, pp. 308–317.

[35] S.A. Attia, V. Azhmyakov, J. Raisch, State jump optimization for a class of hybrid autonomous systems, in: Proceedings of the 2007 IEEE Conference on Control Applications, Singapore, 2007, pp. 1408–1413.

[36] V. Azhmyakov, J. Raisch, Convex control systems and convex optimal control problems, IEEE Transactions on Automatic Control 53 (2008) 993–998.

[37] V. Azhmyakov, A. Poznyak, V.G. Boltyanski, On the dynamic programming approach to multi-model robust optimal control problems, in: Proceedings of the 2008 American Control Conference, Seattle, USA, 2008, pp. 4468–4473.

[38] V. Azhmyakov, A. Poznyak, V.G. Boltyanski, First order optimization techniques for impulsive hybrid dynamical systems, in: Proceedings of the 2008 International Workshop on Variable Structure Systems, Antalya, Turkey, 2008, pp. 173–178.

[39] V. Azhmyakov, V.G. Boltyanski, A. Poznyak, Optimal control of impulsive hybrid systems, Nonlinear Analysis: Hybrid Systems 2 (2008) 1089–1097.

[40] V. Azhmyakov, R. Galvan-Guerra, A. Polyakov, On the method of dynamic programming for linear-quadratic problems of optimal control in hybrid systems, Automation and Remote Control 70 (2009) 787–799.

[41] V. Azhmyakov, R. Galvan-Guerra, M. Egerstedt, Hybrid LQ-optimization using dynamic programming, in: Proceedings of the 2009 American Control Conference, St. Louis, USA, 2009, pp. 3617–3623.

[42] V. Azhmyakov, M. Egerstedt, L. Fridman, A. Poznyak, Continuity properties of nonlinear affine control systems: applications to hybrid and sliding mode dynamics, in: Proceedings of the 3rd IFAC Conference on Analysis and Design of Hybrid Systems, Zaragoza, Spain, 2009, pp. 204–209.

[43] V. Azhmyakov, R. Galvan-Guerra, R. Velazquez, A. Poznyak, On the optimal design of linear networked systems, in: Proceedings of the 2nd IFAC Workshop on Dependable Control of Discrete Systems, Bari, Italy, 2009, pp. 239–244.
[44] V. Azhmyakov, S.N. Morales, Proximal point method for optimal processes governed by ordinary differential equations, Asian Journal of Control 12 (2010) 15–25.
[45] V. Azhmyakov, V.G. Boltyanski, A. Poznyak, The dynamic programming approach to multi-model robust optimization, Nonlinear Analysis: Theory Methods, Applications 72 (2010) 1110–1119.
[46] V. Azhmyakov, J. Raisch, R. Velazquez, An application of the proximal point algorithm to optimal control of affine switched systems, in: Proceedings of the 2010 IEEE Conference on Industrial Technology, Vina del Mar, Chili, 2010, pp. 1733–1738.
[47] V. Azhmyakov, R. Galvan-Guerra, A. Poznyak, On the hybrid LQ-based control design for linear networked systems, Journal of the Franklin Institute 347 (2010) 1214–1226.
[48] V. Azhmyakov, R. Velazquez, On a variational approach to optimization of hybrid mechanical systems, Mathematical Problems in Engineering 2010 (2010) 1–10.
[49] V. Azhmyakov, R. Velazquez, R. Galvan-Guerra, Numerically stable approximations of optimal control processes associated with a class of switched systems, in: Proceedings of the 10th International Workshop on Discrete Event Systems, Berlin, Germany, 2010, pp. 51–56.
[50] V. Azhmyakov, M. Egerstedt, L. Fridman, A. Poznyak, Approximability of nonlinear affine control systems, Nonlinear Analysis: Hybrid Systems 5 (2011) 275–288.
[51] V. Azhmyakov, A gradient type algorithm for a class of optimal control processes governed by hybrid dynamical systems, IMA Journal of Mathematical Control and Information 28 (2011) 291–307.
[52] V. Azhmyakov, M. Basin, The proximal point approach to optimal control of affine switched systems, in: Proceedings of the 18th IFAC World Congress, Milan, Italy, 2011, pp. 10249–10254.
[53] V. Azhmyakov, M. Tulio Angulo, Application of the strong approximability property to a class of affine switched systems and to relaxed differential equations with affine structure, International Journal of Systems Science 42 (2011) 1899–1907.
[54] V. Azhmyakov, M.V. Basin, J. Raisch, Proximal point based approach to optimal control of affine switched systems, Discrete Event Dynamic Systems 22 (2012) 61–81.
[55] V. Azhmyakov, F.A. Miranda Villatoro, Approximability and variational description of the Zeno behavior in affine switched systems, in: Proceedings of the 4th IFAC Conference on Analysis and Design of Hybrid Systems, Eindhoven, The Netherlands, 2012, pp. 307–312.
[56] V. Azhmyakov, M. Basin, A. Gil Garcia, A general approach to optimal control processes associated with a class of discontinuous control systems: applications to the sliding mode dynamics, in: Proceedings of the 2012 Multi-Conference on Systems and Control, Dubrovnik, Croatia, 2012, pp. 1154–1159.
[57] V. Azhmyakov, M.V. Basin, A.E. Gil Garcia, Optimal control processes associated with a class of discontinuous control systems: applications to sliding mode dynamics, Kybernetika 50 (2014) 5–18.
[58] V. Azhmyakov, A. Polyakov, A. Poznyak, Consistent approximations and variational description of some classes of sliding mode control processes, Journal of the Franklin Institute 351 (2014) 1964–1981.
[59] V. Azhmyakov, R. Rodriguez Serrezuela, L.A. Guzman Trujillo, Approximation based control design for a class of switched dynamic systems, in: Proceedings of the 40th Annual Conference of the IEEE Industrial Electronic Society, Dallas, USA, 2014, pp. 90–95.
[60] V. Azhmyakov, R. Rodriguez Serrezuela, A.M. Rios Gallardo, W. Gerardo Vargas, An approximation based approach to optimal control of switched dynamic systems, Mathematical Problems in Engineering 2014 (2014) 1–9.
[61] V. Azhmyakov, J. Cabrera Martinez, A. Poznyak, Optimization of a class of nonlinear switched systems with fixed-levels control inputs, in: Proceedings of the 2015 American Control Conference, Chicago, USA, 2015, pp. 1770–1775.
[62] V. Azhmyakov, R. Juarez, On the projected gradient methods for switched-mode systems optimization, IFAC-PapersOnLine 48 (2015) 181–186.

[63] V. Azhmyakov, R. Juarez, St. Pickl, On the local convexity of singular optimal control problems associated with the switched-mode dynamic systems, IFAC-PapersOnLine 48 (2015) 271–276.
[64] V. Azhmyakov, J. Cabrera, A. Poznyak, Optimal fixed-levels control for non-linear systems with quadratic costs functional, Optimal Control: Applications and Methods 37 (2016) 1035–1055.
[65] V. Azhmyakov, M.E. Bonilla, St. Pickl, L.A. Guzman Trujillo, Constructive approximations of the Zeno dynamics in affine switched systems: the projection based approach, in: Proceedings of the 2016 American Control Conference, Boston, USA, 2016, pp. 5175–5180.
[66] V. Azhmyakov, A. Ahmed, E.I. Verriest, On the optimal control of systems evolving with state suprema, in: Proceedings of the 55th IEEE Conference on Decision and Control, Las Vegas, USA, 2016, pp. 3617–3623.
[67] V. Azhmyakov, L.A. Guzman Trujillo, On the linear quadratic dynamic optimization problems with fixed-levels control functions, Italian Journal of Pure and Applied Mathematics 37 (2017) 219–237.
[68] V. Azhmyakov, R. Juarez, A first-order numerical approach to switched-mode systems optimization, Nonlinear Analysis: Hybrid Systems 25 (2017) 126–137.

B

[69] J. Baillieul, P.J. Antsaklis, Control and communication challenges in networked real-time systems, Proceedings of the IEEE 95 (2007) 9–28.
[70] C. Baiocchi, A. Capelo, Variational and Quasivariational Inequalities. Application to Free Boundary Problems, Wiley, New York, 1984.
[71] A.B. Bakushinskii, Solution methods for monotonous variational inequalities founded on the principle of iterative regularization, U.S.S.R. Computational Mathematics and Mathematical Physics 17 (1977) 12–24.
[72] A.V. Balakrishnan, L.W. Neustadt (Eds.), Conference on Computational Methods in Optimization Problems, Academic Press, New York, 1964.
[73] S.P. Banks, S.A. Khathur, Structure and control of piecewise-linear systems, International Journal of Control 50 (1989) 667–686.
[74] M.V. Basin, A. Ferrara, D. Calderon-Alvarez, Sliding mode regulator solution to optimal control problem, in: Proceedings of the 47th Conference on Decision and Control, Cancun, Mexico, 2008, pp. 2184–2189.
[75] H.H. Bauschke, J.M. Borwein, P.L. Combettes, Bregman monotone optimization algorithms, Preprint, 2002.
[76] H.H. Bauschke, J.M. Borwein, P.L. Combettes, Bregman monotone optimization algorithms, SIAM Journal on Control and Optimization 42 (2000) 596–636.
[77] M. Bebendorf, A note on the Poincaré inequality for convex domains, Zeitschrift für Analysis und ihre Anwendung 22 (2003) 751–756.
[78] R. Bellman, S.E. Dreyfus, Applied Dynamic Programming, Princeton University Press, Princeton, 1962.
[79] J.Y. Bello Cruz, C.W. de Oliveira, On weak and strong convergence of the projected gradient method for convex optimization in Hilbert spaces, arXiv, 2014, pp. 1–18.
[80] M.D. Benedetto, A. Sangiovanni-Vincentelli, Hybrid Systems Computation and Control, Springer, Heidelberg, 2001.
[81] S.C. Bengea, R. Decarlo, Optimal control of switching systems, Automatica 41 (2005) 11–27.
[82] H. Benker, A. Hamel, C. Tammer, A proximal point algorithm for control approximation problems, Mathematical Methods of Operations Research 43 (1996) 261–280.
[83] J. Benoist, J.B. Hiriart-Urruty, What is the subdifferential of the closed convex hull of a function?, SIAM Journal on Mathematical Analysis 27 (1996) 1661–1679.
[84] S.K. Berberian, Fundamentals of Real Analysis, Springer, New York, 1999.
[85] A. Bensoussan, J.L. Menaldi, Hybrid control and dynamic programming, Dynamics and Continuous, Discrete and Impulsive Systems 3 (1997) 395–442.
[86] L.D. Berkovitz, Optimal Control Theory, Springer, New York, 1974.
[87] D.P. Bertsekas, Constrained Optimization and Lagrange Multiplier Method, Academic Press, New York, 1982.

[88] D.P. Bertsekas, Dynamic Programming and Optimal Control, Athena Scientific, Belmont, 1995.
[89] J.T. Betts, Using Sparse Nonlinear Programming to Compute Low Trust Orbit Transfers, Boeing Computer Services Technical Report, 1992.
[90] J.T. Betts, Practical Methods for Optimal Control Using Nonlinear Programming, SIAM, Philadelphia, 2001.
[91] B. Birnir, G. Ponce, N. Svanstedt, The local ill-posedness of the modified KdV equation, Annales de l'Instut Henri Poincare, Anal. Non Lineare 13 (1996) 529–535.
[92] L. Bittner, New conditions for validity of the Lagrange multiplier rule, Mathematische Nachrichten 48 (1971) 353–370.
[93] L. Bittner, Necessary optimality conditions for a model of optimal control processes, in: C. Olech, E. Fidelis (Eds.), Mathematical Control Theory, Polish Scientific Publications, 1976, pp. 25–32.
[94] N.N. Bogoljubov, Sur quelnes methods nouvelles dans le calculus des variations, Annali di Matematica Pura ed Applicata 7 (1930) 249–271.
[95] I. Boiko, Discontinuous Control Systems Frequency-Domain Analysis and Design, Birkhäuser, New York, 2009.
[96] V. Boltyanski, A. Poznyak, The Robust Maximum Principle, Birkhäuser, New York, USA, 2012.
[97] M. Bonilla, N. Alvarez, M. Malabre, V. Azhmyakov, Internal stability of a class o switched systems designed by implicit control techniques, in: Proceedings of the 13th European Control Conference, Strasbourg, France, 2014, pp. 2254–2259.
[98] M. Bonilla, M. Malabre, V. Azhmyakov, Decoupling of internal variable structure for a class of switched systems, in: Proceedings of the 2015 European Control Conference, Linz, Austria, 2015, pp. 1890–1895.
[99] M. Bonilla, M. Malabre, V. Azhmyakov, An implicit systems characterization of a class of impulsive linear switched control processes: part 1: modelling, Nonlinear Analysis: Hybrid Systems 15 (2015) 157–170.
[100] M. Bonilla, M. Malabre, V. Azhmyakov, An implicit systems characterization of a class of impulsive linear switched control processes: part 2: control, Nonlinear Analysis: Hybrid Systems 18 (2015) 15–32.
[101] J.F. Bonnans, On an algorithm for optimal control using Pontryagin's maximum principle, SIAM Journal of Control and Optimization 24 (1986) 579–588.
[102] J.M. Borwein, A.S. Lewis, Convex Analysis and Nonlinear Optimization, Springer, New York, 2000.
[103] M.S. Branicky, V.S. Borkar, S.K. Mitter, A unified framework for hybrid control: model and optimal control theory, IEEE Transactions on Automatic Control 43 (1) (1998) 31–45.
[104] M.S. Branicky, S.M. Phillips, W. Zhang, Stability of networked control systems: explicit analysis of delay, in: Proceedings of the 2000 American Control Conference, Chicago, USA, 2000, pp. 2352–2357.
[105] M.S. Branicky, Introduction to Hybrid Systems, Birkhäuser, Boston, 2005.
[106] J.V. Breakwell, The optimization of trajectories, SIAM Journal 7 (1959) 215–247.
[107] O. Brezhneva, A.A. Tret'yakov, Optimality conditions for degenerate extremum problems with equality constraints, SIAM Journal on Control and Optimization 2 (2003) 729–745.
[108] R.W. Brockett, D. Liberzon, Quantized feedback stabilization of linear systems, IEEE Transactions on Automatic Control 45 (2000) 1279–1289.
[109] F.E. Browder, Convergence of approximants to fixed points of non-expansive non-linear mappings in Banach spaces, Archive for Rational Mechanics and Analysis 24 (1967) 82–90.
[110] A.E. Bryson, Y.C. Ho, Applied Optimal Control, Wiley, New York, 1975.
[111] A.E. Bryson, W.F. Denham, A steepest ascent method for solving optimum programming problems, Journal of Applied Mechanics 29 (1962) 247–257.
[112] R. Bulirsch, F. Montrone, H.J. Pesch, Abort landing in the presence of windshear as a minimax optimal control problem, part 1: necessary conditions, Journal of Optimization Theory and Applications 70 (1991) 1–23.
[113] R. Bulirsch, F. Montrone, H.J. Pesch, Abort landing in the presence of windshear as a minimax optimal control problem, part 2: multiple shooting and homotopy, Journal of Optimization Theory and Applications 70 (1991) 223–254.
[114] R.S. Burachik, L.M. Grana Drummond, A.N. Iusem, B.F. Svaiter, Full convergence of the steepest descent method with inexact line searches, Optimization 32 (1995) 137–146.

[115] R.S. Burachik, A.N. Iusem, A generalized proximal point algorithm for the variational inequality problem in a Hilbert space, SIAM Journal on Optimization 8 (1998) 197–216.
[116] R.S. Burachik, D. Butnariu, A.N. Iusem, Iterative methods for solving stochastic convex feasibility problems and applications, Computational Optimization and Applications 15 (2000) 269–307.
[117] R.S. Burachik, S. Scheimberg, A proximal point method for the variational inequality problem in Banach spaces, SIAM Journal on Control and Optimization 39 (2001) 1633–1649.
[118] C. Büskens, H. Maurer, SQP-methods for solving optimal control problems with control and state constraints: adjoint variables, sensitivity analysis and real-time control, Journal of Computational and Applied Mathematics 120 (2000) 85–108.
[119] D. Butnariu, A.M. Iusem, On a proximal point method for convex optimization in Banach spaces, Numerical Functional Analysis and Optimization 18 (1997) 723–744.
[120] G. Buttazzo, Semicontinuity, Relaxation and Integral Representation in the Calculus of Variations, Pitman Res. Notes Math. Ser., vol. 207, Longman, 1989.
[121] S. Butzek, W.H. Schmidt, Relaxation gaps in optimal control processes with state constraints, in: W.H. Schmidt, K. Heier, L. Bittner, R. Bulirsch (Eds.), Variational Calculus Optimal Control and Applications, Birkhäuser Basel, Basel, 1998, pp. 21–29.

C

[122] P.E. Caines, M.S. Shaikh, Optimality zone algorithms for hybrid systems computation and control: from exponential to linear complexity, in: Proceedings of the 13th Mediterranean Conference on Control and Automation, Limassol, Cyprus, 2005, pp. 1292–1297.
[123] P.E. Caines, M. Egerstedt, R. Malhame, A. Schoellig, A hybrid Bellman equation for bimodal systems, in: Lecture Notes in Computer Science, vol. 4416, Springer, Berlin, 2007, pp. 656–659.
[124] C. Cassandras, D.L. Pepyne, Y. Wardi, Optimal control of a class of hybrid systems, IEEE Transactions on Automatic Control 46 (2001) 398–415.
[125] C. Cassandras, S. Lafortune, Introduction to Discrete Event Systems, Springer, New York, 2008.
[126] Y. Censor, S.A. Zenios, The proximal minimization algorithm with D-functions, Journal of Optimization Theory and Applications 73 (1992) 451–464.
[127] L. Cesari, Optimization Theory and Applications, Springer, New York, 1983.
[128] F.L. Chernousko, A.A. Lyubuschin, Method of successive approximations for solution of optimal control problems, Optimal Control, Applications and Methods 3 (1982) 101–114.
[129] F.H. Clarke, The generalized problem of Bolza, SIAM Journal on Control and Optimization 14 (1976) 683–699.
[130] F.H. Clarke, Optimal solutions to differential inclusions, Journal of Optimization Theory and Applications 19 (1976) 469–478.
[131] F.H. Clarke, Optimization and Nonsmooth Analysis, SIAM, Philadelphia, 1990.
[132] F.H. Clarke, Yu.S. Ledyaev, R.J. Stern, P.R. Wolenski, Nonsmooth Analysis and Control Theory, Springer, New York, 1998.
[133] P. Colanery, R.H. Middleton, Z. Chen, D. Caporale, F. Blanchini, Convexity of the cost functional in an optimal control problem for a class of positive switched systems, Automatica 50 (2014) 1227–1234.
[134] L. Collatz, Differentialgleichungen, Teubner, Stuttgart, 1990.
[135] J. Cullum, An explicit procedure for discretizing continuous optimal control problems, Journal of Optimization Theory and Applications 8 (1979) 15–34.

D

[136] J.W. Daniel, On the convergence of a numerical method in optimal control, Journal of Optimization Theory and Applications 4 (1969) 330–342.

[137] V.F. Demyanov, A.M. Rubinov, Constructive Nonsmooth Analysis, Peter Lang, Frankfurt, 1995.
[138] S. Dharmatti, M. Ramaswamy, Hybrid control systems and viscosity solutions, SIAM Journal on Control and Optimization 44 (2005) 1259–1288.
[139] J. Dieudonné, Foundations of Modern Analysis, Academic Press, New York, 1960.
[140] X.C. Ding, Y. Wardi, D. Taylor, M. Egerstedt, Optimization of switched-mode systems with switching costs, in: Proceedings of the 2008 American Control Conference, Seattle, USA, 2008, pp. 3965–3970.
[141] X.C. Ding, Y. Wardi, M. Egerstedt, On-line optimization of switched-mode dynamical systems, IEEE Transactions on Automatic Control 54 (2009) 2266–2271.
[142] A. Dmitruk, A.M. Kaganovich, Maximum principle for optimal control problems with intermediate constraints, in: Nonlinear Dynamics and Control, vol. 6, Nauka, 2008 (in Russian).
[143] A. Dmitruk, A.M. Kaganovich, The hybrid Maximum Principle is a consequence of Pontryagin Maximum Principle, Systems and Control Letters 57 (2008) 964–970.
[144] A.L. Dontchev, F. Lempio, Difference methods for differential inclusions: a survey, SIAM Review 34 (1992) 263–294.
[145] A. Dontchev, T. Zolezzi, Well Posed Optimization Problems, Springer, Berlin, 1993.
[146] A. Dontchev, W.W. Hager, Lipschitz stability in nonlinear control and optimization, SIAM Journal on Control and Optimization 31 (1993) 569–603.
[147] A.L. Dontchev, Discrete approximations in optimal control, in: B.S. Mordukhovich, H.J. Sussmann (Eds.), Nonsmooth Analysis and Geometric Methods in Deterministic Optimal Control, Springer, New York, 1996, pp. 59–80.
[148] J.C. Dunn, On state constraint representations and mesh-dependent gradient projection convergence rates for optimal control problems, SIAM Journal of Control and Optimization 39 (2000) 1082–1111.
[149] J.C. Dunn, Diagonally modified conditional gradient method for input constrained optimal control problems, SIAM Journal of Control and Optimization 24 (1986) 1177–1191.

E

[150] M. Egerstedt, Y. Wardi, H. Axelsson, Transition-time optimization for switched-mode dynamical systems, IEEE Transactions on Automatic Control 51 (2006) 110–115.
[151] M. Egerstedt, B. Mishra (Eds.), Hybrid Systems: Computation and Control, Lecture Notes in Computer Science, vol. 4981, Springer, Berlin, 2008.
[152] M. Egerstedt, C. Martin, Control Theoretic Splines: Optimal Control, Statistics, and Path Planning, Princeton University Press, Princeton, USA, 2009.
[153] I. Ekeland, R. Temam, Convex Analysis and Variational Problems, North-Holland, Amsterdam, 1976.

F

[154] H.O. Fattorini, Infinite Dimensional Optimization and Control Theory, Cambridge University Press, Cambridge, 1999.
[155] R.P. Fedorenko, Priblizhyonnoye Reshenyie Zadach Optimalnogo Upravlenya, Nauka, Moscow, 1978 (in Russian).
[156] U. Felgenhauer, Stability and local growth near bounded-strong optimal controls, in: E.W. Sachs, R. Tichatschke (Eds.), IFIP TC7 20th Conference on System Modeling and Optimization, Kluwer, Boston, 2003, pp. 213–227.
[157] M.M.A. Ferreira, F.A.C.C. Fontes, R.B. Vinter, Nondegenerate necessary conditions for nonconvex optimal control problems with state constraints, Journal of Mathematical Analysis and Applications 233 (1999) 116–129.
[158] A.V. Fiacco, G. McCormick, Nonlinear Programming: Sequential Unconstrained Minimization Techniques, Wiley, New York, 1968.

[159] A.F. Filippov, On certain questions in the theory of optimal control, Vestnik Moskovskovo Universiteta 2 (1959) 25–32 (in Russian). English translation: SIAM Journal on Control 1 (1962) 76–84.
[160] S. Fitzpatrick, Metric projection and the differentiability of distance functions, Bulletin of the Australian Mathematical Society 22 (1980) 291–312.
[161] W.H. Fleming, R.W. Rishel, Deterministic and Stochastic Optimal Control, Springer Verlag, New York, 1975.
[162] R. Fletcher, Practical Optimization, J. Wiley, Chichester, 1989.
[163] Z. Foroozandeh, M. Shamsi, V. Azhmyakov, M. Shafiee, A modified pseudospectral method for solving trajectory optimization problems with singular arc, Mathematical Methods in the Applied Science 4 (2017) 1783–1793.
[164] M. Fukushima, H. Mline, A generalized proximal point algorithm for certain non-convex minimization problems, International Journal of Systems Science 12 (1981) 989–1000.
[165] M. Fukushima, Y. Yamamoto, A second-order algorithm for continuous-time nonlinear optimal control problems, IEEE Transactions on Automatic Control 31 (1986) 673–676.
[166] A.T. Fuller, Relay control systems optimized for various performance criteria, Automatic and Remote Control 1 (1961) 510–519.

G

[167] R. Gabasov, F. Kirillova, Qualitative Theory of Optimal Processes, Nauka, Moscow, 1971 (in Russian).
[168] H. Gajewski, K. Gröger, K. Zacharias, Nichtlineare Operatorengleichungen und Operatorendifferentialgleichungen, Akademie-Verlag, Berlin, 1974.
[169] R. Galvan-Guerra, V. Azhmyakov, Relations between dynamic programming and the Maximum Principle for impulsive hybrid LQ optimal control problems, in: Proceedings of the 5th International Conference on Electrical Engineering, Computing Science and Automatic Control, Mexico City, Mexico, 2008, pp. 131–136.
[170] R. Galvan-Guerra, V. Azhmyakov, J.E. Velazquez-Velazquez, A. Poznyak, An approach to optimization of linear networked systems based on the hybrid LQ methodology, in: Proceedings of the 6th International Conference on Electrical Engineering, Computing Science and Automatic Control, Toluca, Mexico, 2009, pp. 96–101.
[171] R. Galvan-Guerra, V. Azhmyakov, M. Egerstedt, On the LQ-based optimization techniques for impulsive hybrid control systems, in: Proceedings of the 2010 American Control Conference, Baltimore, USA, 2010, pp. 129–135.
[172] R. Galvan-Guerra, V. Azhmyakov, M. Egerstedt, Optimization of the multiagent systems with increasing state dimensions: hybrid LQ approach, in: Proceedings of the 2011 American Control Conference, San Francisco, USA, 2011, pp. 881–887.
[173] R. Gamkrelidze, Principles of Optimal Control Theory, Plenum Press, London, 1978.
[174] M. Garavello, B. Piccoli, Hybrid necessary principle, SIAM Journal on Control and Optimization 43 (2005) 1867–1887.
[175] P.E. Gill, W. Murray, M.H. Wright, Practical Optimization, Academic Press, New York, USA, 1981.
[176] B. Ginsburg, A. Ioffe, The maximum principle in optimal control of system governed by semilinear equations, in: B.S. Mordukhovich, H.J. Sussmann (Eds.), Nonsmooth Analysis and Geometric Methods in Deterministic Optimal Control, Springer-Verlag, New York, 1996, pp. 81–110.
[177] R.H. Goddard, A Method for Reaching Extreme Altitude, Smithsonian Collection, vol. 2, 1919.
[178] A.A. Goldstein, Convex programming in Hilbert space, Bulletin of the American Mathematical Society 70 (1964) 709–710.
[179] G.C. Goodwin, M.M. Seron, J.A. Dona, Constrained Control and Estimation, Springer, London, UK, 2005.
[180] A. Göpfert, C. Tammer, H. Riahi, Existence and proximal point algorithms for nonlinear monotone complementarity problems, Optimization 45 (1999) 57–68.
[181] O. Güler, On the convergence of the proximal point algorithm for convex minimization, SIAM Journal on Control and Optimization 29 (1991) 403–419.

H

[182] J. Hadamard, Lectures on Cauchy Problems in Linear Partial Differential Equations, Yale University Press, New Haven, 1923.
[183] W.W. Hager, Rate of convergence for discrete approximations to unconstrained control problems, SIAM Journal on Numerical Analysis 13 (1976) 449–471.
[184] W.W. Hager, G.D. Ianculescu, Dual approximations in optimal control, SIAM Journal on Control and Optimization 22 (1990) 1061–1080.
[185] J.K. Hale, S.M.V. Lunel, Introduction to Functional Differential Equations, Springer-Verlag, New York, 1993.
[186] M. Hale, Mode scheduling under dwell time constraints in switched-mode systems, in: Proceedings of the 2014 American Control Conference, Portland, USA, 2014, pp. 3954–3959.
[187] F. Hartung, M. Pituk (Eds.), Recent Advances in Delay Differential and Difference Equations, Springer, Basel, 2014.
[188] M. Hestenes, Conjugate Direction Methods in Optimization, Springer, New York, 1980.
[189] H. Hirano, M. Mukai, T. Azuma, M. Fujita, Optimal control of discrete-time linear systems with network-induced varying delay, in: Proceedings of the 2005 American Control Conference, Portland, USA, 2005, pp. 1419–1424.
[190] J.B. Hiriart-Urruty, C. Lemarechal, Convex Analysis and Minimization Algorithms, Springer, Berlin, 1993.
[191] B. Hofmann, Ill-posedness and local ill-posedness concepts in Hilbert spaces, Optimization 48 (2000) 219–238.

I

[192] A.D. Ioffe, V.M. Tikhomirov, Theory of Extremal Problems, North Holland, Amsterdam, 1979.
[193] A.D. Ioffe, Regular points of Lipschitz functions, Transactions of the American Mathematical Society 251 (1979) 61–69.
[194] E. Isaacson, H.B. Keller, Analysis of Numerical Methods, Dover Publications Inc., New York, 1994.
[195] A.N. Iusem, Inexact version of proximal point and augmented Lagrangian algorithms in Banach spaces, Numerical Functional Analysis and Optimization 22 (2001) 609–640.
[196] A.F. Izmailov, M.V. Solodov, Optimality conditions for irregular inequality-constrained problems, SIAM Journal on Control and Optimization 40 (2001) 1280–1295.
[197] A.F. Izmailov, A.A. Tretjakov, 2-Regular Solutions of Nonlinear Problems, Nauka, Moscow, 1999.

J

[198] J. Jahn, An Introduction to the Theory of Nonlinear Optimization, Springer, Berlin, 1974.
[199] K.H. Johansson, M. Egerstedt, J. Lygeros, S. Sastry, On the regularization of Zeno hybrid automata, Systems and Control Letters 38 (1999) 141–150.

K

[200] L.V. Kantorovich, G.P. Akilov, Functional Analysis, Pergamon Press, Oxford, 1982.
[201] L.V. Kantorovich, G.P. Akilov, Functional Analysis, Nauka, Moscow, 1982 (in Russian).
[202] A. Kaplan, R. Tichatschke, Stable Methods for Ill-Posed Variational Problems – Prox-Regularization of Elliptical Variational Inequalities and Semi-Infinite Optimization Problems, Akademie Verlag, Berlin, 1994.
[203] A. Kaplan, R. Tichatschke, Proximal point approach and approximation of variational inequalities, SIAM Journal on Control and Optimization 39 (2000) 1136–1159.
[204] C.T. Kelley, E.W. Sachs, Mesh independence of the gradient projection method for optimal control problems, SIAM Journal of Control and Optimization 30 (1992) 477–493.

[205] H.K. Khalil, Nonlinear Systems, Prentice Hall, Upper Saddle River, 1996.
[206] D. Kirk, Optimal Control Theory, Dover, New York, USA, 1998.
[207] A. Kojimaa, M. Morari, LQ control for constrained continuous-time systems, Automatica 40 (2004) 1143–1155.
[208] A.N. Kolmogorov, S.V. Fomin, Introductory Real Analysis, Prentice-Hall, Englewood Cliffs, 1970.
[209] A.N. Kolmogorov, S.V. Fomin, Elements of Theory of Functions and Functional Analysis, Nauka, Moscow, 1972 (in Russian).
[210] V. Komornik, Exact Controllability and Stabilization, Masson, Paris, 1994.
[211] G. Köthe, Topological Vector Spaces I, Springer, Berlin, 1983.
[212] K. Kozlowski, Robot Motion and Control 2007, Lecture Notes in Control and Information Sciences, Springer, London, UK, 2007.
[213] A.J. Kurdila, M. Zabarankin, Convex Functional Analysis, Springer, New York, USA, 2006.
[214] M.A. Krasnoselskii, Solutions of equations involving adjoint operators by successive approximations, Uspekhi Mathematicheskikh Nauk 15 (1960) 161–165.
[215] R. Kress, Linear Integral Equations, Springer, Berlin, 1989.
[216] K. Kuratowski, Topology, Academic Press, New York, 1968.

L

[217] U. Ledzewicz, H. Schättler, A hight-order generalization of the Lysternik theorem, Nonlinear Analysis 34 (1998) 793–815.
[218] A. Lamperski, A.D. Ames, Sufficient conditions for Zeno behavior in Lagrangian hybrid systems, Lecture Notes in Computer Science 4981 (2008) 622–625.
[219] A. Lamperski, A.D. Ames, Lyapunov theory for Zeno stability, IEEE Transactions on Automatic Control 58 (1) (2013) 100–112.
[220] N. Lehdili, A. Moudafi, Combining the proximal algorithm and Tikhonov regularization, Optimization 37 (1996) 239–252.
[221] B. Lemaire, On the convergence of some iterative methods for convex minimization, in: P. Gritzmann, R. Horst, E. Sachs, R. Tichatschke (Eds.), Recent Developments in Optimization, Springer, Berlin, 1995, pp. 252–268.
[222] D. Li, Zero duality gap for a class of nonconvex optimization problems, Journal of Optimization Theory and Applications 85 (1995) 309–324.
[223] D. Liberzon, Switching in Systems and Control, Birkhäuser, Boston, 2003.
[224] D. Liberzon, Hybrid feedback stabilization of systems with quantized signals, Automatica 39 (2003) 1543–1554.
[225] B. Lincoln, A. Rantzer, Optimizing linear systems switching, in: Proceedings of the 50th CDC-ECC Conference, IEEE Conference on Decision and Control, Orlando, USA, 2011, pp. 2063–2068.
[226] J.P. Lions, Controle Optimal de Systèmes Gouvernés par des Équations aux Dérivées Partielles, Dunod Gauthier-Villars, Paris, 1968.
[227] L.A. Liusternik, On the conditional extrema of functionals, Matematicheski Sbornik 41 (1934) 390–401 (in Russian).
[228] A. Louis, Inverse und schlecht gestellte Probleme, Teubner-Verlag, Stuttgart, 1989.
[229] D.G. Luenberger, Introduction to Dynamic Systems: Theory, Models and Applications, J. Wiley, New York, 1979.
[230] D.G. Luenberger, Investment Science, Oxford University Press, New York, 1998.
[231] J. Lygeros, Lecture Notes on Hybrid Systems, Cambridge University Press, Cambridge, 2003.
[232] J. Lygeros, An overview of hybrid systems control, in: Handbook of Networked and Embedded Control Systems, Birkhäuser, Boston, 2005, pp. 519–537.

M

[233] K.C.P. Machielsen, Numerical Solution of Optimal Control Problems with State Constraints by Sequential Quadratic Programming in Function Space, Thesis, Technical University of Eindhoven, Eindhoven, 1987.
[234] K. Malanowski, Finite difference approximations to constrained optimal control problems, in: Optimization and Optimal Control, Springer, New York, 1981, pp. 243–254.
[235] K. Malanowski, On normality of Lagrange multipliers for state constrained optimal control problems, Optimization 52 (2002) 75–91.
[236] O.L. Mangasarian, S. Fromovitz, The Fritz John necessary optimality conditions in the presence of equality and inequality constraints, Journal of Mathematical Analysis and Applications 17 (1967) 37–47.
[237] B. Martinet, Regularisation d'inequations variationelles par approximations successives, Revue Francaise d'Automatique, Informatique, Recherche Operationnelle 4 (1970) 154–159.
[238] H. Maurer, Numerical solution of singular control problems using multiple shooting techniques, Journal of Optimization Theory and Applications 18 (1976) 235–257.
[239] D.Q. Mayne, E. Polak, A superlinearly convergent algorithm for constrained optimization problem, Mathematical Programming Study 16 (1982) 45–61.
[240] M.A. Mehrpouya, M. Shamsi, V. Azhmyakov, An efficient solution of Hamiltonian boundary value problems by combined Gauss pseudospectral method with differential continuation approach, Journal of the Franklin Institute 351 (2014) 4765–4785.
[241] H. Minc, Nonnegative Matrices, J. Wiley, New York, 1988.
[242] G. Minty, Monotone (nonlinear) operators in a Hilbert space, Duke Mathematical Journal 29 (1962) 341–348.
[243] A. Mitsos, B. Chachuat, P.I. Barton, McCormic-based relaxation algorithm, SIAM Journal on Optimization 20 (2009) 573–601.
[244] N.N. Moiseev, Numerical Methods in the Theory of Optimal Systems, Nauka, Moscow, 1971 (in Russian).
[245] T. Moor, J. Raisch, Supervisory control of hybrid systems within a behavioural framework, Systems and Control Letters 38 (1999) 157–166.
[246] B.S. Mordukhovich, Metric approximations and necessary optimality conditions for general cases of nonsmooth extremal problems, Soviet Mathematics Doklady 22 (1980) 526–530.
[247] B.S. Mordukhovich, Approximation Methods in Problems of Optimization and Control, Nauka, Moscow, 1988.
[248] B.S. Mordukhovich, Discrete approximations and refined Euler–Lagrange conditions for nonconvex differential inclusions, SIAM Journal on Control and Optimization 33 (1995) 882–915.
[249] B.S. Mordukhovich, Optimization and finite difference approximations of nonconvex differential inclusions with free time, in: B.S. Mordukhovich, H.J. Sussmann (Eds.), Nonsmooth Analysis and Geometric Methods in Deterministic Optimal Control, Springer, New York, 1996, pp. 153–202.
[250] U. Mosco, Convergence of convex sets and solutions of variational inequalities, Advances in Mathematics 3 (1969) 510–585.

N

[251] A. Nagurney, D. Zhang, Projected Dynamical Systems and Variational Inequalities with Applications, Kluwer Academic Publishers, New York, 1996.
[252] J. Nocedal, S. Wright, Numerical Optimization, Springer, New York, 1999.

O

[253] H.J. Oberle, Numerical solution of minimax optimal control problems by multiple shooting technique, Journal of Optimization Theory and Applications 50 (1986) 331–364.
[254] Y. Orlov, Discontinuous Systems: Lyapunov Analysis and Robust Synthesis under Uncertainty Conditions, Springer, New York, 2008.

P

[255] B. Piccoli, Necessary conditions for hybrid optimization, in: Proceedings of the 38th IEEE Conference on Decision and Control, Phoenix, USA, 1999, pp. 410–415.
[256] R.R. Phelps, Convex Functions, Monotone Operators and Differentiability, Springer, Berlin, 1993.
[257] C. Perez, V. Azhmyakov, A. Poznyak, Practical stabilization of a class of switched systems: dwell-time approach, IMA Journal of Mathematical Control and Information 32 (2015) 689–702.
[258] H.J. Pesch, Numerical computation of neighboring optimum feedback control schemes in real time, Applied Mathematics and Optimization 5 (1979) 231–252.
[259] B.T. Polyak, Optimization Software, Inc. Publ. Division, New York, 1987.
[260] B.T. Polyak, Introduction to Optimization, Optimization Software, Inc. Publ. Division, New York, 1987.
[261] E. Polak, Optimization, Springer, New York, 1997.
[262] E. Polak, T.H. Yang, D.Q. Mayne, A method of centers based on barrier functions for solving optimal control problems with continuous state and control constraints, SIAM Journal on Control and Optimization 31 (1993) 159–179.
[263] E. Polak, On the use of consistent approximations in the solution of semi-infinite optimization and optimal control problems, Mathematical Programming 62 (1993) 385–414.
[264] L.S. Pontryagin, V.G. Boltyanski, R.V. Gamkrelidze, E.F. Mischenko, The Mathematical Theory of Optimal Processes, Wiley, New York, 1962.
[265] A. Poznyak, Advanced Mathematical Tools for Automatic Control Engineers, Elsevier, Amsterdam, 2008.
[266] A. Poznyak, V. Azhmyakov, M. Mera, Practical output feedback stabilization for a class of continuous-time dynamic systems under sample-data outputs, International Journal of Control 84 (8) (2011) 1408–1416.
[267] A. Poznyak, A. Polyakov, V. Azhmyakov, Attractive Ellipsoids in Robust Control, Birkhäuser, New York, USA, 2014.
[268] W.H. Press, S.A. Teukolsky, W.T. Vetterling, B.P. Flannery, Numerical Recipes in C, Cambridge University Press, Cambridge, 1992.
[269] R. Pytlak, Numerical Methods for Optimal Control Problems with State Constraints, Springer, Berlin, 1999.

R

[270] A. Rantzer, M. Johansson, Piecewise linear quadratic optimal control, IEEE Transactions on Automatic Control 45 (2000) 629–637.
[271] W. Reddy, Expanding maps on compact metric spaces, Topology and Its Applications 13 (1982) 327–334.
[272] M. Reed, B. Simon, Methods of Modern Mathematical Physics. I Functional Analysis, Academic Press, New York, 1972.
[273] A.P. Robertson, W. Robertson, Topological Vector Spaces, Cambridge University Press, Cambridge, 1966.
[274] S.M. Robinson, Stability theory for systems of inequalities in nonlinear programming, part II: differentiable nonlinear systems, SIAM Journal on Numerical Analysis 13 (1976) 457–513.
[275] R.T. Rockafellar, Monotone operators and the proximal point algorithm, SIAM Journal on Control and Optimization 14 (1976) 877–898.
[276] R.T. Rockafellar, Augmented Lagrange multiplier functions and applications of the proximal point algorithm in convex programming, Mathematics of Operations Research 1 (1976) 97–116.
[277] R.T. Rockafellar, R.J.-B. Wets, Variational Analysis, Springer, Berlin, 1998.
[278] J.B. Rosen, The gradient projection method for nonlinear programming, SIAM Journal 8 (1960) 180–217.
[279] T. Roubicek, Relaxation in Optimization Theory and Variational Calculus, W. de Gruyter, Berlin, 1997.
[280] T. Roubicek, Approximation theory for generalized Young measures, Numerical Functional Analysis and Optimization 16 (1995) 1233–1253.
[281] J.E. Rubio, Control and Optimization: the Linear Treatment of Nonlinear Problems, Manchester University Press, Manchester, U.K., 1986.
[282] W. Rudin, Functional Analysis, McGraw-Hill Book Company, New York, 1973.
[283] M. Ruzicka, Nichtlineare Funktionalanalysis, Springer, Berlin, 2004.

S

[284] Y. Sakawa, Y. Shindo, Y. Hashimoto, Optimal control of a rotary crane, Journal of Optimization Theory and Applications 35 (1981) 535–557.

[285] W.H. Schmidt, Iterative methods for optimal control processes governed by integral equations, International Series of Numerical Mathematics 111 (1993) 69–82.

[286] W.H. Schmidt, Optimalitätsbedingungen für verschiedene Aufgaben von Integralprocessen in Banachräumen und das Iterationsverfahren von Chernousko, Thesis, University of Greifswald, Greifswald, 1988.

[287] J.K. Scott, P.I. Barton, Convex relaxations for nonconvex optimal control problems, in: Proceedings of the 50th IEEE Conference on Decision and Control and European Control Conference, Orlando, USA, 2011, pp. 1042–1047.

[288] J.K. Scott, P.I. Barton, Convex and concave relaxations for the parametric solutions of semi-explicit index-one differential-algebraic equations, Journal of Optimization Theory and Applications 156 (2013) 617–649.

[289] M.S. Shaikh, P.E. Caines, On the hybrid optimal control problem: theory and algorithms, IEEE Transactions on Automatic Control 52 (9) (2007) 1587–1603.

[290] H.L. Smith, Monotone Dynamical Systems: an Introduction to the Theory of Competitive and Cooperative Systems, Mathematical Surveys and Monographs, vol. 41, AMS, 1995.

[291] P. Spellucci, Numerische Verfahren der Nichtlinearen Optimierung, Birkhäuser, Basel, 1993.

[292] S.A. Stanton, B.G. Marchand, Finite set control transcription for optimal control applications, Journal of Spacecraft and Rockets 47 (2010) 457–471.

[293] H.J. Stetter, Analysis of Discretization Methods for Ordinary Differential Equations, Springer, Berlin, 1973.

[294] O. Stryk, Numerische Lösung Optimaler Steuerungsproblems: Diskretisierung, Parameteroptimierung und Berechnung der Adjungierten Variablen, VDI-Verlag, Düsseldorf, 1995.

[295] J. Stoer, R. Bulirsch, Introduction to Numerical Analysis, Springer, New York, USA, 2002.

[296] H. Sussmann, A Maximum Principle for hybrid optimal control problems, in: Proceedings of the 38th IEEE Conference on Decision and Control, Phoenix, USA, 1999, pp. 425–430.

T

[297] W. Takahashi, Nonlinear Functional Analysis, Yokohama Publishers, Yokohama, 2000.

[298] F. Taringoo, P. Caines, The sensitivity of hybrid systems optimal cost, in: Lecture Notes in Computer Science, vol. 5469, Springer, Berlin, Germany, 2009, pp. 475–479.

[299] F. Taringoo, P.E. Caines, On the geometry of switching manifolds for autonomous hybrid systems, in: Proceedings of the 10th International IFAC Workshop on Discrete Event Systems, Berlin, Germany, 2010, pp. 45–50.

[300] K.L. Teo, C.J. Goh, K.H. Wong, A Unified Computational Approach to Optimal Control Problems, Wiley, New York, 1991.

[301] K.L. Teo, C.J. Goh, A computational approach for a class of optimal relaxed control problems, Journal of Optimization Theory and Applications 60 (1989) 117–133.

[302] V.M. Tichomirov, Grundprinzipien der Theorie der Extremalaufgaben, Teubner, Leipzig, 1982.

[303] A. Tikhonov, On the stability of the functional minimization method, USSR Computational Mathematics and Mathematical Physics 6 (1966) 26–33.

[304] A.N. Tikhonov, V.J. Arsenin, Solutions of Ill-Posed Problems, Wiley, New York, 1977.

U

[305] V. Utkin, Sliding Modes in Control and Optimization, Springer, Berlin, 1992.

V

[306] A.J. van der Schaft, J.M. Schumacher, An Introduction to Hybrid Dynamical Systems, Springer, Berlin, 2000.
[307] G. Vanniko, Methods for the Solution of Ill-Posed Problems in Hilbert Space, Tartu University Press, Tartu, 1982.
[308] F.P. Vasil'ev, Methods for Solving the Extremal Problems, Nauka, Moscow, 1981 (in Russian).
[309] V.M. Veliov, Second order discrete approximations to linear differential inclusions, SIAM Journal of Numerical Analysis 29 (1992) 439–451.
[310] E. Verriest, F. Delmotte, M. Egerstedt, Optimal impulsive control of point delay systems with refractory period, in: Proceedings of the 5th IFAC Workshop on Time Delay Systems, Leuven, Belgium, 2004.
[311] E.I. Verriest, Pseudo-continuous multi-dimensional multi-mode systems, Discrete Event Dynamic Systems 22 (2012) 27–59.
[312] E.I. Verriest, G. Dirr, U. Helmke, O. Mitesser, Explicitly solvable bilinear optimal control problems with applications in ecology, in: Proceedings of the 22nd International Symposium on Mathematical Theory of Networks and Systems, Minneapolis, MN, 2016.
[313] E.I. Verriest, V. Azhmyakov, Advances in optimal control of differential systems with the state suprema, in: Proceedings of the 56th IEEE Conference on Decision and Control, Melbourne, Australia, 2017, pp. 739–744.
[314] R. Vinter, Optimal Control, Springer, Basel, 2010.
[315] A. Visintin, Strong convergence results related to strict convexity, Communications in Partial Differential Equations 9 (1984) 439–466.
[316] L. Vu, D. Liberzon, Supervisory control of uncertain systems, International Journal of Adaptive Control and Signal Processing 27 (2012) 739–756.

W

[317] Y. Wardi, M. Egerstedt, Algorithm for optimal mode scheduling in switched systems, arXiv:1107.3099.
[318] Y. Wardi, Optimal control of switched-mode dynamical systems, in: Proceedings of the 11th International Workshop on Discrete Event Systems, Guadalajara, Mexico, 2012, pp. 4–8.
[319] Y. Wardi, M. Egerstedt, P. Twu, A controlled-precision algorithm for mode-switching optimization, in: Proceedings of the 51st IEEE Conference on Decision and Control, Maui, USA, 2012, pp. 713–718.
[320] Y. Wardi, M. Egerstedt, M. Hale, Switched-mode systems: gradient-descent algorithms with Armijo step sizes, Discrete Event Dynamic Systems 25 (4) (2015) 571–599.
[321] J. Warga, Optimal Control of Differential and Functional Equation, Academic Press, New York, 1972.
[322] S.J. Wright, Interior point method for optimal control of discrete-time systems, Journal of Optimization Theory and Applications 77 (1993) 161–187.

X

[323] X. Xu, P. Antsaklis, Optimal control of hybrid autonomous systems with state jumps, in: Proceedings of the American Control Conference, Denver, USA, 2003, pp. 5191–5196.
[324] X. Xu, P. Antsaklis, Quadratic optimal control problems for hybrid linear autonomous systems with state jumps, in: Proceedings of the American Control Conference, Denver, USA, 2003, pp. 3393–3398.

Y

[325] H. Yu, C. Cassandras, Perturbation analysis of communication networks with feedback controls using stochastic hybrid models, Nonlinear Analysis 65 (2006) 1251–1280.
[326] L.C. Young, Lectures on the Calculus of Variations and Optimal Control Theory, Saunders, Philadelphia, 1969.

Z

[327] E.H. Zarantonello, Projection on convex sets in Hilbert space and spectral theory, in: Contributions to Nonlinear Functional Analysis, Academic Press, New York, 1971, pp. 237–424.
[328] E. Zeidler, Nonlinear Functional Analysis and Its Applications II/A. Linear Monotone Operators, Springer, New York, 1990.
[329] E. Zeidler, Nonlinear Functional Analysis and Its Applications II: Nonlinear Monotone Operators, Springer, New York, 1990.
[330] E. Zeidler, Nonlinear Functional Analysis and Its Applications III: Variational Methods and Its Applications, Springer, New York, 1990.
[331] E. Zeidler, Nonlinear Functional Analysis and Its Applications I: Fixed Point Theorems, Springer, New York, 1990.
[332] M.I. Zelikin, V.F. Borisov, Theory of Chattering Control with Applications to Astronautics, Robotics, Economics and Engineering, Birkhäuser, Boston, 1994.
[333] J. Zhang, K.H. Johansson, J. Lygeros, S. Sastry, Zeno hybrid systems, International Journal of Robust and Nonlinear Control 11 (2001) 435–451.
[334] H. Zhang, J. Matthew, Optimal control of hybrid systems and a system of quasi-variational inequalities, SIAM Journal on Control and Optimization 45 (2006) 722–761.
[335] P. Zhao, S. Mohan, R. Vasudevan, Optimal control for nonlinear hybrid systems via convex relaxations, Preprint, arXiv:1702.04310v1, 2018.
[336] T. Zolezzi, Well posed optimal control problems: a perturbation approach, in: B.S. Mordukhovich, H.J. Sussmann (Eds.), Nonsmooth Analysis and Geometric Methods in Deterministic Optimal Control, Springer, New York, 1996, pp. 239–256.
[337] J. Zowe, S. Kurcyusz, Regularity and stability for the mathematical programming in Banach spaces, Applied Mathematics and Optimization 5 (1979) 49–62.

Index

D

E

P

Q

R

Printed in the United States
By Bookmasters